T0295206

Handbook of Air Quality

Handbook of Air Quality

Edited by Frieda Bush

www.statesacademicpress.com

States Academic Press,
109 South 5th Street,
Brooklyn, NY 11249, USA

Visit us on the World Wide Web at:
www.statesacademicpress.com

ISBN: 978-1-63989-754-4

Cataloging-in-Publication Data

Handbook of air quality / edited by Frieda Bush.
p. cm.
Includes bibliographical references and index.
ISBN 978-1-63989-754-4
1. Air quality. 2. Air--Pollution. 3. Air quality management.
4. Air quality--Measurement. 5. Air--Pollution--Measurement.
I. Bush, Frieda.
TD883 .A37 2023
628.53--dc23

Table of Contents

Preface

Air quality refers to the degree to which the air at a particular location is pollution-free. Good air quality means that the air is clear and contains only small amounts of solid particles and chemical pollutants. Contrarily, air quality is considered bad when it is not clean and contains large amounts of solid and chemical particles. Air quality is measured with the help of the air quality index (AQI). Source apportionment (SA) refers to a method for identifying the sources of air pollution and the contribution (amount) of each type of pollutant source in the overall air pollution level. Most commonly used SA methods are classified in three categories, namely, emission reduction impact (ERI) methods, mass-transfer (MT) methods, and incremental (INC) methods. This book elucidates the concepts around prospective developments with respect to air quality. It will prove to be immensely beneficial to students and researchers in the field of air quality engineering.

Various studies have approached the subject by analyzing it with a single perspective, but the present book provides diverse methodologies and techniques to address this field. This book contains theories and applications needed for understanding the subject from different perspectives. The aim is to keep the readers informed about the progresses in the field; therefore, the contributions were carefully examined to compile novel researches by specialists from across the globe.

Indeed, the job of the editor is the most crucial and challenging in compiling all chapters into a single book. In the end, I would extend my sincere thanks to the chapter authors for their profound work. I am also thankful for the support provided by my family and colleagues during the compilation of this book.

Editor

1

Machine Learning-Based Prediction of Air Quality

Yun-Chia Liang [1,*], Yona Maimury [1], Angela Hsiang-Ling Chen [2,*] and Josue Rodolfo Cuevas Juarez [1]

1 Department of Industrial Engineering and Management, Yuan Ze University, Taoyuan City 320, Taiwan; s1065439@mail.yzu.edu.tw (Y.M.); s998911@mail.yzu.edu.tw (J.R.C.J.)

2 Department of Industrial and Systems Engineering, Chung Yuan Christian University, Taoyuan City 320, Taiwan

* Correspondence: ycliang@saturn.yzu.edu.tw (Y.-C.L.); achen@cycu.edu.tw (A.H.-L.C.)

Abstract: Air, an essential natural resource, has been compromised in terms of quality by economic activities. Considerable research has been devoted to predicting instances of poor air quality, but most studies are limited by insufficient longitudinal data, making it difficult to account for seasonal and other factors. Several prediction models have been developed using an 11-year dataset collected by Taiwan's Environmental Protection Administration (EPA). Machine learning methods, including adaptive boosting (AdaBoost), artificial neural network (ANN), random forest, stacking ensemble, and support vector machine (SVM), produce promising results for air quality index (AQI) level predictions. A series of experiments, using datasets for three different regions to obtain the best prediction performance from the stacking ensemble, AdaBoost, and random forest, found the stacking ensemble delivers consistently superior performance for R^2 and RMSE, while AdaBoost provides best results for MAE.

Keywords: air quality monitoring; machine learning; air quality index

---

1. Introduction

Worldwide, air pollution is responsible for around 1.3 million deaths annually according to the World Health Organization (WHO) [1]. The depletion of air quality is just one of harmful effects due to pollutants released into the air. Other detrimental consequences, such as acid rain, global warming, aerosol formation, and photochemical smog, have also increased over the last several decades [2]. The recent rapid spread of COVID-19 has prompted many researchers to investigate underlying pollution-related conditions contributing to COVID-19 pandemics in countries. Several shreds of evidence have shown that air pollution is linked to significantly higher COVID-19 death rates, and patterns in COVID-19 death rates mimic patterns in both high population density and high $PM_{2.5}$ exposure areas [3]. All the above mentioned raises an urgent need to anticipate and plan for pollution fluctuations to help communities and individuals better mitigate the negative impact of air pollution. To do so, air quality evaluation plays a significant role in monitoring and controlling air pollution.

The Environmental Protection Agency (EPA) tracks the commonly known criteria pollutants, i.e., ground-level ozone (O_3), Sulphur dioxide (SO_2), particulates matter (PM_{10} and $PM_{2.5}$), carbon monoxide (CO), carbon dioxide (CO_2), and nitrogen dioxide (NO_2). These substances are in compositions of a common index, called the Air Quality Index (AQI), indicating how clean or polluted the air is currently or forecasted to become in areas. As the AQI increases, a higher percentage of the population is exposed. Different countries have their air quality indices, corresponding to different air quality standards. In the United States, the US Environmental Protection Agency monitors six pollutants at more than 4000 sites: O_3, PM_{10}, $PM_{2.5}$, NO_2, SO_2, and lead. Rybarczyk and Zalakeviciute [4] reviewed a selection of the 46 most relevant journal papers and found more studies with O_3, NO_2, PM_{10} and $PM_{2.5}$, and less on an overall AQI.

Recent researches focus more on advanced statistical learning algorithms for air quality evaluation and air pollution prediction. Raimondo et al. [5], Garcia et al. [6], and Park et al. [7] have used neural networks to build models for predicting the prevalence of individual pollutants, e.g., particulates matter measuring less than 10 microns (PM_{10}). Raimondo et al. [5] used a support vector machine (SVM) and artificial neural network (ANN) to train models. Their best ANN model attained almost 79% for specificity with only a 0.82% false-positive rate, while their best SVM model at a specificity of 80% with a false positive rate of only 0.13%. Yu et al. [8] proposed a random forest approach, named RAQ, for AQI category prediction. Then, Yi et al. [9] applied deep neural networks for AQI category prediction. Veljanovska and Dimoski [10] applied different settings to outperform *k*-nearest neighbor (*k*-NN), decision tree, and SVM for predicting AQI levels. Their ANN model achieved an accuracy of 92.3%, outperforming all other tested algorithms.

The work presented in this paper focuses on the development of AQI prediction models for acute air pollution events 1, 8, and 24 h in advance. The following machine learning (ML) algorithms are investigated, i.e., random forest, adaptive boosting (AdaBoost), support vector machine, artificial neural network, and stacking ensemble methods to train models. As well, this research observes how prediction performance decays over longer time frames, and the precision is measured with three commonly used scale-dependent error indexes: mean absolute error (MAE), root mean squared error (RMSE), and R-squared (R^2).

2. Machine Learning Prediction Methods

Machine learning involves computational methods which learn from complex data to build various models for prediction, classification, and evaluation. The study attempts to build forecasting models capable of efficient pattern recognition and self-learning. In this section, the underlying principle of five machine learning methods as the canonical procedure will be discussed respectively.

2.1. Support Vector Machine

Support vector machine, a supervised learning method for classification, regression, and outlier detection, constructs the hyperplane that acts as a boundary between distinct data points and thus the output can be deduced hereafter [11]. Two distinctive versions of SVM are shown in Figure 1. For classification problem in Figure 1a, data points that lie at the edge of an area closest to the hyperplanes are considered as support vectors. The space between these two regions is the margin between the classes. Hyperplanes will determine the number of classes incurred in the dataset and the output of unseen data will be predicted according to which class holds the most similarity with the new data. As for regression problem in Figure 1b, an approximation of such hyperplane to a non-linear function is constructed at the maximal margin with linear regression. Hence, the additional parameter, known as the ε-insensitive loss is introduced to tolerate some deviations that lie inside the ε region tube [12].

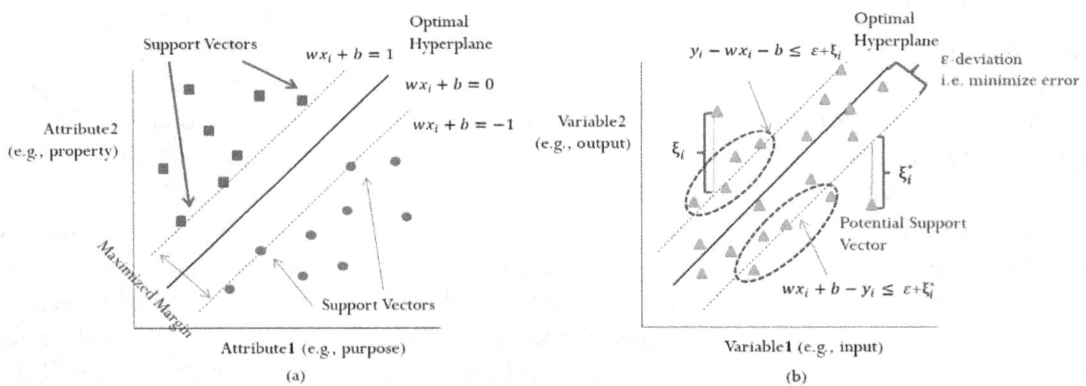

Figure 1. Overview of SVM algorithm: (**a**) SVM for classification; (**b**) SVM for regression.

The boundary lines (dashed lines) across the hyperplane (solid line) in SVR (stands for support vector regression) are defined with regards to parameter ε, in which the resulting lines are the shifted function in the amount of $-\varepsilon$ and $+\varepsilon$ from the hyperplane (assume it is a straight line with an equation of $<w, x_i> +b$). The SVR uses a penalty concept introduced by parameter C (cost factor) for output variables outside the boundaries either above (ξ_i) or below (ξ_i^*). Nevertheless, data points inside the boundaries will be exempted. Since support vectors represent the data points located near these boundary lines (see Figure 1b), if the ε moves further from the hyperplane, the number of support vectors decreases; otherwise, the number of support vectors increases as the ε approximates towards the hyperplane. Finally, since most realistic problems aren't linear, the kernel trick is commonly performed by mapping training data onto the high-dimensional feature space. Kernel functions, e.g., linear, polynomial, radial basis function (RBF), sigmoid, hyperbolic tangent, etc., are used to convert the once inseparable input data into the separable ones.

The parameter ε has brought a couple of advantages, yet is sometimes difficult to tune. Hence, scholars from Australian National University proposed the substitution of parameter ε into parameter ν (hereinafter referred to as ν-SVM) to avoid such a tedious parameter tuning process for regression [13]. Moreover, parameter ν is also applicable for classification, where it becomes the replacement for cost factor C [14]. Values of parameter v with the upper bound of training margin errors and lower bound for the support vectors are recommended from 0 to 1 so that the ν-SVM can offer a more meaningful parameter interpretation [15].

2.2. *Random Forest*

Another prominent machine learning method, random forest, a supervised learning ensemble algorithm, combines multiple decision trees to form a forest and the bagging concept, that latter adds the randomness into the model building. The random selection of features is used to split the individual tree while the random selection of instances is used to create training data subset for each decision tree. At each decision node in every tree, the variable from the random number of features is considered for the best split. If the target attribute is categorical, random forests will choose the most frequent as its prediction. On the other hand, if it's numerical, the average of all predictions will be chosen.

Similar to SVM, the random forest can tackle both classification and regression case. For prediction, each test data point is passed through every decision tree in the forest. The trees then vote on an outcome and the prediction is produced from a majority vote among the models and henceforth resulting in a stronger and more robust single learner. Random forests can overcome the prediction variance that each decision tree has, in the way that the prediction average will approximate the ground truth (classification) or true value (regression). Figure 2 shows the illustration of a random forest that consists of m number of trees.

2.3. *Adaptive Boosting*

The next method, Adaptive Boosting, also came from a branch of ensemble methods where combine several weak learners yet with the sequential arrangement instead of a parallel setting as what random forest does. Boosting trains the base models in sequence one by one and assigns weights to the classifiers based on their accuracy to predict a random set of input instances. By such means, the more accurate classifiers will have more contribution in the final answer. The weights are also attributed to each input item depending on how difficult the instance to be predicted as on average by all classifiers. The higher the weight, the harder it is to estimate the ground truth for the instance and therefore this item will have a higher chance to appear as the training subset in the succeeding iteration. In other words, the boosting process concentrates on the training data that are hard to classify and over-represents them in the training set for the next iteration. The loop will start to be more substantial, as the focus is gathered to solve the difficult-to-predict instances using the stronger classifiers. Classifiers are the base algorithms utilized to perform the prediction, where the common one used in AdaBoost is a decision tree. It also can be constructed from different types of algorithms, e.g., mix of a decision tree, logistic regression, and naïve Bayes (for classification).

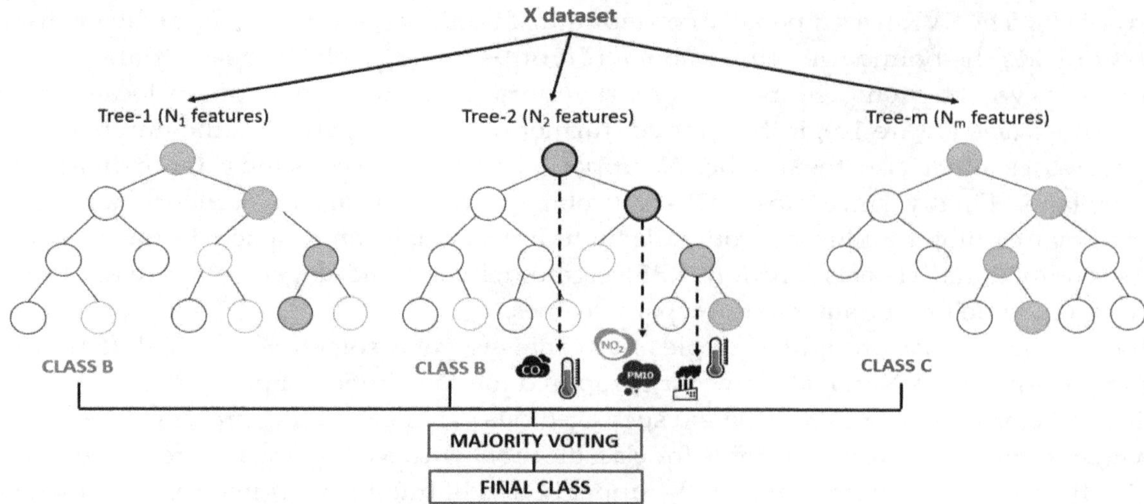

Figure 2. Illustration of a random forest algorithm.

2.4. *Artificial Neural Network*

The next approach preferred in this study is the artificial neural network. Being the earliest algorithm invented among all, ANN is not only seen as the "universal approximator" which can estimate any arbitrary function well [16], but also as the initiator of the most recent progress in the artificial intelligence field as of now, called as deep learning or deep neural network. The neural network simulates the structure and networks of the human brain in the process of information learning. For a human, new things are learned by training the biological neurons in the brain using some examples, where the knowledge extracted will later be stored in the memory. In an ANN, a considerable amount of input data is fed into the artificial neurons where all neurons are trained and the network is adjusted to get a better response, or more specifically output, e.g., in a prediction, or a recognition task. The adjustment of the network is performed by updating the weight ($w_{i,1}$, $w_{i,2}$, $w_{i,3}$, ... , $w_{i,r}$) that each neuron has and biases which are the adder for each summation procedure (see Figure 3). The complexity of the network itself is determined by the number of hidden layers. Furthermore, the net output, denoted by a_i, will be transformed non-linearly by the activation function (f) to form an output y_i that will be forwarded to the next hidden layer. There are numerous types of activation function that are employed to bring the non-linearity property to the input signal as to adapt with a variety of instances and hence results to the highly adaptable network. These are including sigmoid, ReLU, leaky ReLU, hyperbolic tangent (tanh) function, and so on.

2.5. *Linear Regression*

Linear regression is probably the method where most of the academicians started their first machine learning experience. Its main working principle lies behind the fitting of one or more independent variables with the dependent variable into a line in n dimensions. n usually denotes the number of variables within a dataset. This line is supposedly created as it would be minimizing the total errors when trying to fit all the instances into the line. Under machine learning, linear regression is equipped with the capability to learn continuously by optimizing the parameters in the model. These parameters are including w_0, w_1, w_2, ... , w_m (as illustrated in Figure 4). Most commonly, optimization is carried out by a method called gradient descent. It works by partially deriving the loss function and all parameters will be updated by subtracting the previous value with the derivative times a specified learning rate. The learning rate can be tuned by the simplest way, which is rule of thumb (trial and error), or a more sophisticated rule, e.g., meta-heuristic. Another parameter that is left for tuning is the amount of generalization added to the model. Regularization is undergone as an effort to lessen the chance of overfitting and increase the robustness of the model. Two types of regularization used in linear regression are lasso and ridge regression. Lasso regularization will eliminate less important feature by letting the feature's coefficient to zero, and retain another more important one. Ridge regularization on the other hand will not try to eliminate a feature, but instead, tries to shrink the magnitude of coefficients to get a lower variance in the model.

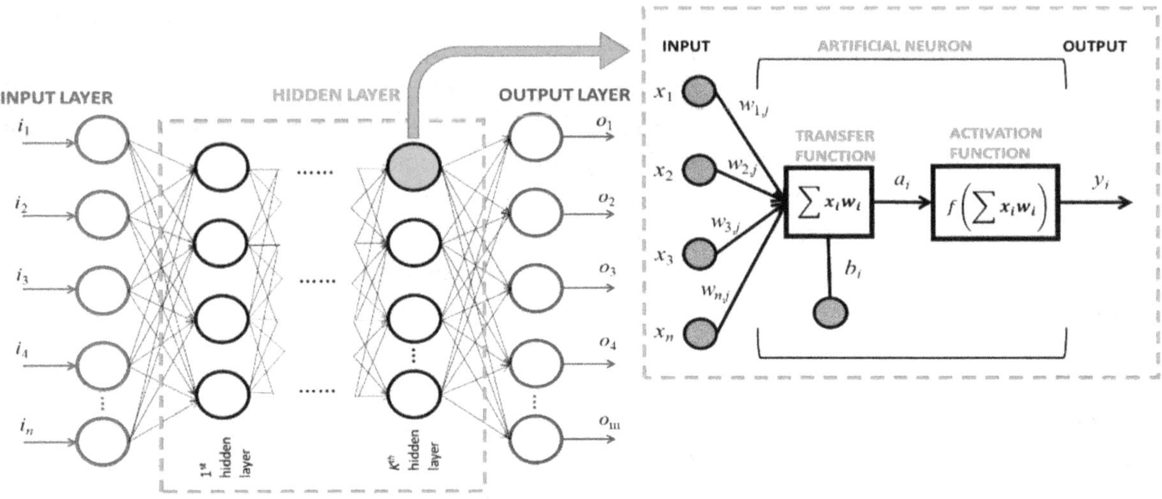

Figure 3. Illustration of artificial neural network.

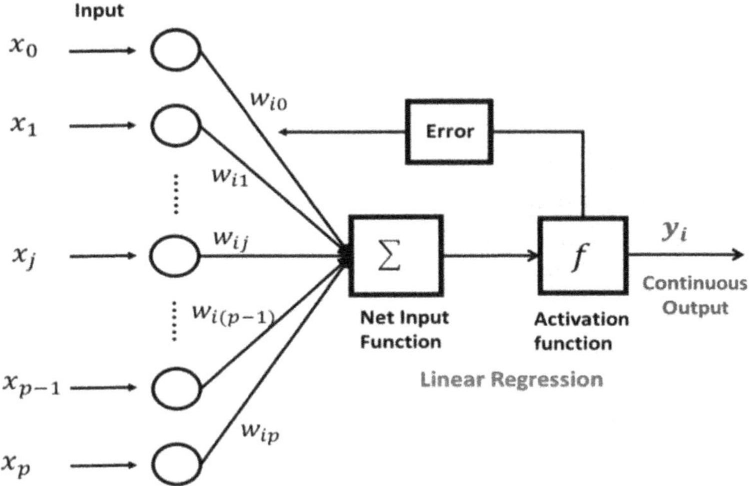

Figure 4. Demonstration of linear regression's learning process.

2.6. Stacking Ensemble

Though coming from the same branch, stacking is quite different from the random forest and boosting strategy in AdaBoost in several ways. In bagging, variance in the final ensemble model is reduced by the random selection of a subset of features as well as instances for each predictor to execute the parallel and independent learning. The outcomes are then aggregated by the averaging method to generate an ensemble prediction. Boosting, on the other hand, will pass the dataset through all the learners which are set sequentially. Each instance and learner are given the attribute, the so-called weight, that is going to be updated on each pass (instance) and each iteration (learner). The weighting procedure results in the uneven contribution of each learner to the final prediction, and uneven prioritization to each instance for the training process - which substitutes the output averaging process mechanism and randomization for training subset in the bagging concept.

For stacking, each base predictor takes the whole dataset without any differentiation on the input and works in the canonical way to produce the result. The special property of this method lies in the aggregation mechanism. After the learning, the outputs from the predictors then become the inputs for the aggregator algorithm to produce the final prediction. The training set in the first learning process occupied by the base predictors is different with the one utilized by the aggregator algorithm because the dataset fed into the predictors has been transformed into the models which are later combined to form the new features. Fitting the aggregator algorithm onto the same instances causes a bias since the inputs are created from these instances. However, splitting two types of datasets raises another problem for a limited amount of data. To overcome this, the common k-folds cross-validation approach is usually adopted to provide more data for training both predictors and aggregators thereby facilitate a more accurate performance measure [17]. In practice, stacking usually considers multiple types of learners to build the prediction, while bagging and boosting are more common to have only homogenous learners. Besides the algorithms used, the design of stacking ensemble can also be altered by the stacking level. If the number of levels is more than 2, the layer in the middle will be filled with multiple aggregators. However, since increasing the number of levels will cost on the time computation, this parameter usually remains in default (i.e., level size = 2).

3. Implementation Methodology

The methodology in this study consists of the following procedures: data collection and preprocessing, feature selection, time windowing, and model building. All the machine learning models exploited in this study will be constructed on the open-source data mining platform, Orange, a software programmed under the python script. In this section, the details of procedures will be discussed respectively.

3.1. Data Collection

The main pollutant emissions in Taiwan are due to energy production industry, traffic, waste incineration and agriculture. In Taiwan, six pollutants (O_3, $PM_{2.5}$, PM_{10}, CO, SO_2, and NO_2) are monitored and controlled based on their concentration time-series. Types of data used as predictors to perform analysis involve AQ: air quality data, MET: meteorological data, and TIME: the day of the month, day of the week, and the hour of the day. From 1 January 2008 to 31 December 2018, air quality data are collected from several monitoring stations across Taiwan and reported via the EPA's website [18]. With the same timeframe, meteorological data are provided in 1-h intervals by Taiwan's Central Weather Bureau (CWB) from three air monitoring stations: Zhongli (Northern Taiwan), Chuanghua (Central Taiwan), and Fengshan (Southern Taiwan). The datasets represent different environmental conditions related to air pollutant concentration.

3.2. Data Pre-Processing

The number of raw data points for the Zhongli, Changhua, and Fengshan monitoring stations includes 91,672, 94,453, and 94,145, respectively. The analysis of these readings begins with a crucial phase – data preprocessing. Various preprocessing operations precede the learning phase. At any particular time, one invalid variable will not affect the whole data group, and thus it will just be either marked blank or, where available, replaced by a value sourced from the CWB, without eliminating the full row. The missing values are treated by imputation to recover the corresponding values. Given the lack of spatial proximity of the readings to the original monitoring stations, the missing values are imputed for relative humidity, temperature, and rainfall, without using wind speed or wind direction. The next imputation process used the k-NN algorithm to substitute the rest of the invalid or missing data that did not qualify for the previous imputation process. Note that the percentage of missing values is lower than 1.3% in all three-station datasets.

Then, input and target data are normalized to eliminate potential biases; thus, variable significance won't be affected by their ranges or their units. All raw data values are normalized to the range of [0, 1].

Inputs with a higher scale than others will tend to dominate the measurement and are consequently given greater priority. Normalization not only improves the model learning rate, but also supports k-NN algorithm performance because the imputation is decided by the distance measure.

3.3. Feature Engineering

In regard to selecting features in the predictive models, the hourly AQI readings with the highest index out of 6 pollutants: O_3, $PM_{2.5}$, PM_{10}, CO, SO_2, and NO_2 are selected. To convert the time-window-specific concentration of 6 pollutants, the AQI Taiwan Guidelines [18] are adopted and the AQI is manually calculated using the following Equations (1) and (2), where index values of O_3, $PM_{2.5}$, and PM_{10} are needed to define AQI in Taiwan, and the lack of one or more of these values will significantly reduce the accurate assessment of current air quality.

$$AQI = \begin{cases} max\{I_{O3}, I_{PM2.5}, I_{PM10}, I_{CO}, I_{SO2}, I_{NO2}\}, I_{O3}, I_{PM2.5}, I_{PM10} \neq \varnothing \\ \varnothing, otherwise \end{cases} \quad (1)$$

Pollutant concentration ($value_i$) is converted to pollutant index (I_i) by the following formula:

$$I_i = LB_j + \frac{value_i - lb_i}{ub_i - lb_i} \times \left(UB_j - LB_j\right) \quad (2)$$

where i = O_3, $PM_{2.5}$, PM_{10}, CO, SO_2, NO_2; j denotes which level in AQI system occupied by the concentration of the specific pollutant using categories of good, moderate, unhealthy which includes specific groups, unhealthy, very unhealthy, and hazardous. The data transformation defines the time-window-specific concentration to calculate I_i values. For example, based on the AQI from Taiwan's EPA website [18], the concentration $value_{O3}$ = 0.06 ppm will fall in the interval with lb_{O3} = 0.055 ppm and ub_{O3} = 0.070 ppm corresponding to the "moderate" pollutant level with $LB_{moderate}$ = 51 and $UB_{moderate}$= 100. The $value_{O3}$ is defined by matching either of two conditions: if the 8-h average concentration is more precautionary for a specific site and is also below 0.2 ppm, then this value is used; otherwise, the 1-h average concentration will be considered. Both $value_{PM2.5}$ and $value_{PM10}$ are the moving average values which consider two time-windows, i.e., the last 12 h and 4 h (see Table 1). Other variables, such as $value_{CO}$ and $value_{NO2}$ only account for a single time window, i.e., last 8 h and 1 h, respectively. Meanwhile, $value_{SO2}$ emphasizes the 24-h average concentration if the 1-h average concentration exceeds 185 ppb; otherwise, the 1-h average value will be used.

The AQI mechanism introduces several new variables to train the prediction model (Table 1). For several pollutants, time windows other than hourly are more sensitive in determining AQI; therefore, the prediction interval related to the accuracy of long-term predictions is under investigation to clarify the time dependency between consecutive data points. As the AQI calculation is already established, the future value of the AQI readings in three different time intervals will be regarded as target variables and are summarized in Table 2.

3.4. Performance Evaluation

According to Isakndaryan et al. [19], the most used metrics are RMSE (root mean squared error) and MAE (mean average error), calculated based on the difference between the prediction result and the true value, while another metric, R^2 (R-squared) is essential to explain the strength of the relationship between predictive models and target variables [20]. These three metrics provide a baseline for comparative analysis across different parameter settings for each model and across different methods. However, performance validation leads to a bias when the data set is split, trained, and tested only one time. This also means the result drawn from the testing dataset may no longer be valid after the testing subset is changed. To overcome this problem, each model is re-built 20 times using different random subsets of training and testing samples. The splitting proportion remains the same (80:20).

All metrics report only a single value from the average performance of 20 identical models validated into 20 different subsets of testing instances.

Table 1. Other features added to the prediction model.

No	Feature	Type	Description
1	O_3 8-h	Numeric	Calculated based on O_3 average of last 8 h
2	PM_{10} moving average	Numeric	Calculated as follows: (0.5× average of PM_{10} in the last 12 h)+(0.5× average of PM_{10} in the last 4 h)
3	$PM_{2.5}$ moving average	Numeric	Calculated using the same rule as the PM_{10} moving average
4	CO 8-h	Numeric	The average concentration for the last 8 h
5	AQI index	Numeric	AQI value based on the maximum index between the AQI pollutants (PM_{10}, $PM_{2.5}$, NO_2, SO_2, O_3, and CO)

Table 2. Description of target variables.

No.	Target	Type	Description
1	F1-AQI	Numeric	AQI index for the next 1 h
2	F8-AQI	Numeric	AQI index for the next 8 h
3	F24-AQI	Numeric	AQI index for the next 24 h

4. Results and Discussion

This section is organized into three parts. First, a general description of the dataset is provided. The datasets are mainly based on geographic distribution across Taiwan. The second part discusses the detailed development of AQI prediction models following their parameter setting and imputation. The last part evaluates the performance of the AQI forecasting models.

4.1. Data Summary

In the Zhongli dataset, moderate is the most frequent AQI level in any given month (Figure 5a). *Unhealthy* occurs more frequently in December through April, indicating that peak pollution usually occurs in winter and spring. The year-based grouping (Figure 5b) clearly shows a general drop in pollution levels from 2014 to 2018, with a small uptick in 2016. In general, the moderate class accounts for 51% cases while good and unhealthy, respectively, account for 28% and 21%.

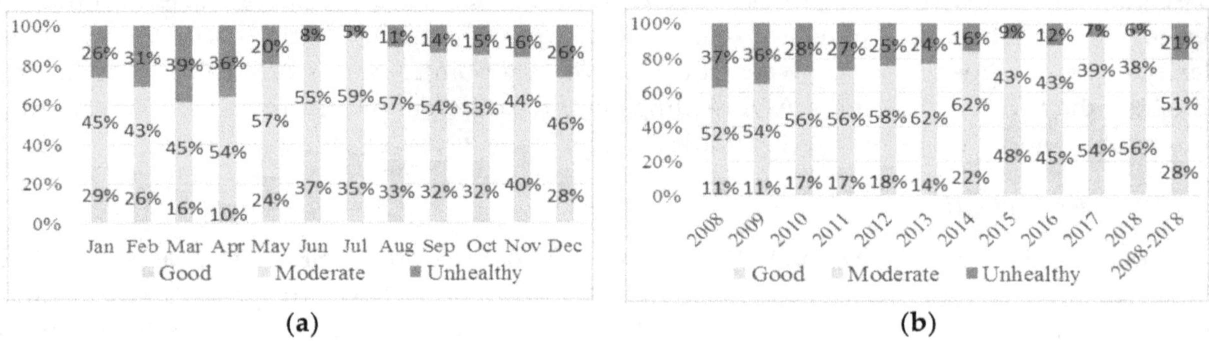

Figure 5. Composition of AQI classes in Zhongli: (**a**) Month-based; (**b**) Year-based and Overall-based.

Similar to the Zhongli AQI pattern, pollution in Changhua peaks in March (Figure 6a). However, the degree of air pollution is more severe in Changhua, with unhealthy accounting for 59% of March readings, as opposed to 39% for Zhongli. Like Zhongli, higher AQI levels in Changhua are also clustered in winter and spring, but September, October, and November also featured significant instances of the unhealthy class (respectively 35%, 38%, and 41%). In general, Changhua has poorer air quality than Zhongli, with more frequent AQI > 100 incidents both monthly and annually. However, the full-year AQI readings in Figure 6b show that air quality has gradually improved over time, with a 34% drop in instances of unhealthy from 2008 to 2018.

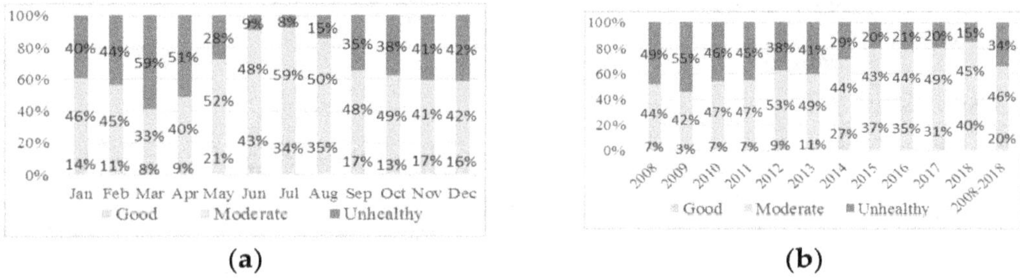

Figure 6. Composition of AQI classes in Changhua: (**a**) Month-based; (**b**) Year-based and Overall-based.

Southern Taiwan, especially Kaohsiung City, is notorious for its poor air quality due not only to emissions from nearby industrial parks but from particulate matter blowing in from China and Southeast Asia. Figure 7a,b shows significant instances of the unhealthy class (red bars) air quality for most of the year, with reduced pollution levels only in May to September. The worst air quality is concentrated in December and January (respectively 78% and 80% unhealthy).

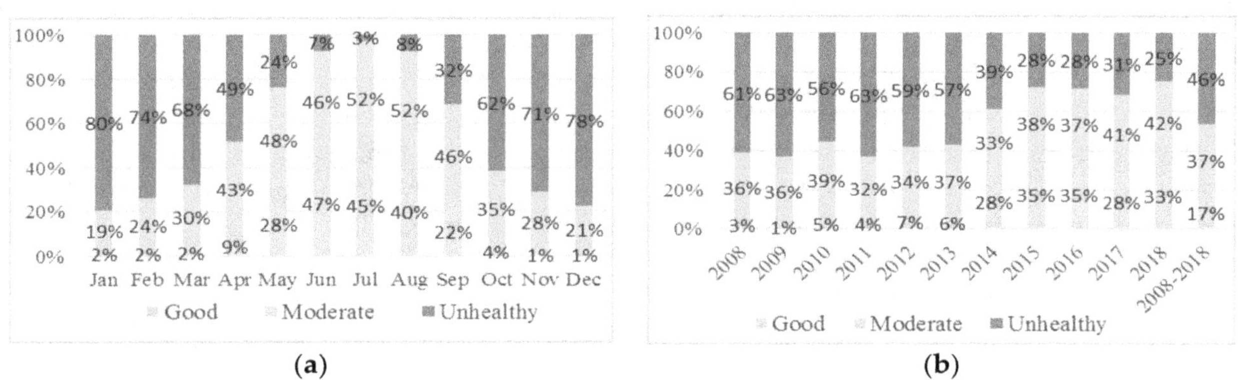

Figure 7. Composition of AQI classes in Fengshan: (**a**) Month-based; (**b**) Year-based and Overall-based.

The winter spike in air pollution is partly due to seasonal atmospheric phenomena that trap air pollution closer to the ground for extended periods. From October to March, Fengshan air quality readings are *good* less than 5% of the time. In terms of year-based AQI class composition, not much improvement is seen until in 2014–2015 with a sharply declined *unhealthy* scores after which levels remain relatively stable. Overall, for the 11 years, the Fengshan dataset is dominated by AQI > 100 (46%) followed by $51 \leq$ AQI ≤ 100 (37%), and AQI ≤ 50 (17%).

4.2. AQI Prediction Model

Table 3 specifies the design of the parameters used to generate the prediction models for all dataset (Zhongli, Changhua, and Fengshan). Note that each particular constant for each dataset supposedly contains three values. However, to ease the documentation, any similar value being used across all datasets or at least across different time steps will be written only once. For example, Changhua dataset which uses the number of trees (i.e., 100) in AdaBoost for all time step categories. Additionally, parameter m in the random forest has only one value in all models. To be able to evaluate the ability of each model in accomplishing the task, 80% of data points will be fed into each training process, while the remaining 20% are spared for the testing purpose.

Table 3. Parameter Design of ML Methods.

Method		Parameter Design (F1/F8/F24)		
		Zhongli	Changhua	Fengshan
Random Forest		No. of Trees = 100/200/200 Min. observation = 6/6/3	No. of Trees = 200 m= 4 Min. observation = 6/3/3	No. of Trees = 200/100/200 Min. observation = 6/3/3
AdaBoost		α = 0.8/0.9999/0.9	# of Trees= 100 α = 0.8/0.9/0.9999	α = 0.8/0.9/0.8
SVM	Linear	C = 3/0.1/0.12, v = 0.5	C = 3/0.12/0.1, v = 0.5	C = 3/0.12/0.1, v = 0.5/0.5/0.9
	Polynomial	C = 3/0.7/0.9, v = 0.5/0.2/0.1, γ = auto, c = 3/5/3, d = 1	C = 3/0.9/0.9, v = 0.5/0.2/0.1, γ = auto, c = 3, d = 1	C =3/0.9/0.9, v = 0.5/0.2/0.9, γ = auto, c = 3, d = 1
	RBF	C = 3/1/1, v = 0.5, γ = auto	C = 3/3/1, v = 0.5, γ = auto	C = 3/3/1, v = 0.5/0.5/0.2, γ = auto
ANN		Max. # of Iterations = 3000 Activation function: Identity; Optimizer: L-BFGS-B; No. of Input neurons = 24; No. of Hidden neurons over layers = 50/50/50; No. of Output neurons = 1		
Stacking Ensemble		α = 0.0001; Max. of Iterations = 300 Regularization: L2 Ridge regression; α = 0.3		

Table 4 describes the evaluation results of Zhongli F1-AQI prediction using 5 methods with and without imputation. It can be inferred that machine learning algorithms performed very well in predicting future AQI levels in Zhongli for the following hour. The linear kernel is shown to be the best input transformation technique for SVM, with R^2 results of 0.953 (without imputation) and 0.963 (with imputation). Imputation allows SVM to produce improvement in all evaluation metrics. Furthermore, in terms of MAE score, SVM-RBF outperforms SVM-Linear, but the opposite is true for the RMSE score. This may be due to RBF having more samples with a larger prediction error despite a smaller average error (larger errors produce a greater penalty for RMSE).

Table 4. Results of ML Algorithms for Zhongli F1-AQI Prediction.

Method	Without Imputation			With Imputation		
	RMSE	MAE	R^2	RMSE	MAE	R^2
SVM-Polynomial	9.836	8.275	0.923	8.145	6.827	0.947
SVM-RBF	9.298	5.119	0.931	8.832	4.617	0.938
SVM-Linear	7.659	6.050	0.953	6.790	5.217	0.963
Random Forest	3.255	2.208	0.992	3.257	2.207	0.992
AdaBoost-Square	3.291	2.187	0.991	3.337	2.185	0.991
AdaBoost-Linear	3.328	2.191	0.991	3.308	2.189	0.991
AdaBoost-Exponential	3.336	2.193	0.991	3.327	2.193	0.991
ANN	3.572	2.438	0.990	3.378	2.396	0.991
Stacking Ensemble	3.236	2.196	0.992	3.243	2.199	0.992

The performance of random forest, AdaBoost, ANN, and stacking ensemble algorithm are all comparable. Random forest and stacking ensemble algorithm obtain slightly better R^2 performance (0.001). Unlike with SVM, imputation does not affect the prediction results for AdaBoost, random forest or the stacking ensemble algorithm, indicating their robustness to missing data. On the other hand, imputation only provides a small degree of improvement on ANN, resulting in tied R^2 values with AdaBoost. Several loss regression functions (square, linear, exponential) are tested on AdaBoost but without a decisive performance outcome due to efforts to avoid bias since the interpretation could be distorted by randomness, especially given very minor degrees of difference.

Table 5 summarizes the results for the 8-h Zhongli AQI prediction. The R^2 value of 0.764 is the best value obtained by the stacking ensemble method. Nonetheless, the performance of SVM becomes worse with an R^2 value less than 0.6 across all kernels. The values of MAE and RMSE are 17 and 23, respectively. However, ANN and random forest perform better than SVM, with R^2 scores exceeding 0.7 and error metrics just slightly lower than those obtained with AdaBoost and stacking ensemble. The results match the expectation since the uncertainty increases with the longer period and leads to higher difficulty in the forecast. The study also finds that the overall values are worse than that of the F1-AQI prediction.

Table 5. Results of ML Algorithms for Zhongli F8-AQI Prediction.

Method	Without Imputation			With Imputation		
	RMSE	MAE	R^2	RMSE	MAE	R^2
SVM-Polynomial	24.308	17.981	0.526	23.244	17.135	0.567
SVM-RBF	23.375	17.283	0.562	23.358	17.278	0.563
SVM-Linear	24.262	18.327	0.528	26.674	20.174	0.430
Random Forest	17.471	12.408	0.755	17.477	12.413	0.755
AdaBoost-Square	17.386	11.801	0.758	17.352	11.788	0.759
AdaBoost-Linear	17.273	11.693	0.761	17.221	11.679	0.762
AdaBoost-Exponential	17.283	11.691	0.761	17.284	11.685	0.761
ANN	18.786	13.502	0.717	18.759	13.486	0.718
Stacking Ensemble	17.167	11.804	0.764	17.178	11.799	0.764

Table 6 shows that no method used for targeting F24-AQI prediction produced an R^2 score above 0.6, with the lowest score of 0.091. Simply put, the yielded predictions fit the dataset poorly. Stacking ensemble still ranks first, but the R^2 gap to the second-best method (AdaBoost-Linear) is larger than in the previous cases. SVM performance tracked far behind the other methods with the highest score for evaluation metrics obtained by RBF kernel. However, the R^2 score is so low that the SVM method is considered not preferable for 24-h prediction.

Table 6. Results of ML Algorithms for Zhongli F24-AQI Prediction.

Method	Without Imputation			With Imputation		
	RMSE	MAE	R^2	RMSE	MAE	R^2
SVM-Polynomial	33.639	24.799	0.098	34.194	25.034	0.068
SVM-RBF	30.635	23.340	0.252	30.335	23.053	0.267
SVM-Linear	37.001	28.904	0.091	36.835	28.595	0.081
Random Forest	24.974	18.648	0.503	25.007	18.667	0.502
AdaBoost-Square	24.219	16.724	0.533	24.226	16.753	0.532
AdaBoost-Linear	24.039	16.586	0.540	24.074	16.614	0.538
AdaBoost-Exponential	24.053	16.574	0.539	24.099	16.620	0.537
ANN	29.150	21.957	0.323	29.113	21.927	0.325
Stacking Ensemble	23.825	16.667	0.548	23.831	16.693	0.548

Predictive model results for F1-AQI Changhua are similar to those for F1-AQI Zhongli. Stacking ensemble, AdaBoost, and random forest provide the best performance for one-hour AQI level prediction (see Table 7). These algorithms perform better for all evaluation metrics in Changhua than in Zhongli. Also, the imputation process reduces the performance of SVM, but not the other algorithms.

When it comes to the F8-AQI prediction (as shown in Table 8), the Changhua prediction again outperforms that of Zhongli. AdaBoost and stacking ensemble both yield R^2 scores exceeding 0.8. Without imputation, stacking ensemble outperforms the other methods. However, with imputation, AdaBoost performance is comparable to that of stacking ensemble. SVM-linear gives the highest MAE and RMSE results, i.e., 23.412 and 31.189, respectively. These error metrics can be further reduced to 19.623 and 25.628 by imputation.

In the Zhongli dataset, the time step selection affects the performance of machine learning methods, and this is consistent with the results for the F24-AQI prediction models in Changhua (Table 9). Declination occurs across all models with a very low R^2. The SVM-Polynomial gives the worst performance for the imputed dataset and an MAE value exceeding 30, and an RMSE value exceeding 40. The best performance is still obtained by the stacking ensemble method, with an R^2 score of 0.605, and MAE and RMSE values respectively below 19 and 26. Among all kernels used by SVM, the radial basis function appears to be the most effective for 24-h AQI predictions. Moreover, AdaBoost-exponential slightly underperforms stacking ensemble in terms of R^2 and RMSE, but consistently provides better MAE results.

Table 7. Results of ML Algorithms for Changhua F1-AQI Prediction.

Method	Without Imputation			With Imputation		
	RMSE	**MAE**	R^2	**RMSE**	**MAE**	R^2
SVM-Polynomial	12.419	11.09	0.907	14.116	12.683	0.880
SVM-RBF	9.672	4.639	0.944	9.596	4.497	0.944
SVM-Linear	9.169	7.055	0.949	10.033	7.638	0.939
Random Forest	3.059	2.055	0.994	3.105	2.066	0.994
AdaBoost-Square	3.093	2.046	0.994	3.126	2.054	0.994
AdaBoost-Linear	3.089	2.043	0.994	3.115	2.048	0.994
AdaBoost-Exponential	3.093	2.046	0.994	3.126	2.054	0.994
ANN	3.914	2.505	0.991	3.870	2.541	0.991
Stacking Ensemble	3.039	2.043	0.994	3.076	2.057	0.994

Table 8. Results of ML Algorithms for Changhua F8-AQI Prediction.

Method	Without Imputation			With Imputation		
	RMSE	**MAE**	R^2	**RMSE**	**MAE**	R^2
SVM-Polynomial	26.225	20.082	0.585	25.818	19.919	0.598
SVM-RBF	25.548	19.422	0.606	25.730	19.597	0.600
SVM-Linear	31.189	23.412	0.413	25.628	19.623	0.603
Random Forest	18.435	13.711	0.795	18.423	13.707	0.795
AdaBoost-Square	17.877	12.747	0.807	17.871	12.734	0.807
AdaBoost-Linear	17.825	12.732	0.808	17.810	12.718	0.809
AdaBoost-Exponential	17.822	12.733	0.808	17.815	12.729	0.808
ANN	20.451	15.329	0.748	20.312	15.213	0.751
Stacking Ensemble	17.801	12.855	0.809	17.792	12.856	0.809

Table 9. Results of ML Algorithms for Changhua F24-AQI Prediction.

Method	Without Imputation			With Imputation		
	RMSE	**MAE**	R^2	**RMSE**	**MAE**	R^2
SVM-Polynomial	40.662	31.689	0.006	41.834	32.727	0.052
SVM-RBF	34.879	26.977	0.269	34.852	26.948	0.270
SVM-Linear	37.451	29.047	0.157	37.092	28.703	0.173
Random Forest	26.765	20.281	0.570	26.786	20.299	0.569
AdaBoost-Square	26.282	18.781	0.585	26.288	18.799	0.585
AdaBoost-Linear	25.786	18.204	0.600	25.817	18.246	0.599
AdaBoost-Exponential	25.747	18.144	0.602	25.773	18.185	0.601
ANN	30.919	23.753	0.426	30.803	23.647	0.430
Stacking Ensemble	25.630	18.255	0.605	25.655	18.294	0.604

Table 10 summarizes the results for the one-hour prediction model without and with *k*-NN imputation step in the Fengshan dataset. Stacking ensemble learning outperforms other techniques in terms of RMSE and R^2, while SVM obtains the worst performance in every prediction case. However, imputation slightly enhances the results, particularly for the RBF and linear kernels, but not for the polynomial kernel which shows a performance decline using the imputed dataset. Also note that while comparing with the results in the other two cities, Zhongli and Changhua, Fengshan shows the best performance in all evaluation measures.

Table 10. Results of ML Algorithms for Fengshan F1-AQI Prediction.

Method	Without Imputation			With Imputation		
	RMSE	MAE	R^2	RMSE	MAE	R^2
SVM-Polynomial	7.072	5.556	0.974	7.275	5.821	0.973
SVM-RBF	9.119	5.542	0.957	8.324	4.702	0.964
SVM-Linear	9.529	7.400	0.953	8.485	6.621	0.963
Random Forest	2.971	1.869	0.995	2.979	1.868	0.995
AdaBoost-Square	3.020	1.771	0.995	2.996	1.766	0.995
AdaBoost-Linear	2.995	1.767	0.995	2.983	1.760	0.995
AdaBoost-Exponential	3.020	1.771	0.995	2.996	1.766	0.995
ANN	3.966	2.544	0.992	3.821	2.585	0.992
Stacking Ensemble	2.925	1.823	0.996	2.921	1.814	0.996

In terms of eight-hour prediction, imputation has a significant impact on SVM-Linear, increasing R^2 from 0.318 to 0.546 (as shown in Table 11). RMSE and MAE are also improved by 10% and shift closer to the performance of other SVM kernels. Of the three locations, application of machine learning algorithms has the biggest impact on 8-h predictions in Fengshan, with stacking ensemble providing the greatest improvement, followed by AdaBoost, random forest, ANN, and SVM. This sequence is consistent for all results.

Table 11. Results of ML Algorithms for Fengshan F8-AQI Prediction.

Method	Without Imputation			With Imputation		
	RMSE	MAE	R^2	RMSE	MAE	R^2
SVM-Polynomial	25.658	19.806	0.659	25.512	19.606	0.663
SVM-RBF	25.810	20.392	0.655	25.665	20.248	0.659
SVM-Linear	36.292	26.859	0.318	29.598	23.028	0.546
Random Forest	16.634	12.111	0.857	16.606	12.100	0.857
AdaBoost-Square	16.440	11.399	0.860	16.498	11.391	0.859
AdaBoost-Linear	16.367	11.364	0.861	16.339	11.367	0.862
AdaBoost-Exponential	16.387	11.373	0.861	16.398	11.367	0.861
ANN	19.112	14.285	0.811	18.975	14.182	0.814
Stacking Ensemble	16.302	11.517	0.862	16.279	11.527	0.863

As summarized in Table 12, for the 24-h predictions in Fengshan, while overall SVM results are not promising, the other methods show quite acceptable evaluation scores. The top three methods (stacking ensemble, AdaBoost, and random forest) obtained R^2 scores exceeding 0.71 for which the MAE and RMSE results are comparable to the F8-AQI prediction for Fengshan. Surprisingly, the stacking ensemble is found to be affected by imputation but, even with imputation, the MAE value is still higher than that of all AdaBoost versions (linear, square, and exponential). AdaBoost and stacking ensemble show consistent results, and AdaBoost generally obtains worse RMSE and R^2 but better MAE.

4.3. Implementation of AQI Forecasting Model

This section describes a simulation-like AQI forecasting using stacking ensemble and AdaBoost (the two best methods from the analyses in Section 4.2) as backend techniques. Each prediction is accompanied by a prediction interval (PI) within a 95% confidence level, which describes a given tolerance for the prediction value such that there is 95% chance that the actual observation could fall within this range. The prediction interval is calculated using the formula below [21]:

$$PI = z_{\alpha/2} \times \sigma(3) \quad (3)$$

where σ represents the standard deviation of the residual errors defined as [22]:

$$\sigma = \sqrt{\frac{1}{n-2}\sum_{i=1}^{n}(y_i - \hat{y}_i)^2} \tag{4}$$

Prediction intervals that reflect the uncertainty of a model's output should be adjusted dynamically as new observations are received every hour, thus ensuring that the prediction interval is always current. The one-month samples (December 2018) from the Zhongli dataset are used to obtain the standard deviation.

Table 12. Results of ML Algorithms for Fengshan F24-AQI Prediction.

Method	Without Imputation			With Imputation		
	RMSE	MAE	R^2	RMSE	MAE	R^2
SVM-Polynomial	35.203	28.025	0.357	35.330	28.140	0.353
SVM-RBF	37.696	30.300	0.263	35.368	28.485	0.351
SVM-Linear	35.954	28.520	0.329	36.511	28.763	0.309
Random Forest	23.388	17.476	0.716	23.384	17.485	0.716
AdaBoost-Square	22.935	15.932	0.727	22.927	15.939	0.727
AdaBoost-Linear	22.663	15.743	0.734	22.654	15.753	0.734
AdaBoost-Exponential	22.708	15.777	0.733	22.723	15.790	0.732
ANN	27.008	20.542	0.622	26.882	20.416	0.625
Stacking Ensemble	22.872	16.372	0.729	22.618	16.105	0.735

As shown in Figure 8a, the higher the prediction time step, the wider the tolerance needed to represent the estimation. AdaBoost and stacking ensemble outperform the other techniques tested in the previous section, obtaining similar predictions and prediction intervals. The predictions here are all based on authentic data, where the best models in each prediction category are reused. Figure 8b shows another forecast constructed during winter, providing an example of poor air quality cases captured in the prediction of F1-AQI, F8-AQI, and F24-AQI using AdaBoost and stacking ensemble.

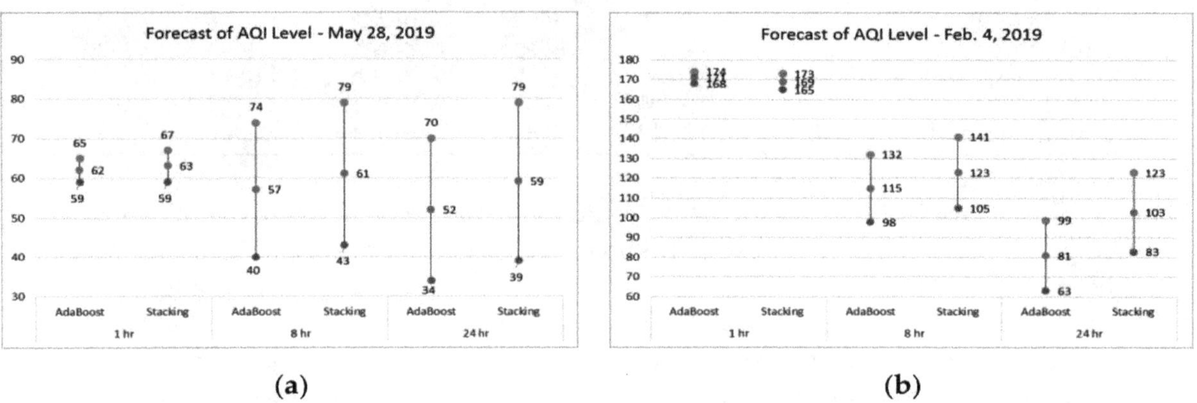

(a) (b)

Figure 8. Forecast of AQI: (**a**) on 28 May 2019, 07:00; (**b**) on 4 February 2019, 03:00.

Figure 9 provides an illustration on how the information will be provided and visualized given a sample of upcoming data for the monitoring and forecasting of the air quality. Noted that as shown by the graph, the higher the time step of prediction the wider the tolerance needed to escort the estimation. AdaBoost and stacking are two methods that outperform other techniques tested in the previous section. Their predictions are close to each other and so are the prediction intervals. The predictions here are based on the real scheme, where the best models of them in each category of prediction were reused again by incorporating the actual values from 24 features.

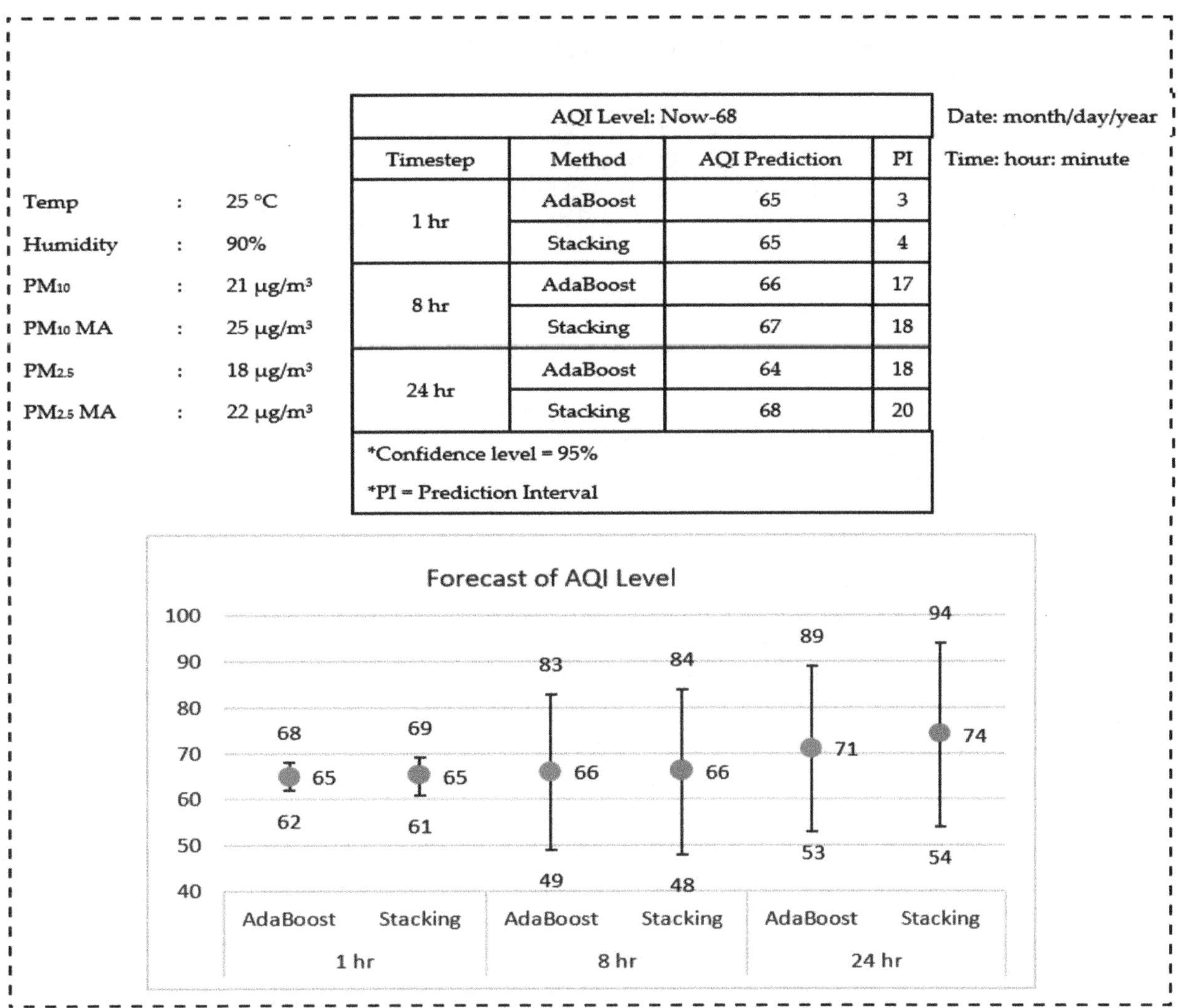

AQI Level: Now-68			
Timestep	Method	AQI Prediction	PI
1 hr	AdaBoost	65	3
	Stacking	65	4
8 hr	AdaBoost	66	17
	Stacking	67	18
24 hr	AdaBoost	64	18
	Stacking	68	20

*Confidence level = 95%

*PI = Prediction Interval

Figure 9. Illustration of Air Quality Monitoring and Forecasting System.

5. Conclusions

Applying artificial intelligence methods provides promising results for AQI forecasting. This study obtained data collected by EPA and CWB of Taiwan over 11 years. Three regions (North: Zhongli, Central: Changhua, South: Fengshan) in Taiwan were considered, including two notorious places (Changhua and Fengshan) for their bad air quality all year round. With good results for R^2, stacking ensemble and AdaBoost offer the best performance of target predictions based on three different datasets. To be more specific, the stacking ensemble delivers the best RMSE results, while AdaBoost provides the best MAE results. All results show that SVM yields the worst results among all methods explored, and only provides meaningful results for 1-h predictions. The results also confirm that the two machine learning methods, AdaBoost and stacking ensemble, employed in this study can outperform popular methods in the literature, such as SVM, random forest, and ANN. In other words, AdaBoost and stacking ensemble can be considered new and superior alternatives for AQI forecast.

This study also indicates that prediction performance varies over different regions in Taiwan. Comparing results from datasets sourced from three different regions displays best results for Fengshan AQI prediction (Southern Taiwan), where performance decay with increased time step is less pronounced than those in Zhongli (north) and Changhua (central). Also, 95% confidence intervals for 1-h, 8-h and 24-h forecast are calculated, respectively. Compared to the single value prediction, the 95% C.I. can provide a better reference to the decision-maker. For example, an event planner can decide if the outdoor activities can go on based on the air quality forecast with better confidence. Future work should focus on improving performance using stacking ensemble, AdaBoost and random forest with hyperparameter optimization, particularly for predictions with larger time steps (F8-AQI and F24-AQI).

Author Contributions: Conceptualization, Y.M., J.R.C.J., and Y.-C.L.; methodology, Y.M. and Y.-C.L.; software, Y.M.; validation, Y.M. and Y.-C.L.; formal analysis, Y.M.; data curation, Y.M.; writing—original draft preparation, Y.M.; writing—review and editing, Y.-C.L. and A.H.-L.C.; visualization, Y.M.; supervision, Y.-C.L.; project administration, Y.-C.L.; funding acquisition, Y.-C.L. All authors have read and agreed to the published version of the manuscript.

References

1. World Health Organization. Air Pollution. Available online: https://www.who.int/health-topics/air-pollution#tab=tab_1/ (accessed on 13 March 2020).
2. Ghorani-Azam, A.; Riahi-Zanjani, B.; Balali-Mood, M. Effects of Air Pollution on Human Health and Practical Measures for Prevention in Iran. *J. Res. Med. Sci.* **2016**, *21*, 1–12.
3. Conticini, E.; Frediani, B.; Caro, D. Can Atmospheric Pollution Be Considered a Co-factor in Extremely High Level of SARS-CoV-2 Lethality in Northern Italy? *Environ. Pollut.* **2020**, *261*, 114465. [CrossRef] [PubMed]
4. Rybarczyk, Y.; Zalakeviciute, R. Machine Learning Approaches for Outdoor Air Quality Modelling: A Systematic Review. *Appl. Sci.* **2018**, *8*, 2570. [CrossRef]
5. Raimondo, G.; Montuori, A.; Moniaci, W.; Pasero, E.; Almkvist, E. A Machine Learning Tool to Forecast PM_{10} Level. In Proceedings of the Fifth Conference on Artificial Intelligence Applications to Environmental Science, San Antonio, TX, USA, 14–18 January 2007; pp. 1–9.
6. Garcia, J.M.; Teodoro, F.; Cerdeira, R.; Coelho, R.M.; Kumar, P.; Carvalho, M.G. Developing a Methodology to Predict PM_{10} Concentrations in Urban Areas Using Generalized Linear Models. *Environ. Technol.* **2016**, *37*, 2316–2325. [CrossRef] [PubMed]
7. Park, S.; Kim, M.; Kim, M.; Namgung, H.-G.; Kim, K.-T.; Cho, K.H.; H, K.; Kwon, S.-B. Predicting PM_{10} Concentration in Seoul Metropolitan Subway Stations Using Artificial Neural Network (ANN). *J. Hazard. Mater.* **2018**, *341*, 75–82. [CrossRef] [PubMed]
8. Yu, R.; Yang, Y.; Yang, L.; Han, G.; Move, O.A. RAQ A Random Forest Approach for Predicting Air Quality in Urban Sensing Systems. *Sensors* **2016**, *16*, 86. [CrossRef] [PubMed]
9. Yi, X.; Zhang, J.; Wang, Z.; Li, T.; Zheng, Y. Deep Distributed Fusion Network for Air Quality Prediction. In Proceedings of the 24th ACM SIGKDD International Conference on Knowledge Discovery and Data Mining, London, UK, 19–23 August 2018; pp. 965–973.
10. Veljanovska, K.; Dimoski, A. Air Quality Index Prediction Using Simple Machine Learning Algorithms. *Int. J. Emerg. Trends Technol. Comput. Sci.* **2018**, *7*, 25–30.
11. Muhammad, I.; Yan, Z. Supervised Machine Learning Approaches: A Survey. *Ictact J. Soft Comput.* **2015**, *5*, 946–952. [CrossRef]
12. Awad, M.; Khanna, R. Support Vector Regression. In *Efficient Learning Machines*; Apress: Berkeley, CA, USA, 2015.
13. Schölkopf, B.; Smola, A.J.; Williamson, R.; Bartlett, P. New Support Vector Algorithms. *Neural Comput.* **2000**, *12*, 1207–1245. [CrossRef] [PubMed]
14. Chang, C.-C.; Lin, C.-J. Training ν-Support Vector Regression: Theory and Algorithms. *Neural Comput.* **2002**, *14*, 1959–1977. [CrossRef] [PubMed]
15. Wu, X.; Srihari, R. New v-Support Vector Machines and Their Sequential Minimal Optimization. In Proceedings of the Twentieth International Conference on Machine Learning, Washington, DC, USA, 21–24 August 2003; AAAI Press: Washington, DC, USA, 2003; pp. 824–831.
16. Yu, L.; Wang, S.; Lai, K.K. Basic Learning Principles of Artificial Neural Networks. In *Foreign-Exchange-Rate Forecasting With Artificial Neural Networks*; Yu, L., Wang, S., Lai, K.K., Eds.; Springer: Boston, MA, USA, 2007; pp. 27–37.
17. Rocca, J. Ensemble Methods: Bagging, Boosting and Stacking. Available online: https://towardsdatascience.com/ensemble-methods-bagging-boosting-and-stacking-c9214a10a205 (accessed on 23 April 2019).

18. Taiwan's Environmental Protection Administration. Taiwan Air Quality Monitoring Network. Available online: https://taqm.epa.gov.tw/taqm/en/b0201.aspx (accessed on 13 March 2020).
19. Iskandaryan, D.; Ramos, F.; Trilles, S. Air Quality Prediction in Smart Cities Using Machine Learning Technologies Based on Sensor Data: A Review. *Appl. Sci.* **2020**, *10*, 2401. [CrossRef]
20. Dufour, J.M. *Coefficients of Determination*; McGill University: Québec, QC, Canada, 2011.
21. Brownlee, J. Prediction Intervals for Machine Learning. Available online: https://machinelearningmastery.com/prediction-intervals-for-machine-learning/ (accessed on 30 May 2018).
22. Shrestha, D.L.; Solomatine, D.P. Machine Learning Approaches for Estimation of Prediction Interval for the Model Output. *Neural Netw.* **2006**, *19*, 225–235. [CrossRef] [PubMed]

A Novel Air Quality Evaluation Paradigm based on the Fuzzy Comprehensive Theory

Xinyue Mo [1,2], Huan Li [3], Lei Zhang [1,2,*] and Zongxi Qu [4]

1 College of Atmospheric Sciences, Lanzhou University, Lanzhou 730000, China; moxy16@lzu.edu.cn
2 Collaborative Innovation Center for Western Ecological Safety, Lanzhou 730000, China
3 School of Information Science and Engineering, Lanzhou University, Lanzhou 730000, China; lih17@lzu.edu.cn
4 School of Management, Lanzhou University, Lanzhou 730000, China; quzx@lzu.edu.cn
* Correspondence: zhanglei@lzu.edu.cn

Abstract: Air pollution is a prominent problem all over the world, seriously endangering human life. To protect the environment and human health, timely and accurate air quality evaluations are imperative. Recently, with the increasing focus on air pollution, an evaluation tool that can offer intuitive air quality information is especially needed. Though the Air Quality Index (AQI) has played this role over the years, its intrinsic limitations discussed in this study (sharp boundary, biased evaluation, conservative strategy and incomplete criterion) are gradually apparent, limiting its air quality evaluation capability. Therefore, a novel paradigm, the Air Quality Fuzzy Comprehensive Evaluation (AQFCE), is proposed. In the preprocessing module, missing and reversal data are handled by a least square piecewise polynomial fitting and linear regression. An improved fuzzy comprehensive evaluation model is adopted to solve the AQI's above limitations in the evaluation module. The early warning module provides a timely alert and recommendation. To validate the performance of the AQFCE, Beijing, Shanghai and Xi'an in China are selected for case studies, and daily and hourly concentration data of six conventional air pollutants from September 2018 to August 2019 are employed. For daily reports, the AQFCE and AQI have a high consistent rate and correlation coefficient regarding chief pollutants and levels, respectively, while examples show the level of the AQFCE is more reasonable. For hourly reports, AQI has antinomies and cannot reflect actual pollution, but the AQFCE is still effective. Current major pollutants, "weekend and holiday effect" and "peak type" of pollution are also revealed by the AQFCE. Experiment results prove that the AQFCE is accurate under different pollution conditions and an important supplement to the AQI. Furthermore, the AQFCE can provide health guideline for the public and assist the government in making environmental decisions and development policies.

Keywords: air quality evaluation; fuzzy math; abnormal data processing; pollution early warning; government decision-making

---

1. Introduction

In recent years, with the improvement of living standards, people have begun to pay more attention to the living environment and health. Air is an indispensable factor for human survival, and air quality is closely related to human health. However, air pollution is always an anxiety for many countries, which has become the focus of the world [1–5]. To understand the air quality information intuitively, air pollutant concentration can be simplified to a dimensionless index according to the impact on human health and the environment by the Air Quality Index (AQI). The US Environmental Protection Agency firstly proposed the Pollution Standards Index (PSI) in 1976 and renamed it the

AQI after revision in 1999 [6]. The European Union (EU) has begun to support the project of Common Information to European Air (CITEAIR) from March 2004, and the Common Air Quality Index (CAQI) was subsequently developed to support European cities to reach the air quality standard [7]. In 2012, the Ministry of Environmental Protection of China also issued the "Ambient Air Quality Standards" and "Technical Regulation on Ambient Air Quality Index (on trial)", which came into force in 2016 [8]. Although the AQI has been adopted by many countries and under implementation for many years, its limitations are gradually prominent with the increasingly higher demand for precision in air quality evaluation and pollution control. For example, actual air quality depends on the integrated effects of all air pollutants but the evaluation result of the AQI is decided only by the pollutant with the highest subindex. Thus, the evaluation based on the most severe pollutant may overestimate total pollution and bring unnecessary social panic and economic loss from severer control measures, which is biased and too conservative.

Therefore, some scholars tried to develop new methods. Plaia et al. [9] proposed a multisite-multipollutant AQI, which considers the conjoint effect of pollutants. Sowlat et al. [10] developed a fuzzy-based AQI, and a fuzzy inference system evaluates the impacts of ten pollutants. Moreover, a traffic AQI was established by Bagieński for assessing air quality near roadways [11]. Others have also made attempts to either revise the AQI or propose a new index [12–18]. However, these new methods have their shortcomings and do not deal with AQI limitations well. Revisions based on the similar principle of AQI cannot avoid its intrinsic limitations. The sharp boundary in classification is not reasonable enough, and a biased and conservative evaluation will overstate real pollution [6,9,10,12,19–21]. The evaluation result is highly sensitive to the calculation grid, and it is not suitable for distinct models to share the same calculation grid. So, designing a calculation grid that is scientific and compatible with the algorithm is necessary, but there are few relevant studies. It is essential to compare the new method with the accepted AQI in theory and practice, but it is scarce in previous studies. Furthermore, previous methods are too complicated to be used in practice.

Based on the above, an improved air quality evaluation method named Air Quality Fuzzy Comprehensive Evaluation (AQFCE) including preprocessing, evaluation and early warning is proposed in this study. In the preprocessing module, the abnormal data of air pollutants are handled by a least square piecewise polynomial fitting and linear regression. In the evaluation module, the limitations of the AQI (sharp boundary, biased evaluation, conservative strategy and incomplete criterion) are concluded based on sufficient comparison and analysis, and then, inspired by fuzzy math theory, a series of targeted solutions are proposed. Concretely, membership degree can deal with the ambiguity of data near the classification boundary. For the biased evaluation and overly conservative strategy, a principal factor prominent model considering the membership degree and factor weight of all pollutants is adopted to ensure both comprehensive evaluation and public health at the same time. A complete calculation grid based on scientific evidence is established to improve the original evaluation criterion. Moreover, according to the evaluation result, the early warning module can release comprehensive information in a timely manner about air pollution and offer suggestions for the public and government.

2. Current AQI

2.1. Calculation Method of AQI

Although the methods of calculating AQI may be slightly different in form among different countries, they are essentially same with linear interpolation between the class borders [19] and the pollutant with the highest subindex decides both the level of air quality and chief pollutant:

$$I_k = \frac{I_H - I_L}{C_H - C_L}(C_k - C_L) + I_L \tag{1}$$

$$AQI = \max\{I_1, I_2, I_3, \ldots, I_n\} \tag{2}$$

where:

I_k: the subindex of the kth pollutant;

C_k: the concentration of the kth pollutant;

$C_H(C_L)$: the concentration breakpoint that is $\geq$ ($\leq$) C_k;

$I_H(I_L)$: the index breakpoint corresponding to $C_H(C_L)$.

2.2. Calculation Criterion of AQI

Criterion, namely the calculation grid, is composed of pollutant concentration limits. It depends on specific requirement of air quality and is in accordance with local economic development and pollution control capacity. Therefore, criteria vary among regions and Table 1 presents their comparison in three representative regions: the European Union (EU), United States of America (USA) and China (CHN) [7,22,23]. As illustrated by Table 1, there are big differences in some aspects including level, period and limit value. The EU has five concise levels, but there are six levels and level VI still has two sublevels in the USA and CHN, which seem to be redundant. Six conventional pollutants are necessary items in air quality evaluation, so three criteria reach an agreement on pollutants but differ greatly in terms of the period. In addition, apart from $PM_{2.5}$ and PM_{10}, the EU has no daily criteria for other pollutants. The USA has neither daily criterion for NO_2 nor hourly criteria for CO, $PM_{2.5}$ and PM_{10}, while CHN has a shortage of hourly limits of particulate matters. Furthermore, the maximum 8 h moving average of CO (EU, USA) and O_3 (USA, CHN) are used for daily evaluations. At present, daily and hourly reports of air quality are primary tasks, but generally incompleteness and misuse exist in all criteria. As to limit value, the USA and CHN are relatively close to each other. At the same level, daily limits of CHN are close to or higher than that of the USA, which is similar for hourly limit. However, the EU has the most stringent criterion where all limits are far lower than that of the USA and CHN, which greatly protects public health and ecological environment.

2.3. Limitations of AQI

AQI has made remarkable contributions to air quality evaluation, which deserves affirmation indeed. However, at the same time its limitations exist objectively:

(1) Like all environmental quality evaluations, air quality evaluations are also full of ambiguities involving the evaluation object, factor, method, criterion and so forth. The sharp boundary adopted by the AQI in the classification scheme may not be reasonable [10,12,24,25]. A tiny difference (slight variation, monitoring error) of concentration near the boundary can lead to distinct evaluation results, and consequently the public and government may have to take totally different measures facing almost the same state of air, which is unnecessary.

(2) Actual air quality is decided by combined effects of various air pollutants, and six conventional pollutants have been included in monitoring. However, in the evaluation process of the AQI, as illustrated by Equation (2), the pollutant with the highest subindex determines air quality level alone, ignoring all other pollutants [6,9,19,20]. Furthermore, Equation (1) clearly shows that the subindex of pollutants is calculated separately according to its own criterion, lacking comparisons of relative hazards among different pollutants, which is insufficient.

Table 1. The comparison of criterion in EU, USA and CHN.

Level	Pollutant Concentration Limit (EU/USA/CHN)												
	SO_2 (ug/m^3)		NO_2 (ug/m^3)		CO (mg/m^3)			O_3 (ug/m^3)		$PM_{2.5}$ (ug/m^3)		PM_{10} (ug/m^3)	
	24 h	1 h	24 h	1 h	24 h	8 h	1 h	8 h	1 h	24 h	1 h	24 h	1 h
I	-/-/50	50/100/150	-/-/40	50/109/100	-/-/2	5/6/-	-/-/5	-/116/100	60/-/160	10/12/35	15/-/-	15/54/50	25/-/-
II	-/-/150	100/214/500	-/-/80	100/205/200	-/-/4	7.5/12/-	-/-/10	-/150/160	120/-/200	20/35.4/75	30/-/-	30/154/150	50/-/-
III	-/-/475	350/529/650	-/-/180	200/739/700	-/-/14	10/16/-	-/-/35	-/182/215	180/351/300	30/55.4/115	55/-/-	50/254/250	90/-/-
IV	-/-/800	500/869/800	-/-/280	400/1333/1200	-/-/24	20/19/-	-/-/60	-/225/265	240/437/400	60/150.4/150	110/-/-	100/354/350	180/-/-
V	-/1726/1600	>500/-/-	-/-/565	>400/2565/2340	-/-/36	>20/38/-	-/-/90	-/429/800	>240/866/800	>60/250.4/250	>110/-/-	>100/424/420	>180/-/-
VI	-/2297/2100	-/-/-	-/-/750	-/3386/3090	-/-/48	-/51/-	-/-/120	-/-/-	-/1080/1000	-/350.4/350	-/-/-	-/504/500	-/-/-
	-/2869/2620	-/-/-	-/-/940	-/4208/3840	-/-/60	-/63/-	-/-/150	-/-/-	-/1294/1200	-/500.4/500	-/-/-	-/604/600	-/-/-

(3) The overly conservative strategy adopted by the AQI does not mean the optimal solution [6,9,10,19,20]. Representing total air quality only by the chief pollutant will overestimate actual pollution, which means a series of stricter and unified prevention measures such as closing schools, vehicle restrictions, stopping or limiting production in industrial enterprise, etc. However, in fact, sensitive groups vary in specific pollutants and the conservative strategy could lead to unnecessary economic loss, social disorder and public panic. Therefore, a more accurate and comprehensive method is needed.

(4) Calculation criterion of the AQI proposed many years ago requires further improvements. Pollutant concentration limits should not be constant forever because they need to match the changing requirements of air quality and pollution control capability. Incomplete criterion cannot meet various tasks of air quality evaluation [26,27].

3. The Air Quality Fuzzy Comprehensive Evaluation (AQFCE) Paradigm

3.1. Preprocessing Module of AQFCE

Here, for concentration data of air pollutants, missing data and reversal data are defined as abnormal data. Missing data are usually due to routine maintenance of monitoring instrument (zero calibration and criterion calibration), communication failure and power failure [28]. Reversal data mean that the concentration of $PM_{2.5}$ which belongs to a part of PM_{10} is higher than that of PM_{10} in the same sampling site and time. There is no uniform rule on automatic monitoring methods of $PM_{2.5}$ and PM_{10} internationally, and thus distinct monitoring methods of $PM_{2.5}$ and PM_{10} could result in this phenomenon especially under higher relative humidity. The integrity and accuracy of data are crucial for air quality evaluation, and abnormal data can lead to questionable and useless results. Nevertheless, there is little attention paid to the abnormal data though they are very common and unavoidable in monitoring data. Therefore, in the preprocessing module, abnormal data processing algorithms are proposed.

(1) For missing data, the deletion method will lose useful information and break the continuity of time series. Interpolation is the most frequently used method, but it is not applicable to data of air pollutant concentration which are highly nonlinear and nonstationary. No matter which interpolation method is selected, its ability of processing such data is limited by the essence of interpolation. Although the interpolation curve is through the known data points, there are problems of bigger errors on unknown data points and the poor capability of extrapolation. In contrast, curve fitting does not go through known data points and approaches the overall trend of data with the minimum error, offering a reasonable solution with better physical meaning. Therefore, for a single pollutant, least square polynomial fitting is recommended to deal with missing data and piecewise fitting is employed during the whole process. The fitting results are acquired based on the least root mean square error and the best determination coefficient and approved by a significant test (95% confidence level).

(2) For reversal data of particulate matter, it is necessary to understand the cause [29]. Take China for example—PM_{10} was included in the ambient air quality monitoring network far earlier than $PM_{2.5}$ and there are differences in monitoring methods. Measurement errors caused by the loss of volatile components in PM_{10} when heating samples are unavoidable for the old Tapered Element Oscillating Microbalance (TEOM) and β-ray methods which have been used until now. Comparatively, new tapered element oscillating microbalance with a Filter Dynamic Measurement System (TEOM + FDMS) and β-ray with a Dynamic Heating System (β-ray + DHS), which have the calibration of temperature and humidity as well as the compensation of volatile components, can ensure accurate results of $PM_{2.5}$. As a part of PM_{10}, $PM_{2.5}$ accounts for 50%~80% according to Air Quality Guidelines of the WHO and PM_{10} is highly related to $PM_{2.5}$. Least square linear regression is recommended for reversal data of particulate matter and piecewise regression is adopted in the whole process. Normal and paired $PM_{2.5}$ and PM_{10} near reversal data are treated

as the independent variable and dependent variable to obtain the regression equation which is then used to correct reversal data. The regression result has a high determination coefficient (>0.8) and is approved by a significance test (95% confidence level).

3.2. Evaluation Module of the AQFCE

Based on the comprehensive consideration about the limitations of AQI and current air pollution, an improved air quality evaluation model is needed. Air quality is a huge and ambiguous system with a large number of uncertain factors, so fuzzy math is chosen as the solution [30]. Fuzzy math can effectively reduce impacts of a sharp boundary on evaluation results, and the combination of membership degree and factor weight can eliminate subjective factors in classification, providing an objective and comprehensive evaluation. In recent years, although some attempts related to fuzzy math have been made in the field of environmental evaluation [31–34], these methods are either too complicated or too simple to be popularized in practice. There is seldom targeted and comprehensive research about air quality evaluations too. In fact, a fuzzy math model with deep customization can not only have more accurate and comprehensive evaluation result but can be as practical and convenient as the AQI, which is exactly what the present study focuses on.

(1) Factor set

As the evaluation object, air quality is influenced by various air pollutants such as sulfur oxides, nitrogen oxides, particulate matters and so forth, which are treated as evaluation factors and selected. Based on the international consensus on conventional air pollutants, SO_2, NO_2, CO, O_3, $PM_{2.5}$ and PM_{10} are chosen for current air quality evaluation, and other pollutants can be added to our model immediately if necessary in future. The factor set composed of six factors is shown as follows:

$$U = \{u_1, u_2, \ldots, u_6\} = \{SO_2, NO_2, CO, O_3, PM_{2.5}, PM_{10}\} \quad (3)$$

(2) Evaluation set

The evaluation set means the classification of air quality, and it has a corresponding calculation criterion which is similar to that of the AQI. The calculation grid ought to accord with the local situation and the algorithm adopted in the evaluation process. It is really significant because an inappropriate calculation grid will lead to questionable evaluation results, and it is also unsuitable for different algorithms to adopt the same calculation grid. Take China as an example—its calculation grid is in need of revisions for following reasons: Firstly, according to the Annual Bulletin on China's Environmental State of the last eight years [35], there have been tremendous changes in many aspects including industrial structure, pollution source, air quality and pollution control capacity, which means the current calculation grid released in 2012 is no longer applicable for present and future air quality evaluation and management. Secondly, there are too many levels concerned with pollution to distinguish in the current criterion: light pollution, moderate pollution, serious pollution and severe pollution (two sublevels). This is confusing for the general public and, with increasing attention and demand on air quality, a clearer and stricter version is imperative. Thirdly, an air quality report consists of a daily report and hourly report, which ought to be separated and follow diverse concentration limits, but there is incompleteness and misuse in the current criterion (Table 1). Hourly reports of $PM_{2.5}$ and PM_{10} have to follow their daily concentration limits, respectively, leading to severely overrated particulate pollution. Similarly, hourly reports of SO_2 will result in an overestimated evaluation result when its concentration is higher than 800 ug/m^3 and an underestimation of a daily report of O_3 when the concentration is higher than 800 ug/m^3. These problems are not accepted and need corrections immediately. Finally, criterion should be compatible with a specific algorithm. Fuzzy math and the AQI have totally different algorithms, and the fuzzy math model (comprehensive strategy) tends to show a lower level of air quality compared with the AQI (conservative strategy) under the same conditions. Therefore, a stricter criterion is also favorable to the smooth transition from the AQI to AQFCE in practice.

To revise the original calculation criterion, authoritative and instructive literature including the Air Quality Guidelines of the WHO, Directive on Ambient Air Quality and Cleaner Air for Europe, Explanation of Ambient Air Quality Standards for China, documents of the AQI for the EU, USA, UK and CHN and Annual Bulletin on China's Environmental State and Monthly Report on Urban Air Quality have been referenced [7,22,23,35–40], which provides tremendous medical and statistical evidence about the effects of air pollutants on human health and regional air quality features. Ratios of limit values in criteria of the EU, USA, UK and CHN have also been referenced, especially in these cases: one pollutant at the same level in different periods, one pollutant at different levels in the same period and $PM_{2.5}$ and PM_{10} at the same level and period. Consequently, a new calculation grid for China (for the fourth limitation of the AQI) which focuses on health and environmental impacts is proposed (Table 2). In the new calculation grid, most values have clear sources and others are obtained by existing values and ratios. The evaluation set comprises five levels and corresponding air quality categories are "Excellent, Good, Moderate, Poor, Hazardous":

$$V = \{v_1, v_2, \ldots v_5\} = \{\text{I, II, III, IV, V}\}.$$

Table 2. A new calculation grid for China.

Level	Pollutant Concentration Limit											
	SO_2 (ug/m^3)		NO_2 (ug/m^3)		CO (mg/m^3)		O_3 (ug/m^3)		$PM_{2.5}$ (ug/m^3)		PM_{10} (ug/m^3)	
	24 h	1 h	24 h	1 h	24 h	1 h	8 h	1 h	24 h	1 h	24 h	1 h
I	50	100	40	100	2	5	100	150	25	37.5	50	75
II	125	350	80	200	4	10	160	200	75	100	150	200
III	200	500	134	376	8	20	200	245	110	150	220	300
IV	300	600	190	494	16	40	240	320	150	210	250	350
V	400	800	380	940	24	60	400	520	210	280	300	400

(3) Fuzzy matrix

Membership degree is the basic idea of fuzzy math and is objectively existent. It can be denoted by r_{ij}, which means the membership degree of factor u_i aiming at the comment v_j. Membership degree is an effective solution to the problem of data near boundary (for the first limitation of the AQI). For air quality evaluation, a pollutant concentration value can belong to both possible levels to various extents rather than be defined as only one certain level, and further membership degree will be combined with factor weight to make a comprehensive evaluation. The membership function can calculate the membership degree of pollutant concentration to the evaluation level. Commonly employed membership functions include the rectangular distribution function, Cauchy distribution function, Gauss distribution function, ridge distribution function, K-order parabolic distribution function, etc. In view of this evaluation object, factor and criterion, the trapezoid-triangle distribution function is adopted in this study. Details are presented as Equations (4)–(6) and Figure 1.

$$r_{ij} = \begin{cases} 1 & u_i \le v_{ij} \\ (v_{i(j+1)} - u_i)/(v_{i(j+1)} - v_{ij}) & v_{ij} < u_i < v_{i(j+1)} \\ 0 & u_i \ge v_{i(j+1)} \end{cases} \quad j = 1 \tag{4}$$

$$r_{ij} = \begin{cases} \left(u_i - v_{i(j-1)}\right)/\left(v_{ij} - v_{i(j-1)}\right) & v_{i(j-1)} < u_i < v_{ij} \\ (v_{i(j+1)} - u_i)/(v_{i(j+1)} - v_{ij}) & v_{ij} < u_i < v_{i(j+1)} \\ 0 & u_i \ge v_{i(j+1)} \end{cases} \quad j = 2,3,4 \tag{5}$$

$$r_{ij} = \begin{cases} 0 & u_i \leq v_{i(j-1)} \\ \left(u_i - v_{i(j-1)}\right) / \left(v_{ij} - v_{i(j-1)}\right) & v_{i(j-1)} < u_i < v_{ij} \qquad j = 5 \\ 1 & u_i \geq v_{ij} \end{cases} \tag{6}$$

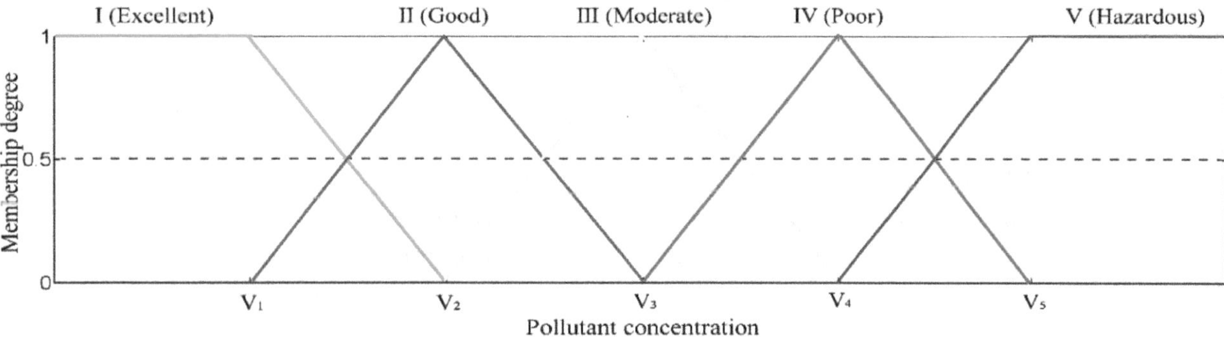

Figure 1. Membership function of air pollutant concentration.

The fuzzy matrix can be expressed by the matrix R:

$$R = \left(r_{ij}\right)_{m \times n} = \begin{bmatrix} r_{11} & r_{12} & \cdots & r_{1n} \\ r_{21} & r_{22} & \cdots & r_{2n} \\ \cdots & \cdots & \cdots & \cdots \\ r_{m1} & r_{i2} & \cdots & r_{mn} \end{bmatrix} \tag{7}$$

(4) Factor weight

Hazard of each evaluation factor needs to be quantified and compared, and a proper weighting method can calculate the weight of factor, namely its relative hazard on air quality. The multiple super-scale weighting method is a frequently used method in environment quality evaluations [33], but it can be further improved. Generally, there are some concentration limits (mainly higher value) without definite scientific bases in the calculation grid, which will significantly influence the evaluation result. Thus, the consideration of all limits in the weighting algorithm cannot accurately reflect the fatalness of each pollutant, while limits of the second level are more reliable. In general, the second level is not only the demarcation between pollution and nonpollution but also the target to protect human health, and it is usually employed by the government to evaluate whether air quality is up to standard. Most of all, they have more recommended limits from world authorities such as the WHO. Therefore, an improved method named the Normalized Secondary-Level Limit Weighting (NSLLW) algorithm is proposed as Equation (8). Additionally, due to the fact that factor weight will be combined with a fuzzy matrix to obtain an overall evaluation result of air quality, it can serve as an independent tool to obtain a great deal of useful information including chief pollutant, warning pollutant ($u_i / v_{i2} > 1$), relative hazard and rank (for the second limitation of the AQI). Table 3 shows an example of evaluation results of NSLLW. In contrary, all evaluation results of the AQI are all decided by pollutant with the highest subindex, which is biased and insufficient.

$$w_i = \left[u_i / (v_{i2})\right] / \sum_{i=1}^{m} \left[u_i / (v_{i2})\right] \tag{8}$$

Table 3. Evaluation results of Normalized Secondary-Level Limit Weighting (NSLLW) on 17 April 2019 in Beijing.

	SO_2	NO_2	CO	O_3	$PM_{2.5}$	PM_{10}
Concentration	9 ug/m^3	54 ug/m^3	1 mg/m^3	189 ug/m^3	105 ug/m^3	137 ug/m^3
Relative hazard (weight)	0.02	0.15	0.06	0.26	0.31	0.20
Rank	6	4	5	2	1	3
Warning pollutant				✓	✓	
Chief pollutant					✓	

(5) Fuzzy operator

The final evaluation result will be obtained by synthesizing the factor weight and fuzzy matrix with an appropriate composite operator which can be defined by the researcher according to the actual evaluation task. The fuzzy operator plays a significant role and determines the nature of a model. Common fuzzy math models can be divided into three categories:

Principal factor decision model: $M(\wedge, \vee)$

$$b_j = \bigvee_{i=1}^{m} \left(w_i \wedge r_{ij}\right) (j = 1, 2, \ldots, n) \tag{9}$$

Principal factor prominent model: $M(\wedge, \oplus)$

$$b_j = \bigoplus_{i=1}^{m} \left(w_i \wedge r_{ij}\right) = \sum_{i=1}^{m} \left(w_i \wedge r_{ij}\right) (j = 1, 2, \ldots, n) \tag{10}$$

Weighted average model: $M(\bullet, +)$

$$b_j = \sum_{i=1}^{m} \left(w_i \bullet r_{ij}\right) \ (j = 1, 2, \ldots, n) \tag{11}$$

$M(\bullet, +)$ is the first choice for environment quality evaluations because this model is favorable for taking all factors into account and obtaining an integrated effect. This weighted average model seems to be reasonable, but diluting the influence of the chief pollutant makes its evaluation level far lower than that of the AQI (conservative strategy) even under the same conditions, which may affect the people who are sensitive to the chief pollutant. Therefore, we choose $M(\bullet, \vee)$ which considers the integrated effect of all pollutants and meanwhile highlights the influence of the chief pollutant to ensure a smooth transition from the AQI to AQFCE (for the third limitation of the AQI). Its equation is as follows:

$$b_j = \bigvee_{i=1}^{m} \left(w_i \bullet r_{ij}\right) \ (j = 1, 2, \ldots, n) \tag{12}$$

(6) Evaluation result

According to the principle of maximum membership degree, the maximum value of B reflects the final evaluation result (level) of air quality.

$$B = (w_1, w_2, \ldots w_m) \ (\bullet, \vee) \begin{bmatrix} r_{11} & r_{12} & \cdots & r_{1n} \\ r_{21} & r_{22} & \cdots & r_{2n} \\ \cdots & \cdots & \cdots & \cdots \\ r_{m1} & r_{m2} & \cdots & r_{mn} \end{bmatrix} \tag{13}$$

3.3. Early Warning Module of AQFCE

Besides air quality evaluations at different time scale, forecasts and early warnings can also be achieved by the AQFCE based on forecast data. In order to prevent air pollution and provide maximum protection for public health in a timely manner, the air pollution early warning module was designed as Table 4, which references documents of the AQI from many countries [7,22,23,39] and emergency plan for severe air pollution of many provinces in China [41–44]. As an alert and recommendation, it can not only guide people's daily activities but provide decision-making support for governments, such as evaluating whether the air quality of a city reaches the standard, taking mandatory measures to address potential air pollution and so forth.

Table 4. Air pollution early warning.

Level	Category	Color (R G B)	Impact	Recommended Measure	
				The Public	Government
I	Excellent	green (0 255 0)	Air quality is perfect with little pollution.	Outdoor activities are suggested for general people.	None
II	Good	blue (0 0 255)	Air quality is basically satisfactory and there may be some pollutants affecting the very few people with abnormal sensitivity.	The people who are abnormally sensitive to pollutants should reduce outdoor activities.	None
III	Moderate	yellow (255 255 0)	Air quality is generally acceptable but sensitive people may present mild symptoms.	Sensitive people, including children, the elderly and patients with respiratory tract, cardiovascular and cerebrovascular diseases, should reduce outdoor activities.	None
IV	Poor	red (255 0 0)	Air pollution may aggravate sensitive people's symptoms and damage general people's health.	Outdoor activities should be suspended for sensitive people and schools. General people should reduce outdoor time and wear masks.	Advocate public transportation and enhance road cleaning and washing. Suspend large-scale open air activities.
V	Hazardous	purple (128 0 128)	Air pollution may seriously aggravate sensitive people's symptoms and symptoms are common in healthy people.	All people should avoid outdoor activities and, outdoors, people must wear masks. Idling and low speed driving of vehicles need to be avoided.	Besides the above, adopt flexible mechanisms in schools and enterprises. Stop or limit production in industrial enterprises and vehicle restrictions. Stop demolition and transportation work in construction sites.

SO_2: Sulphur dioxide is mainly from the combustion of sulphur-containing fuels and can be the precursor of a secondary particle. It can damage the respiratory system of humans and animals, hinder the growth and development of plants and corrode metal materials.

NO_2: Nitrogen dioxide is mainly from thermal power plants, transportation and industrial sources and can be the precursor of a secondary particle. Bronchus and lungs will be affected under high concentration.

CO: Carbon monoxide is from the incomplete combustion of carbon-containing fuel and can cause serious harm to the nervous and cardiovascular system of humans and animals.

O_3: Ozone is from photochemical reaction of nitrogen oxide and volatile organic compound produced by transportation and industry. It is a strong oxidant which can cause coughing, chest discomfort and inflammation of the respiratory tract. Plant and building materials are also injured by it.

PM_{10} and $PM_{2.5}$: Particulate matter is from natural (volcano eruption, sandstorm) and anthropogenic (industrial processes, traffic emission) sources. It can be divided into primary (emitted directly) and secondary (atmospheric chemistry reaction of precursors) particles. PM_{10} ($d \leq 10$ um) can enter into the upper respiratory tract and has a wide range of health effects involving ones on the cardiovascular system, respiratory system, child growth and mortality. It also reduces visibility by scattering and absorbing light. $PM_{2.5}$ (d ≤ 2.5 um) can enter into lungs and has a longer suspension time and transmission distance in air, which means greater harm.

4. Experimental Result and Discussion

4.1. Data and Method

To prove the feasibility of the AQFCE in practice, it has been fully tested under various conditions in China. According to "Three-year Action Plan for Winning the Blue Sky Defense War" issued by the China's State Council in June 2018, the Fenwei Plain has replaced Pearl River Delta to be the key area of air pollution control together with the Jing-jin-ji region and Yangtze River Delta. As representative cities, Beijing, Shanghai and Xi'an (Figure 2), with different environments, industrial structures and pollution features, are selected. Unreasonable industrial and energy structures as well as continuously rapid growth of vehicles have caused huge emissions of local pollution, which is common in the three metropolises. In addition, Beijing and Xi'an are, respectively, located in the northeast and southwest of "Jing-jin-ji air pollution transmission channel", so they are often affected by regional pollution transport. Adjacent to sand sources, they also suffer from sand- and dust-storms in spring. Daily and hourly concentration data of six conventional air pollutants from September 2018 to August 2019 are from the Ministry of Ecology and Environment of China (http://www.mee.gov.cn). In view of the excessive hourly data, the representative month of every season (October, January, April and July) are used in hourly evaluations. The AQI is the most widely used and recognized air quality evaluation method at present, so comparative analysis with the AQI in practice is necessary for any new method. Therefore, based on the same monitor data, comparative analysis between the AQI and AQFCE is carried out on different spaces (Beijing, Shanghai, Xi'an), time (spring, summer, autumn, winter), tasks (daily report, hourly report) and purposes (air quality level, chief pollutant).

4.2. Case Study and Analysis

(1) Beijing

Figure 3 presents the concentration variation of six pollutants and corresponding air quality evaluation results (chief pollutant and level) of Beijing in four seasons. Combined pollution is common in Beijing and O_3 is dominant in late spring, summer and early autumn. Regional pollution transport and local traffic pollution make $PM_{2.5}$ and NO_2 major air pollutants throughout the year, while PM_{10} is more notable in spring mainly from dust transport upstream. In general (Figure 3), there is an increase in daily concentration during weekends and holidays, which can be attributed to the strict traffic restriction on weekdays and celebration fireworks during festivals. For hourly concentration (Figure 4A–D), ozone has one peak in the period from 14:00 to 16:00 every day, while two peaks at 8:00 and 20:00 are apparent for other pollutants, which are in accordance with traffic peaks. Referring to pollutant concentration (Figure 3), it is clear from evaluation results that the AQFCE accurately reflects pollution features and trends under different pollution conditions (lighter pollution, heavier pollution, pollution process), and it is sensitive to sudden and significant changes of concentration. As a statistical indicator, the Consistent Rate (CR) and Kendall Correlation Coefficient (T) are applied to compare the AQI with AQFCE on chief pollutants and air quality level (ordered categorical variable). Furthermore, specific examples will provide empirical support for the comparative analysis.

For daily reports (Figure 3), the CR between the AQI and AQFCE are 82.47% and 58.08% on chief pollutant and air quality level, respectively. The chief pollutant has an important purpose and significantly influences the subsequent evaluation on air quality level. So, accurate identification of the chief pollutant is the premise of air quality evaluation, and both methods have high consistency on this point. In view of their middle consistency of level, differences are further analyzed by specific examples. On 2 March 2019 (Figure 3), concentration values of SO_2, NO_2, CO, O_3, $PM_{2.5}$ and PM_{10} were 3, 14, 0.4, 77, 40 and 157, which means values of five pollutants are far below or close to the limit of I and one pollutant's value is close to the limit of II. The AQI and AQFCE determine the same chief pollutant PM_{10} but a different level III (light pollution) and II (good), respectively, which shows that a sharp boundary and biased evaluation make the AQI overestimate actual pollution. Experimental results indicate almost all their divergences are due to the fact that the level of AQI with a conservative strategy is higher than that of the AQFCE with a comprehensive strategy by one level, but they have a similar trend and high correlation coefficient 0.78. In conclusion, though the AQI and AQFCE are two completely different methods, namely their distinct algorithms, criteria and strategies, their effectiveness in daily reports are proved by each other and the AQFCE is more comprehensive.

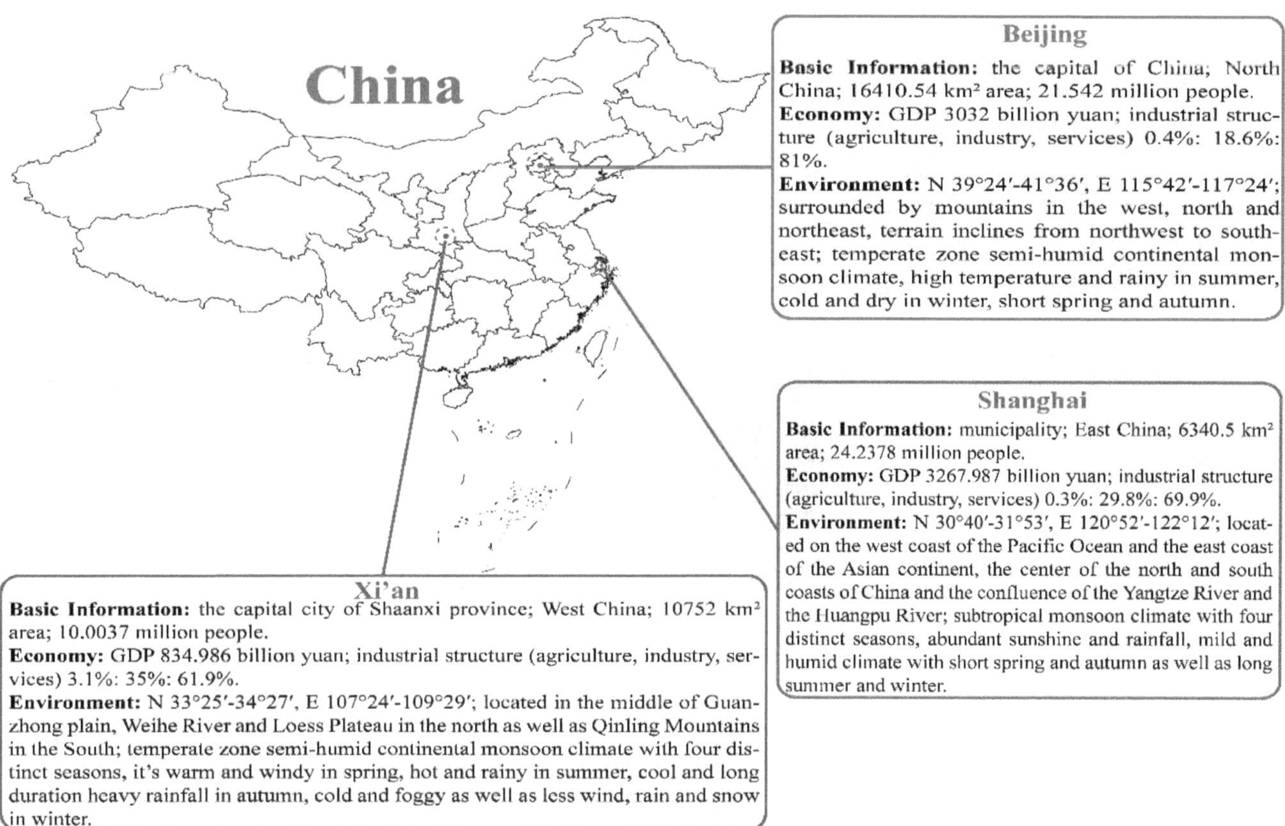

Figure 2. The overview of study areas.

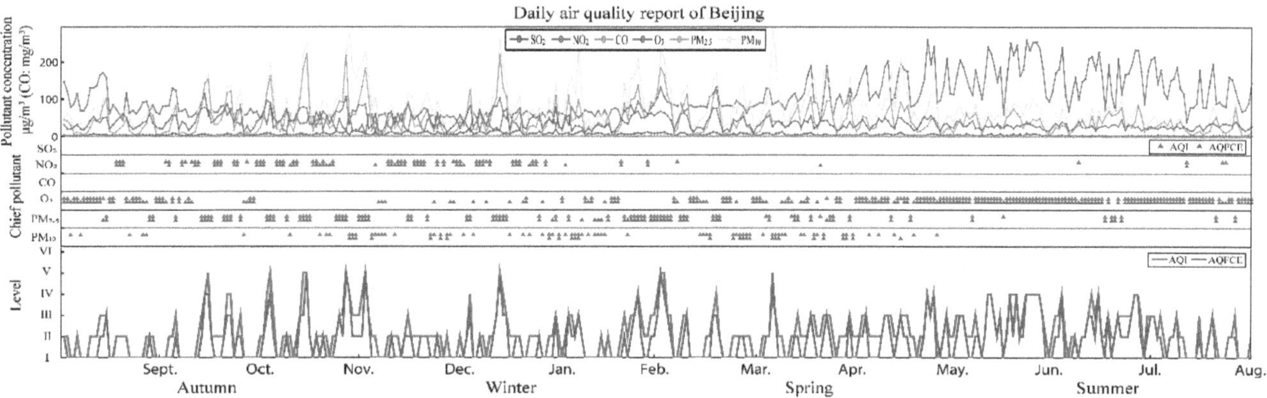

Figure 3. Daily pollutant concentration variation and air quality evaluation result of chief pollutant and level of Beijing.

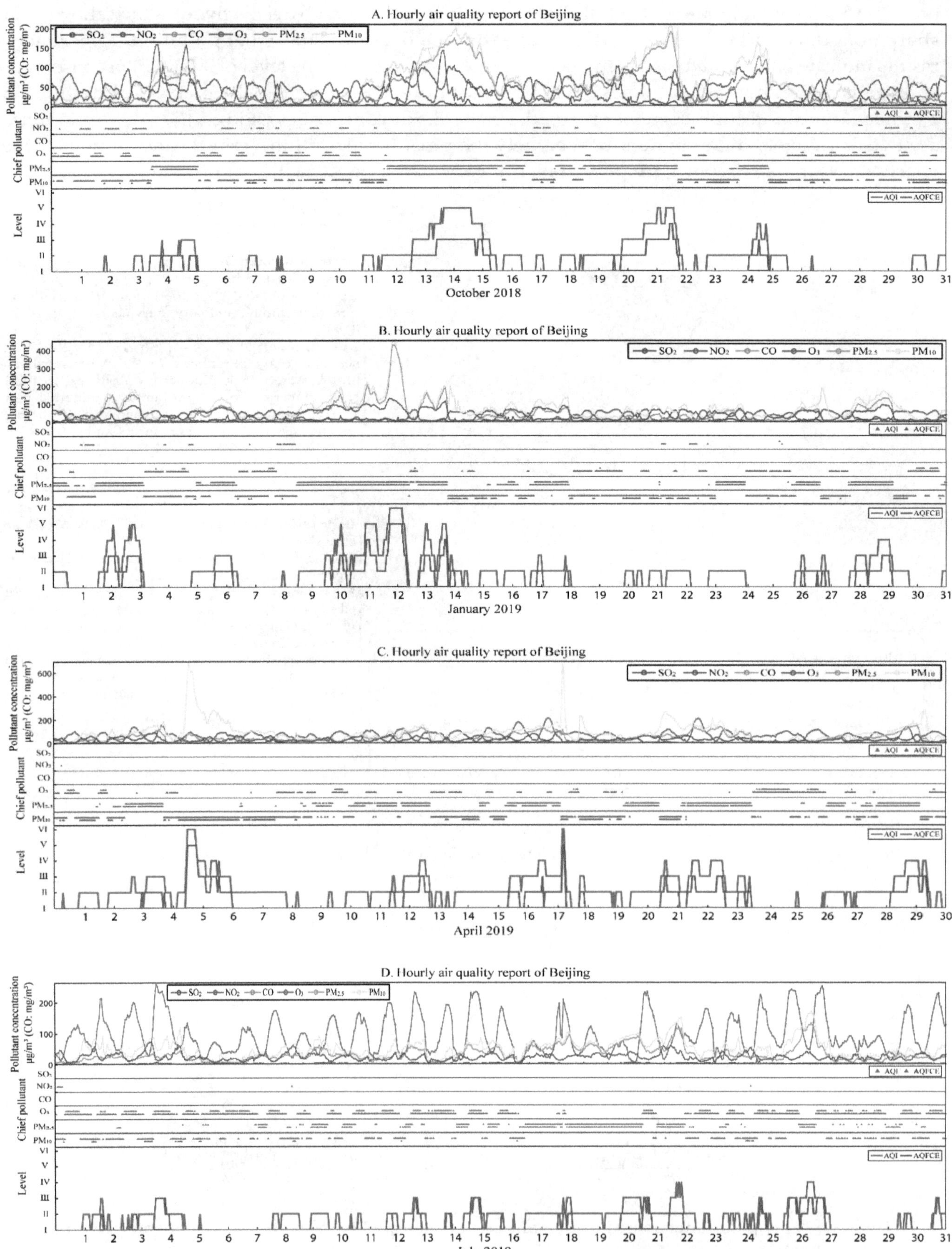

Figure 4. Hourly pollutant concentration variation and air quality evaluation result of chief pollutant and level of Beijing.

For hourly reports (Figure 4A–D), the incomplete criterion of the AQI is really a problem for not only the comparison but also AQI's own evaluation. This is because in the hourly report of the AQI $PM_{2.5}$ and PM_{10} use daily criterion while other pollutants use hourly criterion, which is unreasonable. This leads to an exaggerated hazard of $PM_{2.5}$ and PM_{10}, and they are more likely to be chosen as chief pollutants even if they are not the biggest hazard. The AQI's level is decided by the chief pollutant alone, so a questionable chief pollutant may give a meaningless air quality level. Experimental results prove this and there are antinomies in the results of the AQI. For example, on 23 October 2018, the chief pollutant of the hourly report was PM_{10} in every hour but the daily report's chief pollutant was NO_2, which is contradictory. The AQFCE is complete in hourly reports, and there is a marked drop of CR compared with the daily report as expected—63.04% and 45.49% for the chief pollutant and level. $PM_{2.5}$ and PM_{10} without hourly criterion really disturb the comparison, but CR increase significantly to 89.74% and 97.39%, respectively, after eliminating their influence, which means AQI with appropriate concentration limits is similar to the AQFCE. Actually, an hourly report of the AQI is severely affected by its criterion, and the AQFCE is robust for use in various tasks.

(2) Shanghai and Xi'an

From Figures 5 and 6, a mobile pollution source is prominent in Shanghai, which promotes NO_2 and further helps $PM_{2.5}$ occupy the majority of the particle pollution. Excessive emission of Volatile Organic Compounds (VOCs) and a high temperature leads to the dominance of O_3 during the year, except in winter. The "Weekend effect" and "holiday effect" for daily concentrations (Figure 5) and "Peak Type" for hourly concentrations (Figure 6A–D) appear in Shanghai too, which is common in a metropolis. Two methods have similar performances in terms of their daily reports (Figure 5), and their CRs (T) are 85.48% and 48.77% (0.62) for the chief pollutant and level. From Figure 6A–D, exaggerated hazard of particulate matter by the AQI often leads to a questionable chief pollutant (mainly $PM_{2.5}$ and PM_{10}) and level. Even though when particulate matter is the actual chief pollutant, the level of the AQI is seriously overvalued too. On the contrary, based on the actual pollutant concentration, results of the AQFCE in hourly report are more convincing. Excluding the influence of particulate matter, the CR has a sharp rise about 40% of both the chief pollutant (47.59%–84.76%) and level (58.74%–96.99%).

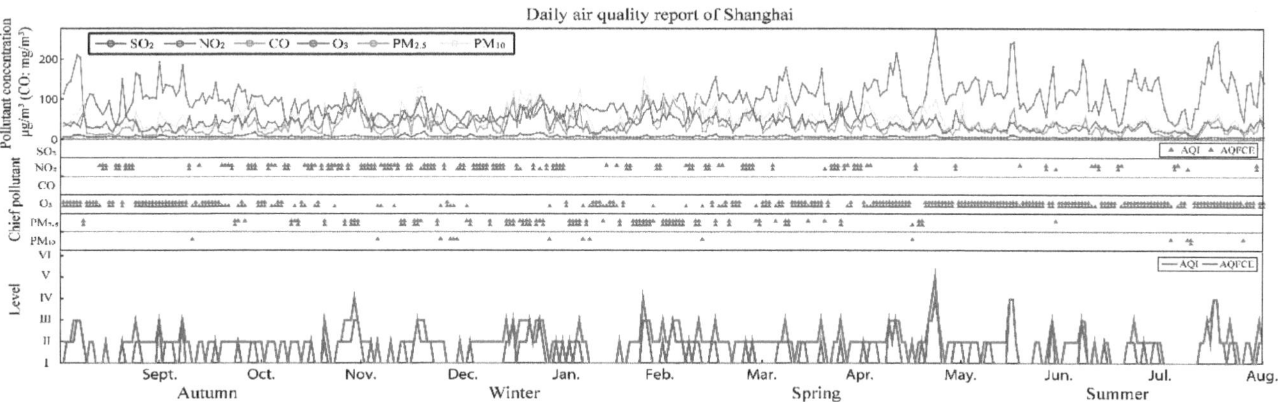

Figure 5. Daily pollutant concentration variation and air quality evaluation result of chief pollutant and level of Shanghai.

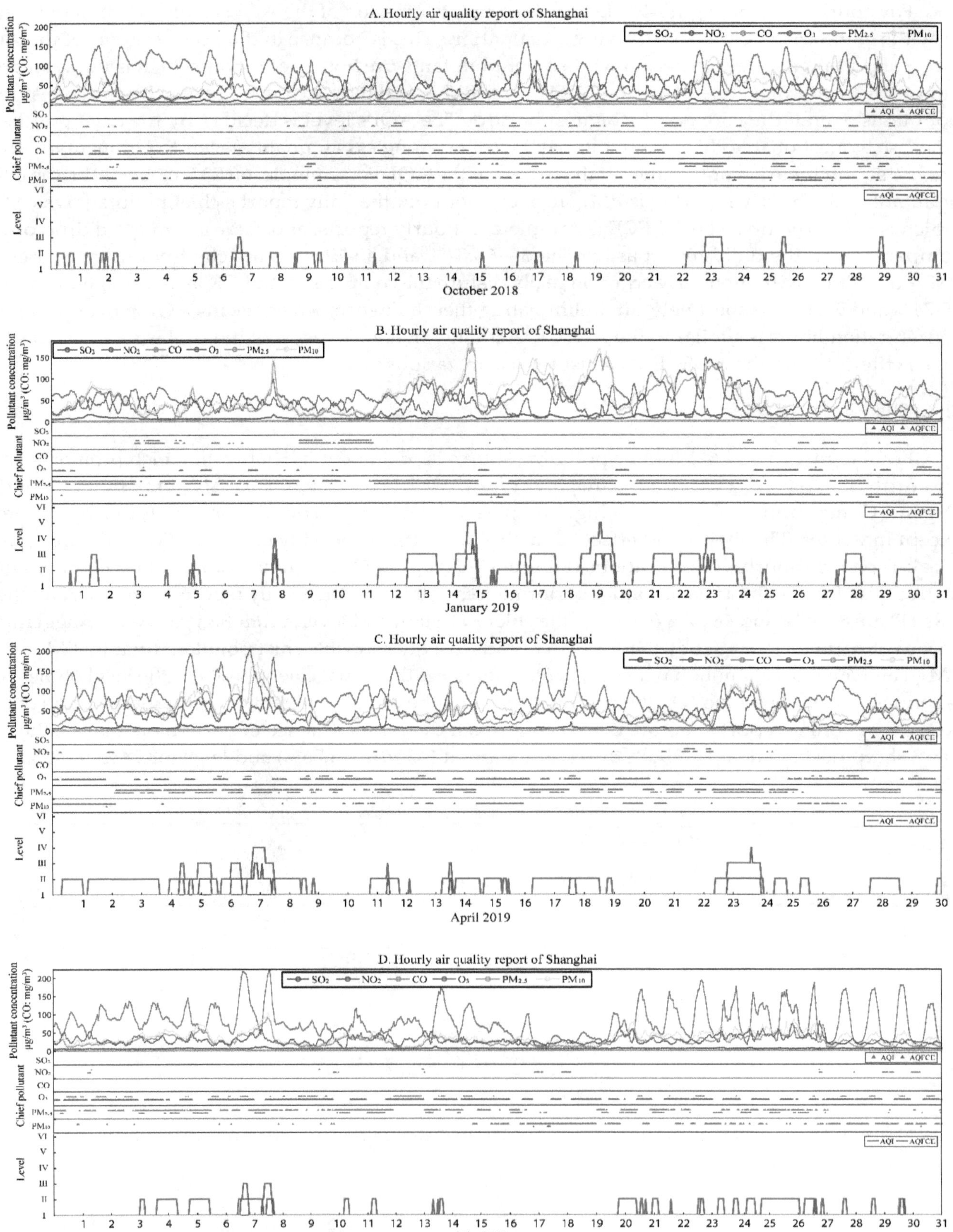

Figure 6. Hourly pollutant concentration variation and air quality evaluation result of chief pollutant and level of Shanghai.

It is clear from Figure 7 that ozone pollution is also a tough problem in the summer of Xi'an. Perennial pollution transport from the northeast and local pollution emission as well as sand dust transport from the northwest cause severe particle pollution in autumn, winter (mainly $PM_{2.5}$) and spring (mainly PM_{10}). From Figure 7, comparison results of daily and hourly reports are similar to Beijing and Shanghai. The AQFCE and AQI have an approximate performance in chief pollutant (83.84%) and level (49.9%/0.78) for daily reports (Figure 7), while huge divergences on the chief pollutant in summer (Figure 8D) and level in winter (Figure 8B) stem from magnified particle pollution by the AQI for hourly reports. Table 5 presents the comparison of evaluation results between the AQI and AQFCE by consistency and similarity.

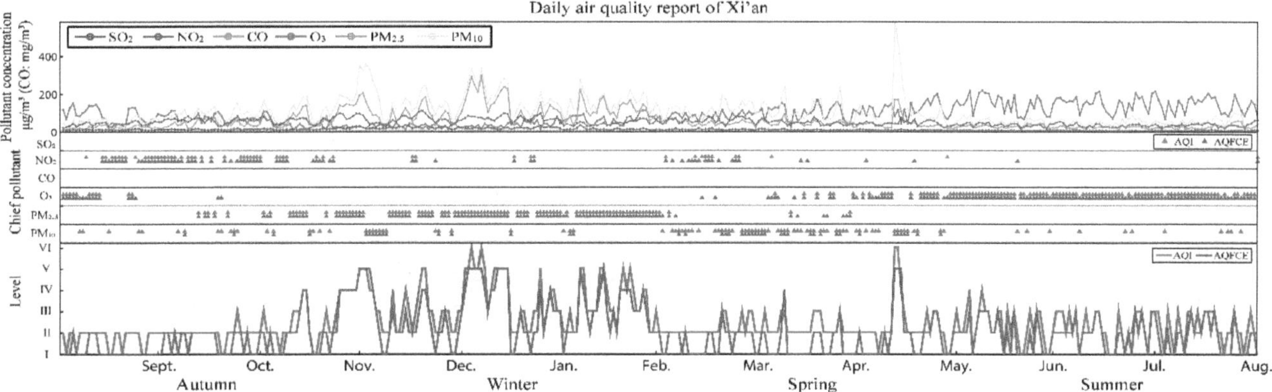

Figure 7. Daily pollutant concentration variation and air quality evaluation result of chief pollutant and level of Xi'an.

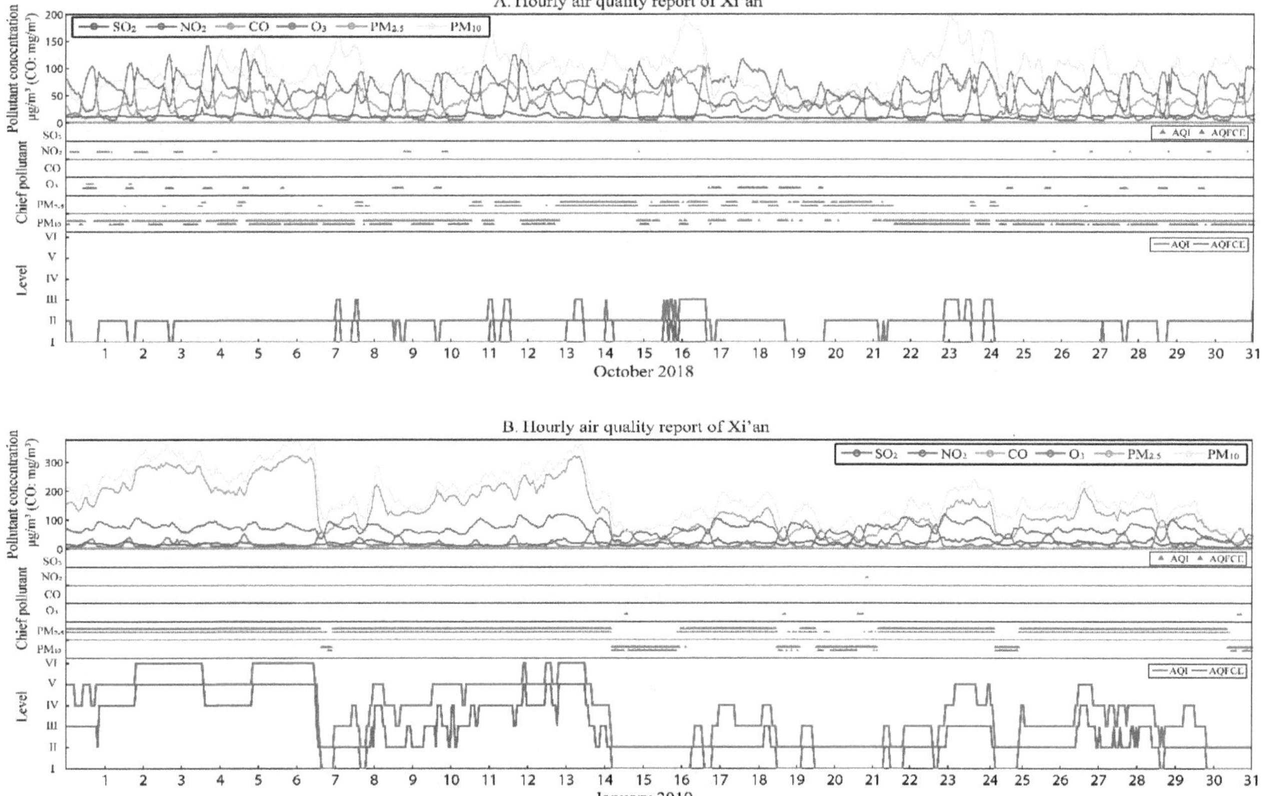

Figure 8. *Cont.*

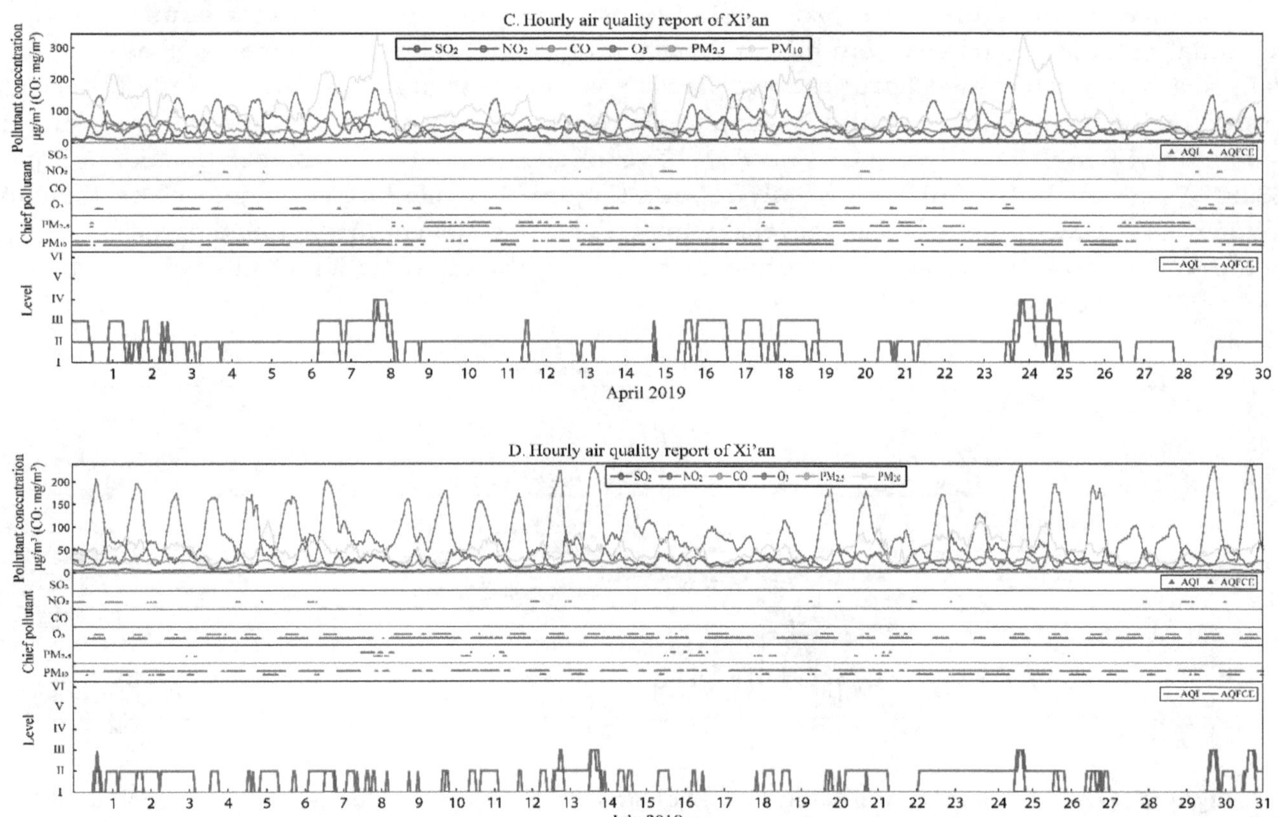

Figure 8. Hourly pollutant concentration variation and air quality evaluation result of chief pollutant and level of Xi'an.

Table 5. The comparison of evaluation result between the Air Quality Index (AQI) and Air Quality Fuzzy Comprehensive Evaluation (AQFCE) by consistent rate (CR) (T).

	Beijing				**Shanghai**				**Xi'an**			
	DR		**HR**		**DR**		**HR**		**DR**		**HR**	
	CP	**Level**	**CP**	**Level**	**CP**	**Level**	**CP**	**Level**	**CP**	**Level**	**CP**	**Level**
AQI and AQFCE	82.47%	58.08% (0.78)	63.04%	45.49%	85.48%	48.77% (0.62)	47.59%	58.74%	83.84%	50.96% (0.78)	69.88%	25.68%
AQI* and AQFCE	-	-	89.74%	97.39%	-	-	84.76%	96.99%	-	-	89.94%	94.99%

CR (T): Consistent Rate (Kendall Correlation Coefficient), DR/HR: Daily Report/Hourly Report, CP: Chief Pollutant, AQI*: exclude $PM_{2.5}$ and PM_{10} in HR.

5. Conclusions

The AQI has been used around the world for many years, yet its intrinsic limitations are increasingly apparent. A sharp boundary leads to completely different evaluation results and corresponding measures facing almost the same air quality. A biased evaluation is not objective and comprehensive enough, and a conservative strategy may exaggerate actual air pollution and bring about unnecessary economic loss and public concern. In addition, incomplete criteria have caused questionable evaluation results. Now people are beginning to realize the inconsistency between their feelings and reports of AQI, and meanwhile governments are also considering how to balance air pollution control and economic development. Therefore, based on the comparative analysis of AQI in theory, we designed a novel air quality evaluation paradigm, the AQFCE, which consists of preprocessing, evaluation and early warning. To ensure the integrity and accuracy of data, least square piecewise polynomial fitting and linear regression was used for missing and reversal data of air pollutant concentration in the preprocessing module. In the evaluation module, membership degree is adopted to deal with the problem of a sharp boundary. Compared with the biased evaluation and conservative strategy of the AQI, a principal factor prominent model $M(\bullet, \vee)$ based on the membership degree and factor weight of all pollutants can give more objective evaluation results and balance human health protection and pollution control cost. A complete and targeted calculation grid based on scientific evidence is proposed too. Moreover, the early warning module can provide timely alerts and recommended measures according to the evaluation results of air quality. In general, the AQFCE can achieve an air quality evaluation and forecast at different time scale. Comparative analysis of the AQI and AQFCE in practice was carried out in China under various conditions including different space (Beijing, Shanghai and Xi'an), time (autumn, winter, spring and summer), task (daily report, hourly report) and purpose (chief pollutant, level). Experimental results offer some meaningful information:

(1) Referring to actual pollutant concentration, the AQFCE accurately reflects the trend of air pollution under different pollution conditions (lighter pollution, heavier pollution, pollution process) and is sensitive to sudden and significant changes in concentration. In daily reports, the AQFCE and AQI have high consistent rates and correlation coefficients regarding the chief pollutant and level, respectively, in Beijing, Shanghai and Xi'an, while examples prove the AQFCE with a comprehensive strategy is more reasonable than the AQI with a conservative strategy. In hourly reports, examples indicate that the AQI, influenced by incomplete criterion, has questionable evaluation results and even antinomies, while the AQFCE is still as effective as daily reports.

(2) The AQFCE successfully reveals that O_3, $PM_{2.5}$, PM_{10} and NO_2 are major air pollutants in China at present, while SO_2 and CO have faded out. The "Weekend effect" and "holiday effect" for daily concentration and "single peak type" of O_3 as well as "double peak type" of the other pollutants for hourly concentration are found. In addition, with common ozone pollution in summer and fine particle pollution in winter, pollution features vary among the Jing-jin-ji region, Yangtze River Delta and Fenwei Plain for different pollution sources and diffusion conditions.

In conclusion, comparative analysis in theory and practice prove the feasibility of the AQFCE. The AQFFCE solves the limitations of the AQI and is an important improvement and supplement to the AQI. In the future, we will further improve the AQFCE and hope that it can provide health guidelines for the public and support in short-term environmental decision-making as well as a long-term development plan for government.

Author Contributions: Conceptualization, X.M. and Z.Q.; Methodology, X.M.; software, H.L.; validation, X.M., H.L., writing—original draft preparation, X.M., H.L.; writing—review and editing, X.M., H.L., L.Z. and Z.Q.; supervision, L.Z.; project administration, L.Z.; funding acquisition, L.Z. All authors have read and agreed to the published version of the manuscript.

References

1. Hadley, M.B.; Vedanthan, R.; Fuster, V. Air pollution and cardiovascular disease: A window of opportunity. *Nat. Rev. Cardiol.* **2018**, *15*, 193–194. [CrossRef]
2. Wang, H.; Zhang, Y.; Zhao, H.; Lu, X.; Zhang, Y.; Zhu, W.; Nielsen, C.P.; Li, X.; Zhang, Q.; Bi, J.; et al. Trade-driven relocation of air pollution and health impacts in China. *Nat. Commun.* **2017**, *8*, 1–7. [CrossRef]
3. Waller, L.A. Estimate suggests many infant deaths in sub-Saharan Africa attributable to air pollution. *Nat. Cell Biol.* **2018**, *559*, 188–189. [CrossRef]
4. Costa, A.F.; Hoek, G.; Brunekreef, B.; De Leon, A.C.P. Air Pollution and Deaths among Elderly Residents of São Paulo, Brazil: An Analysis of Mortality Displacement. *Environ. Heal. Perspect.* **2017**, *125*, 349–354. [CrossRef]
5. Sofia, D.; Gioiella, F.; Lotrecchiano, N.; Giuliano, A. Cost-benefit analysis to support decarbonization scenario for 2030: A case study in Italy. *Energy Policy* **2020**, *137*, 111137. [CrossRef]
6. Cheng, W.-L.; Chen, Y.-S.; Zhang, J.; Lyons, T.; Pai, J.-L.; Chang, S.-H. Comparison of the Revised Air Quality Index with the PSI and AQI indices. *Sci. Total. Environ.* **2007**, *382*, 191–198. [CrossRef]

7. Elshout, S.V.D.; Léger, K.; Heich, H. CAQI Common Air Quality Index — Update with $PM_{2.5}$ and sensitivity analysis. *Sci. Total. Environ.* **2014**, *489*, 461–468. [CrossRef]
8. Chen, W.; Tang, H.; Zhao, H. Urban air quality evaluations under two versions of the national ambient air quality standards of China. *Atmos. Pollut. Res.* **2016**, *7*, 49–57. [CrossRef]
9. Plaia, A.; Di Salvo, F.; Ruggieri, M.; Agró, G. A Multisite-Multipollutant Air Quality Index. *Atmospheric Environ.* **2013**, *70*, 387–391. [CrossRef]
10. Sowlat, M.H.; Gharibi, H.; Yunesian, M.; Mahmoudi, M.T.; Lotfi, S. A novel, fuzzy-based air quality index (FAQI) for air quality assessment. *Atmos. Environ.* **2011**, *45*, 2050–2059. [CrossRef]
11. Bagieński, Z. Traffic air quality index. *Sci. Total. Environ.* **2015**, *505*, 606–614. [CrossRef]
12. Olvera-García, M.Á.; Carbajal-Hernández, J.J.; Sánchez-Fernández, L.P.; Hernández-Bautista, I. Air quality assessment using a weighted Fuzzy Inference System. *Ecol. Inform.* **2016**, *33*, 57–74. [CrossRef]
13. Li, Z.Y.; Zhang, Z.J.; Wang, J.Y. Universal Index Formulae of Air Quality Evaluation by Transformed Values of Indexes. *Environ. Sci. Technol.* **2012**, *35*, 179–184. (In Chinese)
14. Gorai, A.K.; Kanchan; Upadhyay, A.; Tuluri, F.; Goyal, P.; Tchounwou, P.B. An innovative approach for determination of air quality health index. *Sci. Total. Environ.* **2015**, *533*, 495–505. [CrossRef] [PubMed]
15. Adams, M.D.; Kanaroglou, P. A criticality index for air pollution monitors. *Atmos. Pollut. Res.* **2016**, *7*, 482–487. [CrossRef]
16. Thach, T.Q.; Tsang, H.; Cao, P.; Ho, L.-M. A novel method to construct an air quality index based on air pollution profiles. *Int. J. Hyg. Environ. Heal.* **2018**, *221*, 17–26. [CrossRef]
17. Wang, Y.; Fu, X.-K.; Jiang, W.; Wang, T.; Tsou, M.-H.; Ye, X. Inferring urban air quality based on social media. *Comput. Environ. Urban Syst.* **2017**, *66*, 110–116. [CrossRef]
18. Chen, Y.; Wang, L.; Li, F.; Du, B.; Choo, K.-K.R.; Hassan, H.; Qin, W. Air quality data clustering using EPLS method. *Inf. Fusion* **2017**, *36*, 225–232. [CrossRef]
19. Ruggieri, M.; Plaia, A. An aggregate AQI: Comparing different standardizations and introducing a variability index. *Sci. Total. Environ.* **2012**, *420*, 263–272. [CrossRef]
20. Plaia, A.; Ruggieri, M. Air quality indices: A review. *Rev. Environ. Sci. Bio/Technol.* **2010**, *10*, 165–179. [CrossRef]
21. Sofia, D.; Lotrecchiano, N.; Giuliano, A.; Barletta, D.; Poletto, M. Optimization of Number and Location of Sampling Points of an Air Quality Monitoring Network in an Urban Contest. *Chem. Eng. Trans.* **2019**, *74*, 277.
22. Environmental Protection Agency of USA. National Ambient Air Quality Standards for Ozone [EB/OL]. Available online: https://www.govinfo.gov/content/pkg/FR-2015-10-26/pdf/2015-26594.pdf (accessed on 1 December 2020).
23. Ministry of Environmental Protection of China. Technical Regulation Ambient Air Quality Index (Trial) (HJ 633-2012) [EB/OL]. Available online: http://www.cnemc.cn/jcgf/dqhj/201706/t20170606_647274.shtml (accessed on 1 December 2020).
24. Gorai, A.K.; Kanchan; Upadhyay, A.; Goyal, P. Design of fuzzy synthetic evaluation model for air quality assessment. *Environ. Syst. Decis.* **2014**, *34*, 456–469. [CrossRef]
25. Suo, C.; Li, Y.; Sun, J.; Yin, S. An air quality index-based multistage type-2-fuzzy interval-stochastic programming model for energy and environmental systems management under multiple uncertainties. *Environ. Res.* **2018**, *167*, 98–114. [CrossRef]
26. Wang, S.; Du, L.; Wang, R. Comparison of Air Quality Index between China and Foreign Countries. *Environ. Monit. China* **2013**, *29*, 58–65. (In Chinese)
27. Gao, Q.-X.; Liu, J.-R.; Li, W.-T.; Gao, W.-K. Comparative Analysis and Inspiration of Air Quality Index between China and America. *Environ. Sci.* **2015**, *36*, 1141–1147. (In Chinese)
28. China National Environmental Monitoring Center. Release Notes of Urban Air Quality Real Time Release Platform of China [DB/OL]. Available online: http://106.37.208.233:20035/ (accessed on 1 December 2020).
29. Pan, B.F.; Zheng, H.H.; Li, L.N.; Wang, W. The Characteristic and Reason about the Reversal between PM2.5 and PM10 in Ambient Air Quality Automatic Monitoring. *Environ. Monit. China* **2014**, *30*, 90–95. (In Chinese)
30. Chen, G.; Pham, A.T.T. *Introduction to Fuzzy Sets, Fuzzy Logic, and Fuzzy Control Systems*; CRC Press: Boca Raton, FL, USA, 2001.
31. Ping, J.; Chen, B.; Husain, T. Risk Assessment of Ambient Air Quality by Stochastic-Based Fuzzy Approaches. *Environ. Eng. Sci.* **2010**, *27*, 233–246. [CrossRef]

32. Vadiati, M.; Moghaddam, A.A.; Nakhaei, M.; Adamowski, J.; Akbarzadeh, A. A fuzzy-logic based decision-making approach for identification of groundwater quality based on groundwater quality indices. *J. Environ. Manag.* **2016**, *184*, 255–270. [CrossRef]
33. Xu, Y.; Yang, W.; Wang, J. Air quality early-warning system for cities in China. *Atmos. Environ.* **2017**, *148*, 239–257. [CrossRef]
34. Yang, Z.; Wang, J. A new air quality monitoring and early warning system: Air quality assessment and air pollutant concentration prediction. *Environ. Res.* **2017**, *158*, 105–117. [CrossRef]
35. China National Environmental Monitoring Center. Bulletin on China's Environmental State [EB/OL]. Available online: http://www.cnemc.cn/jcbg/zghjzkgb/ (accessed on 1 December 2020).
36. Jindal, S.K. Regional Office for Europe. Air quality guidelines: Global update 2005. Particulate matter, ozone, nitrogen dioxide and sulfur dioxide. *Indian J. Med. Res.* **2007**, *4*, 492–493.
37. Directive 2008/50/EC of the European Parliament and of the Council of 21 May 2008 on Ambient Air Quality and Cleaner Air for Europe. Available online: http://eur-lex.europa.eu/LexUriServ/LexUriServ.do?uri=OJ:L:2008:152:0001:0044:EN:PDF (accessed on 1 December 2020).
38. Ministry of Ecology and Environment of the People's Republic of China. Explanation of Ambient Air Quality Standards [EB/OL]. Available online: http://www.mee.gov.cn/gkml/hbb/bgth/201011/t20101130_198128.htm (accessed on 1 December 2020).
39. Holgate, S. *Review of the UK Air Quality Index. A Report by the Committee on the Medical Effects of Air Pollutants*; Health Protection Agency: London, UK, 2011.
40. Ministry of Ecology and Environment of the People's Republic of China. Monthly Report on Urban Air Quality [EB/OL]. Available online: http://www.mee.gov.cn/hjzl/dqhj/cskqzlzkyb/index.shtml (accessed on 1 December 2020).
41. The People's Government of Beijing Municipality. Emergency Plan for Severe Air Pollution of Beijing [EB/OL]. Available online: http://www.beijing.gov.cn/zhengce/zfwj/zfwj2016/szfwj/201905/t20190522_61613.html (accessed on 1 December 2020).
42. The People's Government of Hongkou District, Shanghai. Emergency Plan for Severe Air Pollution of Shanghai [EB/OL]. Available online: http://xxgk.shhk.gov.cn/hkxxgk/depart/showinfo.aspx?infoid=488327c6-0a30-48da-9267-c83142ae78c6&categorynum=002004005001 (accessed on 1 December 2020).
43. The People's Government of Xi'an. Emergency Plan for Severe Air Pollution of Xi'an [EB/OL]. Available online: http://www.xa.gov.cn/gk/zhsgjy/yjyj/5d490976fd850833ac58c3ff.html (accessed on 1 December 2020).
44. Tianjin Municipal People's Government. Emergency Plan for Severe Air Pollution of Tianjin [EB/OL]. Available online: http://www.tj.gov.cn/zwgk/szfwj/tjsrmzfbgt/202005/t20200519_2370645.html (accessed on 1 December 2020).

3

Changes in Air Quality during the COVID-19 Pandemic and Associated Health Benefits in Korea

Ji Hoon Seo [1,2], Ji Soo Kim [3], Jinho Yang [3], Hyunjun Yun [3], Minwoo Roh [4], Jung Woon Kim [1], Songhun Yu [1], Na Na Jeong [1], Hyun Woo Jeon [3], Joung Sook Choi [3] and Jong Ryeul Sohn [1,2,3,*]

1 Department of Health & Environmental Science, Korea University, Seoul 02841, Korea; wlgns304@korea.ac.kr (J.H.S.); holysanta@korea.ac.kr (J.W.K.); thdgns0212@korea.ac.kr (S.Y.); nana8931@naver.com (N.N.J.)

2 Department of Public Health Sciences, BK21 FOUR R&E Center for Learning Health Systems, Graduate School, Korea University, Seoul 02841, Korea

3 Department of Health and Safety Convergence Science, Korea University, Seoul 02841, Korea; zs1414@korea.ac.kr (J.S.K.); iamjinho0@naver.com (J.Y.); hyunjun0116@korea.ac.kr (H.Y.); 2012250541@korea.ac.kr (H.W.J.); pooh03788@korea.ac.kr (J.S.C.)

4 Department of Environmental Science and Ecological Engineering, Korea University, Seoul 02841, Korea; mw0406toto@korea.ac.kr

* Correspondence: sohn1956@korea.ac.kr

Abstract: The COVID-19 pandemic was caused by a highly contagious coronavirus that has triggered worldwide control actions such as social distancing and lockdowns. COVID-19 control actions have resulted in improved air quality locally and around the world in the short-term by limiting human activity. We analyzed the impacts of social distancing and transboundary pollutants on air quality changes using open data and examined the corresponding health benefits focusing on two domestic cities (Seoul and Daegu) in Korea where the spread of coronavirus was severe. During the COVID-19 pandemic, $PM_{2.5}$, PM_{10}, and NO_2 concentrations decreased significantly by 31%, 61%, and 33%, respectively, compared to the previous three years. In particular, the $PM_{2.5}/PM_{10}$ ratio fell 24.5% after the implementation of social distancing, suggesting a decrease in anthropogenic emissions. Moreover, we found that the air quality index (AQI) also improved significantly, with a focus on reducing exposure to sensitive groups. In Seoul and Daegu, improved air quality prevented 250 and 78 premature deaths, and health costs were USD 884 million and USD 278 million, respectively. On the other hand, health loss due to COVID-19 deaths was in sharp contrast to USD 7.1 million and USD 543.6 million. Our findings indicate a significant association between COVID-19 prevalence patterns and health outcomes.

Keywords: COVID-19; air quality; health benefit; $PM_{2.5}$; PM_{10}

---

1. Introduction

The novel coronavirus (COVID-19) was reported in Wuhan, China at the end of 2019 and spread rapidly around the world, leading the World Health Organization to declare a pandemic in March 2020 [1]. By August, the number of confirmed cases worldwide exceeded 29.5 million, with 9.33 million deaths [2]. South Korea has experienced two coronavirus incidents in the past decade, middle east respiratory syndrome coronavirus (MERS) and COVID-19. Although COVID-19 has a lower fatality rate than MERS or severe acute respiratory syndrome (SARS) [3], the highest reproduction number was estimated to be 3.54 in the early stages of the epidemic in Korea [4], making it more contagious than any other coronavirus. Many countries where COVID-19 has been considered an international concern due to the spread of the virus have taken steps ranging from urging residents to stay at home to social distancing efforts and even lockdowns [5–12].

Starting with the first case on January 20, Korea's spread rate rapidly increased around the metropolitan areas, reporting the second-largest number of confirmed cases in the world until February. Social distancing to prevent the spread of coronavirus began on February 29, and at the end of April, the daily new confirmed cases remained below 10, entering a period of stabilization. However, in August, large-scale cluster infections were confirmed across the country and the second pandemic began, compelling the government to issue stronger social distancing guidelines. The increase in the cumulative number of confirmed cases and social distancing period in Korea are shown in Figure 1.

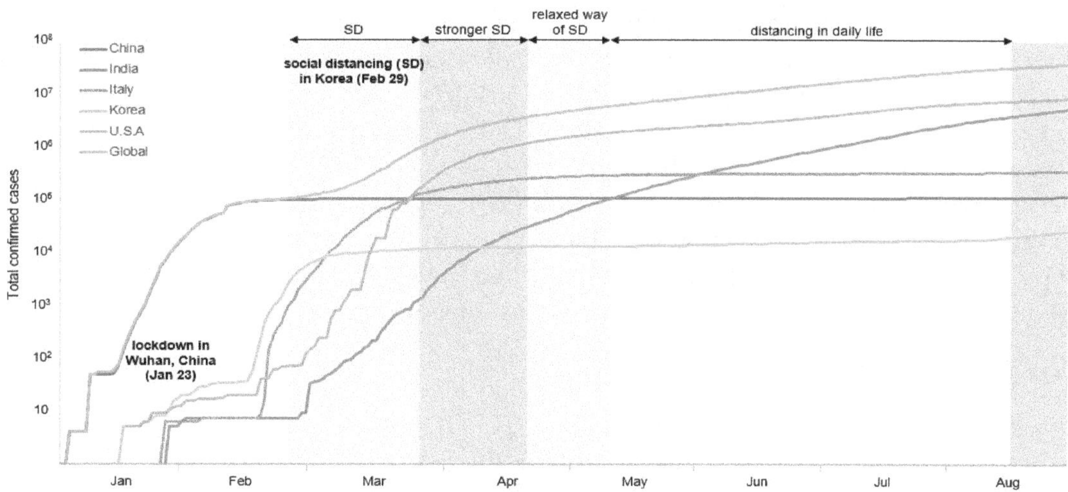

Figure 1. The cumulative number of COVID-19 cases worldwide. The highlighted shadow represents the level of social distancing (SD) in Korea. The data were collected from the WHO Coronavirus Disease (COVID-19) Dashboard [2].

Many recent studies and media have reported that COVID-19 control actions have not only slowed the spread of the virus [13–15], but also brought local and short-term changes in air quality [12,16–21]. Reduced human and industrial activities have decreased air pollutant emissions, and significant improvements in global air quality are being observed in 2020 [22–24]. In particular, in central China where the first COVID-19 confirmed case was reported, a significant decrease in particulate matter (PM) was observed immediately after lockdown. Compared to the same period in the previous three years, from January to March 2020, fine particles ($PM_{2.5}$) and respiratory suspended particles (PM_{10}) decreased by 30.1 to 40.5%, respectively, and the concentration of nitrogen dioxide (NO_2) was significantly reduced by 27.9% [24].

A decrease in air pollutant levels was also reported in Korea, which adopted social distancing, a regulation somewhat weaker than a COVID-19 lockdown [12,25,26]. The reduction in traffic-related emissions due to social distancing and reductions in transboundary pollutants from the neighboring countries lowered the $PM_{2.5}$, PM_{10}, and NO_2 concentrations by 45%, 36%, and 20%, respectively, in March compared to the previous three years [25]. The improved air quality in March was very unusual because, traditionally in Korea, the air quality in spring deteriorates appreciably due to the influence of Asian dust [27].

Meanwhile, in Sao Paulo, a representative megacity in Brazil where the confirmed cases increased rapidly, the concentration of $PM_{2.5}$, PM_{10}, and NO_2 also decreased by 45%, 46%, and 58%, respectively. This improvement in air quality prevented 802 premature deaths and saved about USD 720 million in health costs [11]. Similarly, in Korea, a study focused on Seoul to evaluate the health costs associated with $PM_{2.5}$ decreases, but only the overall change over four months was observed. Thus, detailed evaluations according to short-term changes in air quality were not performed [26]. Also, no research has evaluated only $PM_{2.5}$ but also air pollutants representative of Korea, such as PM_{10} and NO_2.

Moreover, in prior studies, air pollution levels were evaluated and health burden was assessed after COVID-19 control actions in major cities around the world through satellite data analysis, but only a slight difference in CO levels was observed in the Republic of Korea and other pollutants could not be accessed due to the lack of data [28].

Meanwhile, the WHO has reported 7.2 million deaths each year from air pollution, the fifth most serious health-related factor that causes disease [29]. Specifically, in Korea, it was predicted that the number of early deaths by air pollution would increase from 17,000 in 2010 to 30,000 in 2030 and 54,000 in 2060 [30]. Particulate matter (PM) and nitrogen dioxide (NO_2) are well-known air pollutants that cause numerous adverse health effects including cardiovascular diseases in Korea [31,32].

Reductions in air pollution caused a substantial and instant impact on health. In particular, within a few weeks, respiratory and irritation symptoms, cardiovascular diseases, and all-cause mortality were significantly reduced. Health improvements related to air quality led to significant health benefits and contributed to avoidable health risk [33].

The improvements in air quality since COVID-19 have provided an opportunity to evaluate health costs due to air pollutant reductions, including PM [25]. The purpose of our research was to observe the changes in air quality using open data in two representative cities with the rapid spread of COVID-19 in Korea. In addition, the potential health benefit was also investigated by analyzing deaths prevented by social distancing. This study is an applied research that assesses the short-term effect on health and air quality changes due to social measurement during the COVID-19 pandemic.

2. Materials and Methods

To access the health benefits resulting from COVID-19 control actions, the concentration of pollutants, exposure, environmental factors, mortality, and COVID-19 data were comprehensively analyzed as follows.

2.1. Study Site

After the first confirmed case was reported on January 20, local social infections spread widely around Daegu. By February 29, the cumulative confirmed cases accounted for 71% of the nation's confirmed cases (2236/3150), and even though the rate was reduced to 35% on August 31, the city still had the largest number of confirmed cases in Korea (7047/19947) [34].

Seoul is the most densely populated city in the OECD [35], with a population of about 10 million. The spread in Seoul began later than in Daegu, but since mid-August, there have been more than 100 newly confirmed cases per day, with 19% of the confirmed cases reported in Seoul on August 31 (3867/19947) [32]. Therefore, this study focused on the metropolitan city where coronavirus was most prevalent to observe changes in air quality before and after COVID-19 occurrence and evaluate the health benefits.

2.2. Air Pollutant Levels and Meteorological Conditions

The concentrations of $PM_{2.5}$, PM_{10}, and NO_2 were obtained from Air Korea (www.airkorea.or.kr/web) [36], and all data were applied to the Korea Registry of Environment (KMOE)'s Quality Assurance/Quality Control (QA/QC) procedure. There are 25 and 15 The Air Quality Monitoring Stations (AQMS) in Seoul and Daegu, respectively (Figure 2). We obtained the air pollutant levels on an hourly basis and calculated the daily and weekly averages to compare them to the previous three years (2017–2019). In this study, air pollution levels were analyzed for a total of 18 weeks. Observations were made from January 1, 2020, until May 5, when social distancing (SD) was implemented.

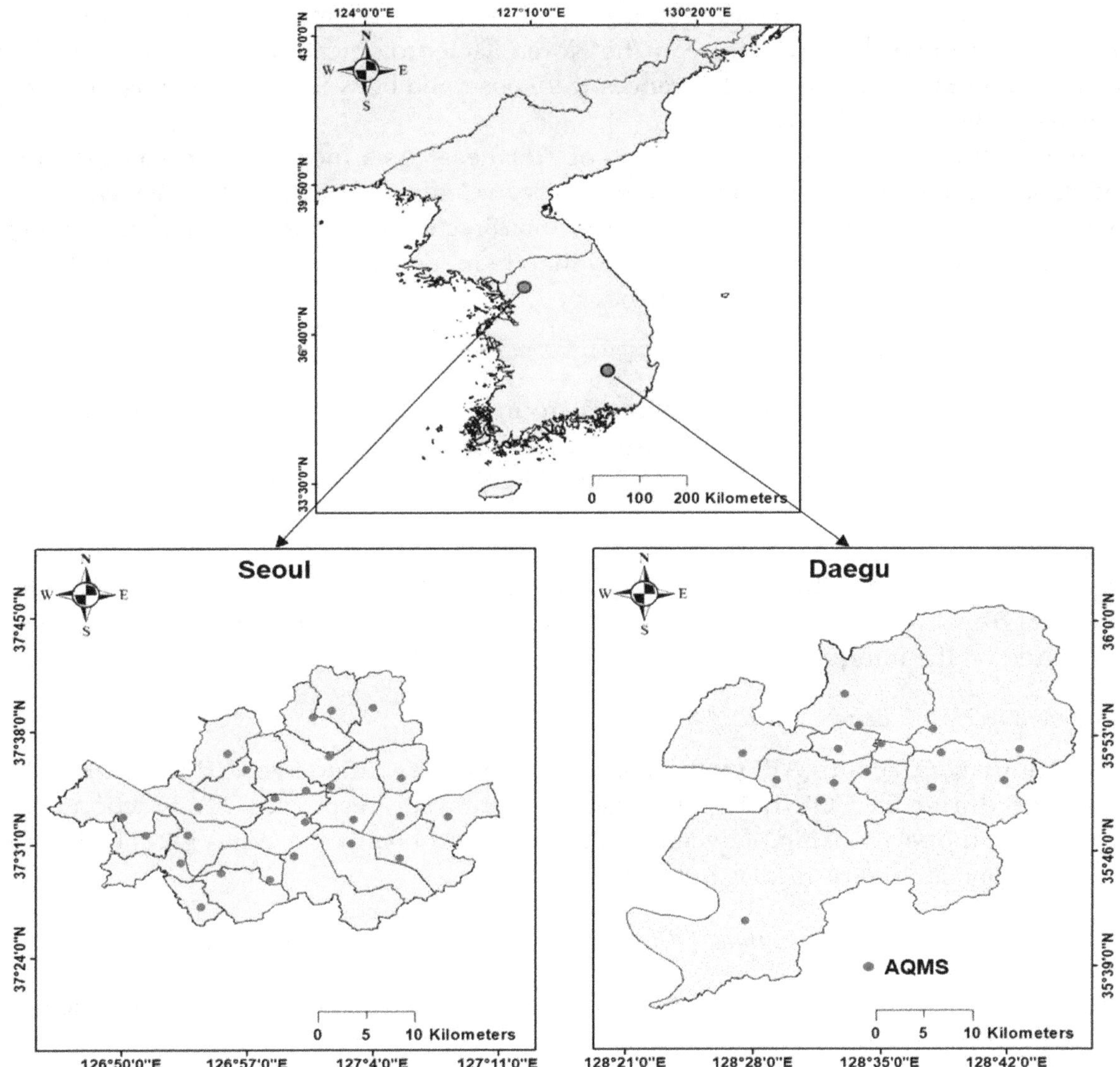

Figure 2. Location of Seoul (left) and Daegu (right) in Korea. Red dots indicate the location of the air quality monitoring stations (AQMS).

In addition, we evaluated the AQI to efficiently access air quality related to health concerns. The AQI is widely used as a tool to easily communicate how polluted the air currently is and the level of health effects on the public. The AQI was classified into six categories and expressed in colors representing different levels of health concern. The AQI of the air pollutants was calculated as interpolation Equation (1) [37]:

$$I = \frac{I_{high} - I_{low}}{C_{high} - C_{low}} * (C - C_{low}) + I_{low} \quad (1)$$

where

I: Air Quality Index
C: pollutant concentration
C_{low}: the concentration breakpoint $\leq C$
C_{high}: the concentration breakpoint $\geq C$
I_{low}: the index breakpoint corresponding to C_{low}
I_{high}: the index breakpoint corresponding to C_{high}

Meteorological conditions including precipitation, wind speed, temperature, and humidity were analyzed through open data available on the Korea Meteorological Administration website [38]. As shown in Table S1, no significant differences were observed between the previous three years and 2020 in meteorological conditions.

To analyze the wind speed and direction of Northeast Asia including Korea, 'ERA5 monthly averaged data on single levels' provided by the European Centre for Medium-Range Weather Forecasts (ECMWF) [39] was used to investigate speed and wind direction in the East Asian region from January to March 2020. Using this data, we made a wind map in meters per second by using ArcGIS 10.6 at a height of 100 m above the surface of the Earth.

2.3. COVID-19 and Mortality Data

The cumulative number of COVID-19 cases around the world was expressed by modifying the statistical data of the WHO Coronavirus Disease Dashboard [2]. Daily confirmed cases and death statistics related to COVID-19 were obtained through the Korea Disease Control and Prevention Agency (KCDA) [32].

In Korea, mortality data is published to the public through Statistics Korea (KOSTAT) [40], and since it is usually accessible after 1.5 years, the mortality rate in 2020 was estimated based on statistical data from 2016 to 2018. To evaluate all-cause mortality, we used non-accidental (A00-R00) mortality based on the International Classification of Disease [41].

2.4. Assessment of Health Benefits Related to Air Quality

The attributable fraction (AF) method was adopted to evaluate avoided deaths due to $PM_{2.5}$, PM_{10}, and NO_2 during the COVID-19 pandemic. Concentration-response (C-R) functions are widely used to evaluate the relationship between air pollution and health outcomes [42–44]. In this study, log-linear C-R functions were used as follows:

$$Avoidable\ (preventable)\ deaths = \left(1 - exp^{-\beta \times \Delta x}\right) \times number\ of\ deaths \qquad (2)$$

where β represents C-R coefficients, Δx is the decrease in the level of air pollutants, and the number of deaths was based on mortality over the same period in past years (2016–2018).

In this study, the C-R coefficients of each air pollutant adopted values derived from well-designed previous studies (Table 1). In addition, we used AirQ+ software tool that was developed by WHO to quantify the health impacts of air pollution.

Table 1. The concentration-response (C-R) coefficients for $PM_{2.5}$-, PM_{10}-, and NO_2-induced mortality.

Air Pollutants	Estimated C-R Regression Coefficients	Epidemiological References
$PM_{2.5}$	0.00405	[11,45]
PM_{10}	0.0008	[46]
NO_2	0.00135	[47]

Finally, we evaluated health outcomes to investigate the cost of mortality at the society level. The value of statistical life (VSL) represents how much individuals are willing to pay (WTP) to reduce the risk of death [48], and has been widely used in studies dealing with air quality and health [49–53]. The economic cost of a mortality impact was calculated through VSL and multiplied by premature deaths (COVID-19) or avoided deaths. In our study, we evaluated economic outcomes related to health by adopting USD 3.53 million proposed by the OECD as a representative VSL in Korea [54].

3. Results and Discussion

3.1. Comparision of Air Pollutant Levels

The time-series of $PM_{2.5}$, PM_{10} and NO_2 concentrations in Seoul and Daegu for the control period (2017–2019) and COVID-19 period in 2020 are shown in Figures 3–5, respectively. The concentration of $PM_{2.5}$ during the observation was significantly lower than that of the control period, with an average reduction of 30.7 μg/m^3. For $PM_{2.5}$, it was reduced by 44.2 μg/m^3 and 42.6 μg/m^3 in Seoul and Daegu, immediately after implementing SD (week 8) respectively (Figure 3). Significant reductions in $PM_{2.5}$ were also observed even before SD (weeks 1–7) due to changes in transboundary $PM_{2.5}$ concentrations [25]. Especially in China, the plant operations temporarily stopped in early January due to the New Year holiday, reducing industrial emissions. Also at the end of January, strict COVID-19 lockdown measures centered on Hubei Province were implemented, which curbed $PM_{2.5}$ man-made emissions due to traffic volume and industrial development [24]. The impact lasted until March, leading to a decrease in $PM_{2.5}$ concentrations in South Korea. Korea also showed a sharp drop in traffic after SD was implemented and a huge reduction in $PM_{2.5}$ concentrations in the early SD period as a result of decreased anthropogenic emissions as human activity decreased [12]. Since March 22nd, SD efforts were further strengthened (week 14) and another significant $PM_{2.5}$ decrease was observed. In particular, Seoul showed its greatest reduction during the observation period, with a decrease of nearly 80% from 57.1 μg/m^3 to 11.6 μg/m^3. $PM_{2.5}$ concentration changes were immediately apparent in the first SD efforts and two weeks after the implementation of strict SD (week 16), probably related to reductions in regional atmospheric congestion.

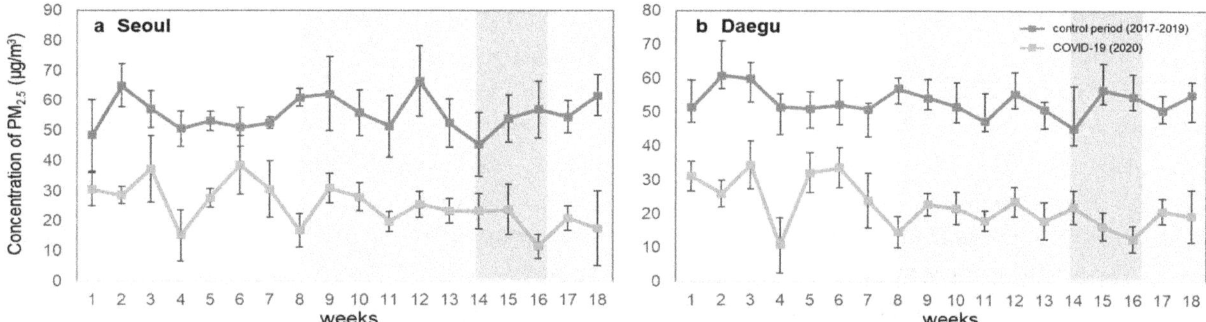

Figure 3. Comparison of average $PM_{2.5}$ concentrations by week between 2020 and the previous three years in (**a**) Seoul and (**b**) Daegu. Highlighted shadow represents the level of social distancing. The error bar means standard deviation.

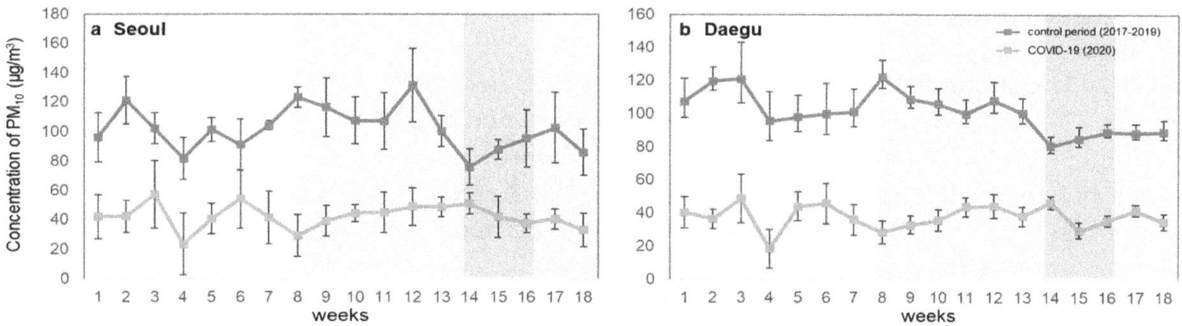

Figure 4. Comparison of average PM_{10} concentrations by week between 2020 and the previous three years in (**a**) Seoul and (**b**) Daegu. Highlighted shadow represents the level of social distancing. The error bar means standard deviation.

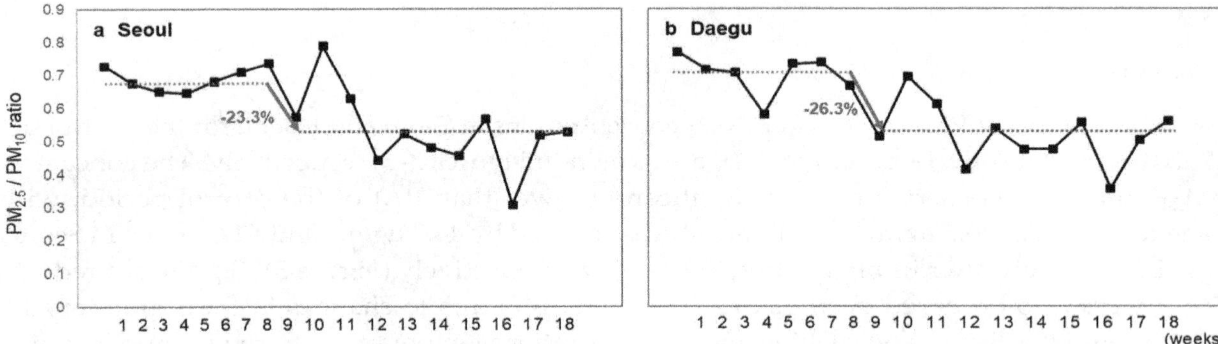

Figure 5. Changes in weekly $PM_{2.5}/PM_{10}$ ratios in (**a**) Seoul and (**b**) Daegu. The black dotted line represents the average ratio before SD, and the red dotted line represents the average ratio after SD. The reduction in the $PM_{2.5}/PM_{10}$ ratio is highlighted.

The changes in PM_{10} concentrations were similar to those of $PM_{2.5}$, as shown in Figure 4. Immediately after the implementation of SD measures, PM_{10} concentrations were reduced by 59.5 μg/ m^3 and 63.3 μg/m^3 in Seoul and Daegu, respectively. As described earlier, this was interpreted as the result of reductions in transboundary PM from the neighboring countries and decreases in PM due to SD. In particular, after the implementation of SD, PM_{10} concentrations in both cities were reduced by 76% compared to the control period. The effect of the enhanced SD is weak compared to that observed in $PM_{2.5}$ but still led to a continued decrease in PM_{10} concentrations.

We estimated the contribution of particles to the atmospheric PM by their size using the $PM_{2.5}/PM_{10}$ ratio (Figure 5). The ratio is widely used to identify the source of PM. A high ratio indicates a high contribution of $PM_{2.5}$, which represents artificial emission generation, and a low ratio represents the contribution of coarse particulates from natural sources such as Asian dust or wildfires [55,56]. Before and after SD, the $PM_{2.5}/PM_{10}$ ratio decreased significantly, which was related to reductions in anthropogenic $PM_{2.5}$ due to the restriction of human activities. Interestingly, notable changes in the $PM_{2.5}/PM_{10}$ ratio were also observed in the first two weeks following COVID-19 control actions. This delayed effect was mainly seen around 15 to 16 weeks when the wind speed was relatively slow (1.0–2.2 m/s) and the atmosphere was relatively stagnant.

The change in NO_2 concentrations was minute compared to PM but was significantly reduced compared to the control period as shown in Figure 6. The sharp decline in traffic since SD has led to a decrease in traffic-related NO_2 in both cities. Particularly in week 8, the average reduction was 38.7 μg/m^3 and a delayed reduction effect was shown two weeks after the implementation of enhanced SD (week 16). The average NO_2 concentration before SD was lower than that of the control period, but higher concentrations were partially observed, resulting in a relatively weaker effect of reduced transboundary pollutants compared to PM. According to a Korea-China-Japan cooperative study (Long-range Transboundary Air Pollutants in North East Asia: LTP), Chinese sources contributed about 40% of Korea's NOx, while the transboundary concentration of PM was up to 80% [57]. This supports our findings that the effects of transboundary NO_2 before SD were weaker than those of PM.

Even before the implementation of SD in Korea, Korea's air pollution levels were affected by the neighboring countries, which had already implemented COVID-19 control actions, resulting in significant reductions in concentrations along with the limiting effect of domestic sources due to SD.

In general, due to the nature of the Northeast Asian environment, which is dominated by western winds, South Korea is significantly affected by air pollutants from neighboring countries including China and North Korea [58]. According to KMOE, atmospheric PM in Korea from winter to spring is affected 28% to 82% from neighboring countries [59]. The Northeast Asia cooperative study also reported that the transboundary $PM_{2.5}$ from China was 32%, and the effect increased to 70% in the period of high concentration. NOx concentrations were also shown to have a transboundary effect of about 40% [55]. These previous research results support that air quality improvements before week

8, which were not related to social distancing in Korea, were related to strong COVID-19 lockdown measures in China.

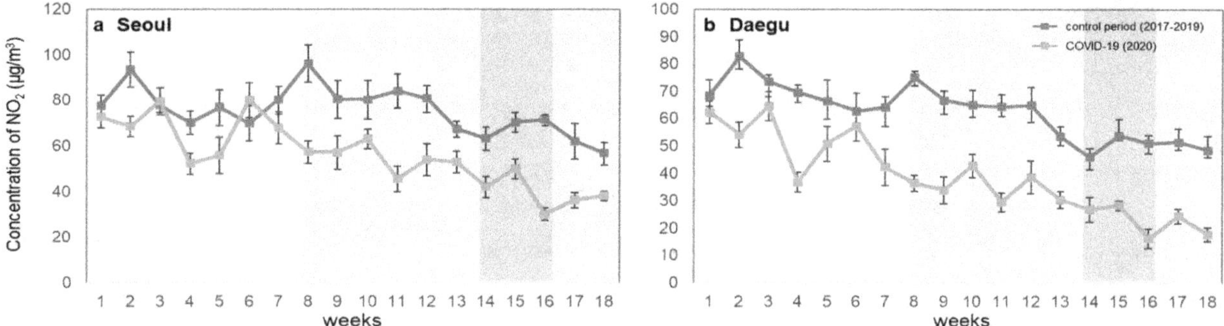

Figure 6. Comparison of average NO_2 concentrations by week between 2020 and the previous three years in (**a**) Seoul and (**b**) Daegu. Highlighted shadow represents the level of social distancing. The error bar means standard deviation.

The average wind speed and direction during the observation period are shown in Figure 7. As expected, the western wind was quite dominant in January and affected measurements in February and March. The results of wind direction and wind speed analysis during the pre- and post-SD periods support the impact of improved air quality in neighboring countries across the borders, especially in Northeast Asia.

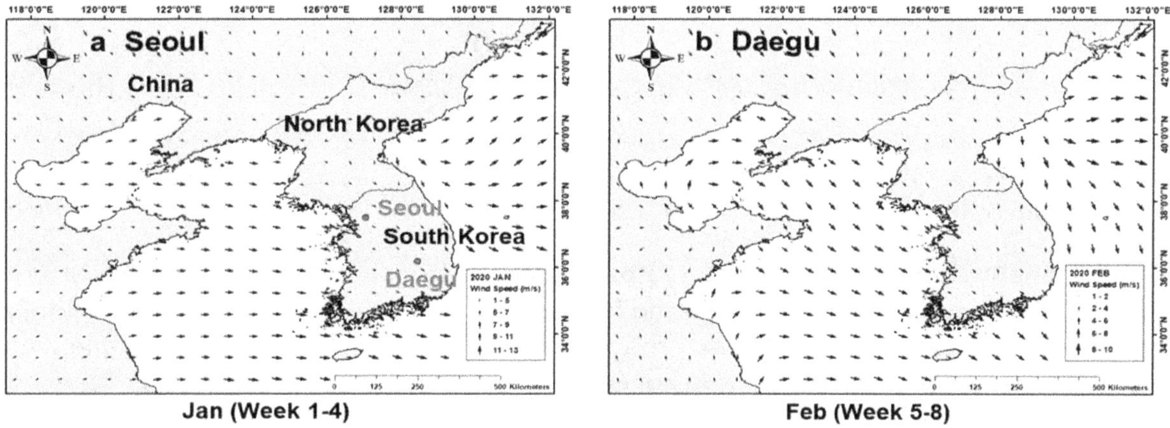

Figure 7. Map of wind speed and direction analysis in the Northeast Asia region in January and February. ((**a**): Seoul and (**b**): Daegu)

3.2. Distribution of Air Quality Index (AQI)

The daily AQI based on the concentrations of $PM_{2.5}$ and PM_{10} in Seoul and Daegu is shown in the Figure 8. NO_2 could only be calculated from an AQI above 200, but no day exceeded 200 in this study. The AQI results, shown in contrasting colors between the COVID-19 pandemic period and the control period showed a dramatic improvement in air quality. Interestingly, the AQI for $PM_{2.5}$ has not been classified as unhealthy (red) since the COVID-19 outbreak in either city. During the control period, the values unhealthy for the sensitive group (SG) accounted for the majority, with 91.2% and 92.8% in Seoul and Daegu, respectively, whereas in 2020, the proportions were moderate at 68.3% and 65.9%, respectively.

The AQI for PM_{10} also showed a significant reduction. In both cities, it was moderate during the control period at 86.4–98.4%, while good at 75.4–86.5% in 2020.

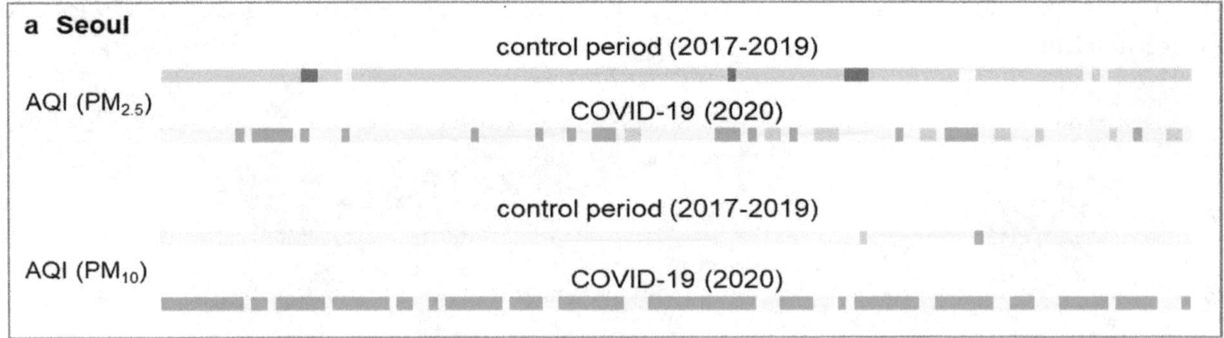

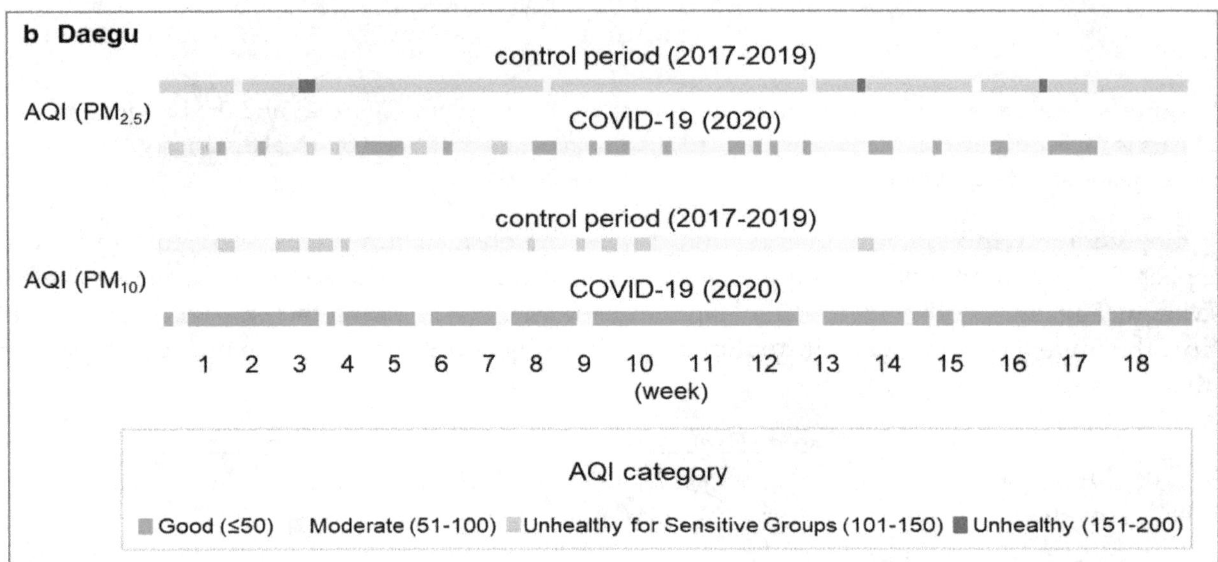

Figure 8. Comparison of daily AQI distribution of $PM_{2.5}$ and PM_{10}. Each AQI color indicates a different level of health concern. ((**a**): Seoul and (**b**): Daegu)

The AQI analysis showed a significant improvement in air quality centered on PM. In particular, it was rare that air quality deteriorated to the level of unhealthy for the SG, and significant health benefits would have been seen in at-risk groups (e.g., the elderly, children, and people with lung disease) due to the effects of air quality improvements in 2020.

In previous studies, a decrease in the AQI level of 14.9–32.2% was observed in central China where the COVID-19 lockdown was implemented [24]. In addition, significant improvement in air quality was also found in cities that implemented measures to prevent the spread of coronavirus [11], although not lockdowns, consistent with our study. These improvements in AQI have been observed worldwide, including the United States [44,60], Italy [61], and India [8,23] experiencing COVID-19.

3.3. Corresponding Health Outcomes

Based on the results of the concentration compared to the control period, the relative risks (RRs) and attributable factors (AFs) for each pollutant from January 2020 were calculated and are shown in Tables S2 and S3. In Seoul, the RR was 1.05–1.20 for $PM_{2.5}$, 1.02–1.08 for PM_{10}, and 0.99–1.06 for NO_2. In Daegu, the RR was 1.08–1.19 for $PM_{2.5}$, 1.03–1.08 for PM_{10}, and 1.01–1.05 for NO_2. In week 8, when the largest reduction in air pollutants was observed, the RR was the highest, and the AF values derived from the RRs were in the order of $PM_{2.5}$ (0.12) > PM_{10} (0.05) > NO_2 (0.03) in both cities. Noticeably, compared to the AFs observed in Sao Paulo, Brazil [11], a megacity similar to Seoul, the $PM_{2.5}$ and PM_{10} had greater AFs in Seoul, whereas NO_2 had greater AFs in Sao Paulo. This difference seems to have occurred because the main indicator of air quality is NO_2 in Sao Paulo whereas, in Korea, $PM_{2.5}$ is the contributory pollutant.

Based on the AF values, we assessed the preventable deaths and monetary values in each city (Table 2). The analysis showed many more preventable deaths due to reduced air pollution in Seoul, which has a large population. The decrease in PM_{10} and NO_2 saved the lives of at least 55 people, and especially $PM_{2.5}$ prevented about 250 premature deaths. The effects in Daegu were not as appreciable as those in Seoul, but due to a significant drop in $PM_{2.5}$, the number of avoidable deaths was estimated to reach about 78. The number of deaths from COVID-19 differed greatly from two in Seoul to 154 in Daegu during the investigation period.

Table 2. Assessment of weekly avoided deaths due to reductions in $PM_{2.5}$, PM_{10}, and NO_2.

Estimated Avoided Deaths						
Location	**Seoul**			**Daegu**		
Weeks	**$PM_{2.5}$**	**PM_{10}**	**NO_2**	**$PM_{2.5}$**	**PM_{10}**	**NO_2**
1	8.2	5.0	0.8	3.3	2.2	0.3
2	16.6	7.4	4.0	5.2	2.5	1.5
3	9.5	4.3	−0.3	3.5	2.0	0.4
4	16.9	5.8	3.0	5.7	2.3	1.6
5	11.5	5.6	3.3	2.7	1.5	0.8
6	6.0	3.5	−1.7	2.7	1.6	0.3
7	10.2	5.9	2.0	3.6	1.8	1.0
8	19.7	8.7	6.1	5.1	2.3	1.6
9	13.5	6.8	3.5	4.4	2.2	1.6
10	11.9	5.5	2.5	4.2	2.0	1.1
11	13.3	5.4	5.6	3.9	1.5	1.6
12	16.1	6.7	3.7	4.2	1.7	1.2
13	12.2	4.4	2.1	4.5	1.8	1.1
14	9.4	2.2	3.1	3.1	0.9	0.9
15	12.5	3.9	2.9	5.0	1.4	1.1
16	18.4	4.9	5.9	5.1	1.4	1.5
17	13.6	5.2	3.7	3.9	1.3	1.2
18	18.0	4.5	2.7	4.7	1.5	1.4

Health benefits based on the VSL approach were USD 884 million and USD 278 million in Seoul and Daegu, respectively (Figure 9). Interestingly, Seoul had two COVID-19 deaths, with economic losses of USD 7.1 million, while Daegu had 154 deaths and USD 543.6 million in health costs, showing sharply contrasting results. In Seoul, despite its dense population, social distancing was effective, whereas Daegu saw the rapid spread of secondary, tertiary, and other infections from cluster infection cases at the early stages of COVID-19. The results suggest that the initial spread rate of infectious diseases can have a significant effect on subsequent health outcomes.

The number of prevented deaths related to reductions in transboundary $PM_{2.5}$ before SD was 78 and 26 in Seoul and Daegu, respectively. After 8 weeks, since the effects of reductions in transboundary $PM_{2.5}$ were synergistic with the SD effect, it could not be evaluated separately, but the COVID-19 control actions of the neighboring countries probably continued to affect Korea. Thus, the premature deaths avoided before the 8th week of SD suggests that changes in air pollution in the neighboring countries could affect not only air quality but also health outcomes in other countries.

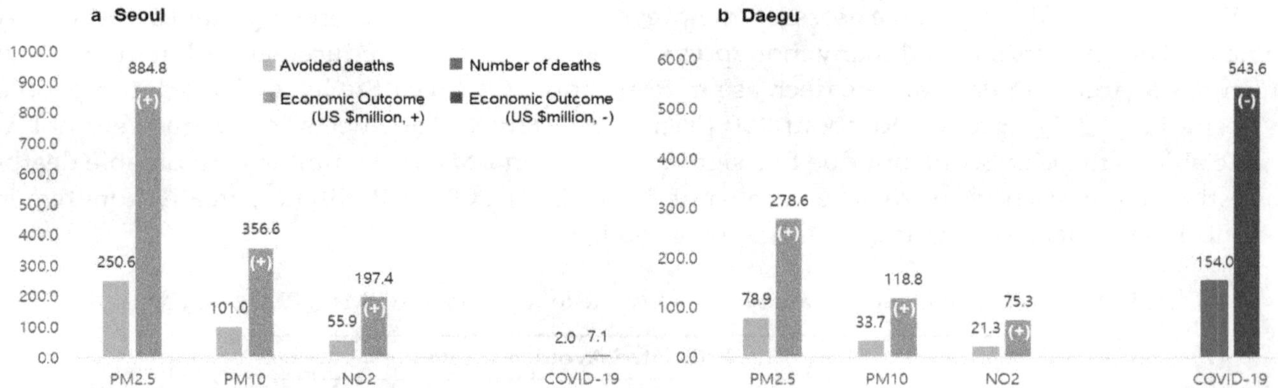

Figure 9. Avoided deaths from $PM_{2.5}$, PM_{10}, and NO_2 reductions over 18 weeks, and associated economic outcomes (+). The economic outcome from COVID-19 deaths was evaluated as (-). ((**a**): Seoul and (**b**): Daegu)

4. Conclusions

We investigated the impact of COVID-19 control actions on air quality and the resulting health benefits. Dramatic air quality improvements were observed in both cities. In the case of $PM_{2.5}$, it was reduced by 30.6–30.8% and PM_{10} was significantly reduced by 59.5–63.3%. NO_2 also decreased by 26.1–39.6%, supporting the positive effects of human activities.

The $PM_{2.5}/PM_{10}$ ratio significantly decreased from 0.69–0.70 before SD to 0.52–0.53 after SD. This proves that while the effect of improving air quality before SD was due to reductions in transboundary pollutants, the reductions in anthropogenic emissions in Korea after SD also affected the results.

The AQI of $PM_{2.5}$ and PM_{10} showed air quality improvement more clearly. South Korea generally experiences severe air pollution from winter to spring, but it was never rated as unhealthy in 2020. In addition, the number of days assessed as unhealthy for sensitive groups was considerably reduced, and it is thought that there would be a significant health effect on at-risk groups. However, the $PM_{2.5}/PM_{10}$ ratio still exceeded 0.5 and the AQI distribution often exceeded 100, confirming that $PM_{2.5}$ was still a contributory pollutant to Korean air quality.

We were able to assess the deaths prevented from reduced air pollutant levels. $PM_{2.5}$ was still a key pollutant in the health benefit evaluation, and there were 250 and 78 preventable deaths in Seoul and Daegu, respectively. The resulting health benefits were estimated at USD 884 million and USD 278 million, respectively. This was a conservative approach and could be evaluated as a greater health benefit when the effects of PM_{10} and NO_2 are considered together.

The number of deaths caused by COVID-19 showed a large difference, with two and 154 in Seoul and Daegu, and USD 7.1 million and USD 543.6 million in economic losses, respectively, showing a prime example of the remarkable health effects from the initial spread of coronavirus.

Several limitations exist in this study. First, the evaluation of air pollutant levels reflecting meteorological conditions was not performed. Precipitation, wind speed, and temperature can affect atmospheric diffusion, and more accurate concentration comparisons are needed in consideration of meteorological conditions. Second, since 2020 mortality data were not available, the past average mortality rate was used. Differences in mortality rates due to COVID-19 may occur, and the mortality statistics for 2020 can be used to access health benefits in more detail.

Nevertheless, we confirmed that COVID-19 control actions caused changes in air quality and contributed to preventing deaths from air pollution. Moreover, the results implied that changes in air quality in the neighboring countries caused effects across borders. Meanwhile, we assessed health benefits due to improved air quality, but it was evaluated that the global disease burden caused by COVID-19 is even greater.

The results of this study indicate the direction we should take to lead a sustainable and healthy life in preparation for the post-COVID-era and emphasize the necessity of national and global efforts.

Supplementary Materials:
Table S1: Meteorological conditions during COVID-19 pandemic and control periods, Table S2: Relative risks (RRs) and attributable factors (AFs) for $PM_{2.5}$, PM_{10}, and NO_2 (Seoul), and Table S3: Relative risks (RRs) and attributable factors (AFs) for $PM_{2.5}$, PM_{10}, and NO_2 (Daegu).

Author Contributions: Conceptualization, J.H.S. and J.R.S.; methodology, J.H.S., J.S.K., J.Y. and H.Y.; software, H.W.J. and M.R.; validation, J.W.K., S.Y. and J.Y.; formal analysis, J.H.S., J.W.K., S.Y., J.S.K. and H.Y.; investigation, J.H.S., H.W.J. and J.S.C.; data curation, N.N.J., J.Y. and J.S.C.; writing—original draft preparation, J.H.S. and J.R.S.; writing-review and editing, J.H.S. and J.R.S.; visualization, J.H.S. and M.R. All authors read and agreed to the published version of the manuscript.

Acknowledgments: The authors acknowledge the free use of the Air Korea data. The authors thank G. Byun and D. A. Huh for their insightful comments in writing the manuscript.

References

1. Cucinotta, D.; Vanelli, M. WHO declares COVID-19 a pandemic. *Acta Bio-Medica: Atenei Parmensis* **2020**, *91*, 157–160.
2. World Health Organization (WHO) Health Emergency Dashboard. Available online: https://covid19.who.int (accessed on 30 September 2020).
3. Petrosillo, N.; Viceconte, G.; Ergonul, O.; Ippolito, G.; Petersen, E. COVID-19, SARS and MERS: Are they closely related? *Clin. Microbiol. Infect.* **2020**, *26*, 729–734. [CrossRef] [PubMed]
4. Choi, S.; Ki, M. Estimating the reproductive number and the outbreak size of COVID-19 in Korea. *Epidemiol. Health* **2020**, *42*. [CrossRef] [PubMed]
5. Collivignarelli, M.C.; Abbà, A.; Bertanza, G.; Pedrazzani, R.; Ricciardi, P.; Miino, M.C. Lockdown for CoViD-2019 in Milan: What are the effects on air quality? *Sci. Total Environ.* **2020**, *732*, 139280. [CrossRef] [PubMed]
6. Dantas, G.; Siciliano, B.; França, B.B.; da Silva, C.M.; Arbilla, G. The impact of COVID-19 partial lockdown on the air quality of the city of Rio de Janeiro, Brazil. *Sci. Total Environ.* **2020**, *729*, 139085. [CrossRef]
7. Hashim, B.M.; Al-Naseri, S.K.; Al-Maliki, A.; Al-Ansari, N. Impact of COVID-19 lockdown on NO2, O3, PM2. 5 and PM10 concentrations and assessing air quality changes in Baghdad, Iraq. *Sci. Total Environ.* **2020**, 141978.
8. Kumari, P.; Toshniwal, D. Impact of lockdown measures during COVID-19 on air quality–A case study of India. *Int. J. Environ. Health Res.* **2020**, 1–8. [CrossRef]
9. Nakada, L.Y.K.; Urban, R.C. COVID-19 pandemic: Impacts on the air quality during the partial lockdown in São Paulo state, Brazil. *Sci. Total Environ.* **2020**, *730*, 139087. [CrossRef]
10. Singh, R.P.; Chauhan, A. Impact of lockdown on air quality in India during COVID-19 pandemic. *Air Qual. Atmos. Health* **2020**, *13*, 921–928. [CrossRef]
11. Debone, D.; da Costa, M.V.; Miraglia, S.G. 90 days of COVID-19 social distancing and its impacts on air quality and health in Sao Paulo, Brazil. *Sustainability* **2020**, *12*, 7440. [CrossRef]
12. Han, B.-S.; Park, K.; Kwak, K.-H.; Park, S.-B.; Jin, H.-G.; Moon, S.; Kim, J.-W.; Baik, J.-J. Air quality change in Seoul, South Korea under COVID-19 social distancing: Focusing on PM2. 5. *Int. J. Environ. Res. Public Health* **2020**, *17*, 6208. [CrossRef]
13. Anderson, R.M.; Heesterbeek, H.; Klinkenberg, D.; Hollingsworth, T.D. How will country-based mitigation measures influence the course of the COVID-19 epidemic? *Lancet* **2020**, *395*, 931–934. [CrossRef]
14. Chen, S.; Yang, J.; Yang, W.; Wang, C.; Bärnighausen, T. COVID-19 control in China during mass population movements at New Year. *Lancet* **2020**, *395*, 764–766. [CrossRef]
15. Lau, H.; Khosrawipour, V.; Kocbach, P.; Mikolajczyk, A.; Schubert, J.; Bania, J.; Khosrawipour, T. The positive impact of lockdown in Wuhan on containing the COVID-19 outbreak in China. *J. Travel Med.* **2020**, *27*, taaa037. [CrossRef] [PubMed]

16. He, G.; Pan, Y.; Tanaka, T. The short-term impacts of COVID-19 lockdown on urban air pollution in China. *Nat. Sustain.* **2020**, 1–7. [CrossRef]
17. Menut, L.; Bessagnet, B.; Siour, G.; Mailler, S.; Pennel, R.; Cholakian, A. Impact of lockdown measures to combat Covid-19 on air quality over western Europe. *Sci. Total Environ.* **2020**, *741*, 140426. [CrossRef] [PubMed]
18. Kroll, J.H.; Heald, C.L.; Cappa, C.D.; Farmer, D.K.; Fry, J.L.; Murphy, J.G.; Steiner, A.L. The complex chemical effects of COVID-19 shutdowns on air quality. *Nat. Chem.* **2020**, *12*, 777–779. [CrossRef]
19. Stratoulias, D.; Nuthammachot, N. Air quality development during the COVID-19 pandemic over a medium-sized urban area in Thailand. *Sci. Total Environ.* **2020**, *746*, 141320. [CrossRef]
20. Shehzad, K.; Sarfraz, M.; Shah, S.G.M. The impact of COVID-19 as a necessary evil on air pollution in India during the lockdown. *Environ. Pollut.* **2020**, *266*, 115080. [CrossRef]
21. Rodríguez-Urrego, D.; Rodríguez-Urrego, L. Air quality during the COVID-19: PM2. 5 analysis in the 50 most polluted capital cities in the world. *Environ. Pollut.* **2020**, 115042. [CrossRef]
22. Li, L.; Li, Q.; Huang, L.; Wang, Q.; Zhu, A.; Xu, J.; Liu, Z.; Li, H.; Shi, L.; Li, R. Air quality changes during the COVID-19 lockdown over the Yangtze River Delta Region: An insight into the impact of human activity pattern changes on air pollution variation. *Sci. Total Environ.* **2020**, 139282. [CrossRef]
23. Sharma, S.; Zhang, M.; Gao, J.; Zhang, H.; Kota, S.H. Effect of restricted emissions during COVID-19 on air quality in India. *Sci. Total Environ.* **2020**, *728*, 138878. [CrossRef] [PubMed]
24. Xu, K.; Cui, K.; Young, L.-H.; Wang, Y.-F.; Hsieh, Y.-K.; Wan, S.; Zhang, J. Air quality index, indicatory air pollutants and impact of COVID-19 event on the air quality near central China. *Aerosol Air Qual. Res.* **2020**, *20*, 1204–1221. [CrossRef]
25. Ju, M.J.; Oh, J.; Choi, Y.-H. Changes in air pollution levels after COVID-19 outbreak in Korea. *Sci. Total Environ.* **2020**, *750*, 141521. [CrossRef] [PubMed]
26. Han, C.; Hong, Y.-C. Decrease in ambient fine particulate matter during COVID-19 crisis and corresponding health benefits in Seoul, Korea. *Int. J. Environ. Res. Public Health* **2020**, *17*, 5279. [CrossRef] [PubMed]
27. Sharma, A.P.; Kim, K.H.; Ahn, J.w.; Shon, Z.H.; Sohn, J.R.; Lee, J.H.; Ma, C.J.; Brown, R.J. Ambient particulate matter (PM10) concentrations in major urban areas of Korea during 1996–2010. *Atmos. Pollut. Res.* **2014**, *5*, 161–169. [CrossRef]
28. Tseng, C.-H.; Wang, H.-C.; Xiao, N.-Y.; Chang, Y.-M. Examining the feasibility of prediction models by monitoring data and management data for bioaerosols inside office buildings. *Build. Environ.* **2011**, *46*, 2578–2589. [CrossRef]
29. World Health Organization (WHO). World Health Statistics. 2015. Available online: https://www.who.int/gho/publications/world_health_statistics/2015/en/ (accessed on 30 September 2020).
30. OCDE, O. *Economic Consequences of Outdoor Air Pollution*; Organisation for Economic Co-operation and Development: Paris, France, 2016.
31. Son, J.-Y.; Bell, M.L. The relationships between short-term exposure to particulate matter and mortality in Korea: Impact of particulate matter exposure metrics for sub-daily exposures. *Environ. Res. Lett.* **2013**, *8*, 014015. [CrossRef]
32. Kim, S.-Y.; Kim, H.; Lee, J.-T. Health Effects of air-quality regulations in Seoul Metropolitan area: Applying synthetic control method to controlled-interrupted time-series analysis. *Atmosphere* **2020**, *11*, 868. [CrossRef]
33. Schraufnagel, D.E.; Balmes, J.R.; De Matteis, S.; Hoffman, B.; Kim, W.J.; Perez-Padilla, R.; Rice, M.; Sood, A.; Vanker, A.; Wuebbles, D.J. Health benefits of air pollution reduction. *Ann. Am. Thorac. Soc.* **2019**, *16*, 1478–1487. [CrossRef]
34. Ministry of Health and Welfare (MOHW). COVID-19. Available online: http://ncov.mohw.go.kr/bdBoardList_Real.do (accessed on 30 September 2020).
35. Organization for Economic Cooperation and Development (OECD). OECD Better Life Index. Available online: http://www.oecdbetterlifeindex.org/countries/korea/ (accessed on 30 September 2020).
36. Korea Ministry of Environment (KMOE). AirKorea. Available online: https://www.airkorea.or.kr/index (accessed on 30 September 2020).
37. Mintz, D. *Technical Assistance Document for the Reporting of Daily Air Quality-The Air Quality Index (AQI)*; US Environmental Protection Agency: Research Triangle Park, NC, USA, 2009.
38. Korea Meteorological Administration (KMA). Weather Information. Available online: http://www.kma.go.kr/eng/index.jsp (accessed on 30 September 2020).

39. Copernicus Climate Change Service (C3S). ERA5: Fifth Generation of ECMWF Atmospheric Reanalyses of the Global Climate. 2017. Available online: https://www.ecmwf.int/sites/default/files/elibrary/2017/17312-news-c3s-era5.pdf (accessed on 30 September 2020).
40. Statistics Korea (KOSTAT). MicroData Intergrated Service (MDIS). Available online: https://mdis.kostat.go.kr/index.do (accessed on 30 September 2020).
41. World Health Organization. *International Statistical Classification of Diseases and Related Health Problems: Instruction Manual*; World Health Organization: Geneva, Switzerland, 2004; Volume 2.
42. Pinichka, C.; Makka, N.; Sukkumnoed, D.; Chariyalertsak, S.; Inchai, P.; Bundhamcharoen, K. Burden of disease attributed to ambient air pollution in Thailand: A GIS-based approach. *PLoS ONE* **2017**, *12*, e0189909. [CrossRef]
43. Pope, C.A., III; Cropper, M.; Coggins, J.; Cohen, A. Health benefits of air pollution abatement policy: Role of the shape of the concentration–response function. *J. Air Waste Manag. Assoc.* **2015**, *65*, 516–522. [CrossRef] [PubMed]
44. Pan, S.; Jung, J.; Li, Z.; Hou, X.; Roy, A.; Choi, Y.; Gao, H.O. Air quality implications of COVID-19 in California. *Sustainability* **2020**, *12*, 7067. [CrossRef]
45. US Environmental Protection Agency. *Quantitative Health Risk Assessment for Particulate Matter*; Office of Air and Radiation, Office of Air Quality Planning and Standards: Triangle Park, NC, USA, 2010.
46. Chalvatzaki, E.; Chatoutsidou, S.E.; Lehtomäki, H.; Almeida, S.M.; Eleftheriadis, K.; Hänninen, O.; Lazaridis, M. Characterization of human health risks from particulate air pollution in selected European cities. *Atmosphere* **2019**, *10*, 96. [CrossRef]
47. Zhong, S.; Yu, Z.; Zhu, W. Study of the effects of air pollutants on human health based on Baidu indices of disease symptoms and air quality monitoring data in Beijing, China. *Int. J. Environ. Res. Public Health* **2019**, *16*, 1014. [CrossRef] [PubMed]
48. Ligus, M. Measuring the willingness to pay for improved air quality: A contingent valuation survey. *Pol. J. Environ. Stud.* **2018**, *27*, 763–771. [CrossRef]
49. Hubbell, B.J. Evaluating the health benefits of air pollution reductions: Recent developments at the US EPA. In Proceedings of the Prepared for the Symposium on The Measurement and Economic Valuation of Health Effects of Air Pollution, London, UK, 19–20 February 2001; Institute of Materials: London, UK, 2001; pp. 3–5.
50. Amann, M.; Holland, M.; Maas, R.; Vandyck, T.; Saveyn, B. *Costs, Benefits and Economic Impacts of the EU Clean Air Strategy and Their Implications on Innovation and Competitiveness*; IIASA Report; IIASA: Laxenburg, Austria, 2017.
51. World Health Organization. *Economic Cost of the Health Impact of Air Pollution in Europe: Clean Air, Health and Wealth*; WHO Regional Office for Europe: Copenhagen, Denmark, 2017.
52. Markandya, A.; Sampedro, J.; Smith, S.J.; Van Dingenen, R.; Pizarro-Irizar, C.; Arto, I.; González-Eguino, M. Health co-benefits from air pollution and mitigation costs of the Paris Agreement: A modelling study. *Lancet Planet. Health* **2018**, *2*, e126–e133. [CrossRef]
53. Martinez, G.S.; Spadaro, J.V.; Chapizanis, D.; Kendrovski, V.; Kochubovski, M.; Mudu, P. Health impacts and economic costs of air pollution in the metropolitan area of Skopje. *Int. J. Environ. Res. Public Health* **2018**, *15*, 626. [CrossRef]
54. Roy, R.; Braathen, N.A. *The Rising Cost of Ambient Air Pollution thus far in the 21st Century: Results from the BRIICS and the OECD Countries*; OECD: Paris, France, 2017.
55. Zhao, D.; Chen, H.; Yu, E.; Luo, T. PM2. 5/PM10 ratios in eight economic regions and their relationship with meteorology in China. *Adv. Meteorol.* **2019**, *2019*. [CrossRef]
56. Xu, G.; Jiao, L.; Zhang, B.; Zhao, S.; Yuan, M.; Gu, Y.; Liu, J.; Tang, X. Spatial and temporal variability of the PM2. 5/PM10 ratio in Wuhan, Central China. *Aerosol Air Qual. Res.* **2016**, *17*, 741–751. [CrossRef]
57. Joint Research Project for Long–range Transboundary Air Pollutants in Northeast Asia, Summary Report of the 4th stage (2013–2017) LTP Project. 2019. Available online: http://www.neaspec.org/sites/default/files/3-2%20Joint%20Research%20Project%20on%20Long-Range%20Transboundary%20Air%20Pollutants.pdf (accessed on 30 September 2020).
58. Kim, M.J. The effects of transboundary air pollution from China on ambient air quality in South Korea. *Heliyon* **2019**, *5*, e02953. [CrossRef]

59. National Institute of Environmental Reaserch. Analysis of fine dust occurrence cases. Available online: https://www.nier.go.kr/NIER/cop/bbs/selectNoLoginBoardArticle.do?menuNo=14003&bbsId=BBSMSTR_000000000022&nttId=24685&Command=READ (accessed on 30 September 2020).
60. Berman, J.D.; Ebisu, K. Changes in US air pollution during the COVID-19 pandemic. *Sci. Total Environ.* **2020**, *739*, 139864. [CrossRef] [PubMed]
61. Zoran, M.A.; Savastru, R.S.; Savastru, D.M.; Tautan, M.N. Assessing the relationship between ground levels of ozone (O_3) and nitrogen dioxide (NO_2) with coronavirus (COVID-19) in Milan, Italy. *Sci. Total Environ.* **2020**, *740*, 140005. [CrossRef] [PubMed]

4

Indoor Air Pollution from Residential Stoves: Examining the Flooding of Particulate Matter into Homes during Real-World use

Rohit Chakraborty [1,*], James Heydon [2], Martin Mayfield [1] and Lyudmila Mihaylova [3]

[1] Department of Civil and Structural Engineering, The University of Sheffield, Sheffield S1 3JD, UK; martin.mayfield@sheffield.ac.uk
[2] School of Sociology and Social Policy, The University of Nottingham, Nottingham NG7 2RD, UK; james.heydon@nottingham.ac.uk
[3] Department of Automatic Control and Systems Engineering, The University of Sheffield, Sheffield S1 3JD, UK; l.s.mihaylova@sheffield.ac.uk
* Correspondence: rohit.chakraborty@sheffield.ac.uk

Abstract: This study concerns the levels of particulate matter ($PM_{2.5}$ and PM_1) released by residential stoves inside the home during 'real world' use. Focusing on stoves that were certified by the UK's Department of Environment, Food, and Rural Affairs (DEFRA), PM sensors were placed in the vicinity of 20 different stoves over four weeks, recording 260 uses. The participants completed a research diary in order to provide information on time lit, amount and type of fuel used, and duration of use, among other details. Multivariate statistical tools were used in order to analyse indoor PM concentrations, averages, intensities, and their relationship to aspects of stove management. The study has four core findings. First, the daily average indoor PM concentrations when a stove was used were higher for $PM_{2.5}$ by 196.23% and PM_1 by 227.80% than those of the non-use control group. Second, hourly peak averages are higher for $PM_{2.5}$ by 123.91% and for PM_1 by 133.09% than daily averages, showing that PM is 'flooding' into indoor areas through normal use. Third, the peaks that are derived from these 'flooding' incidents are associated with the number of fuel pieces used and length of the burn period. This points to the opening of the stove door as a primary mechanism for introducing PM into the home. Finally, it demonstrates that the indoor air pollution being witnessed is not originating from outside the home. Taken together, the study demonstrates that people inside homes with a residential stove are at risk of exposure to high intensities of $PM_{2.5}$ and PM_1 within a short period of time through normal use. It is recommended that this risk be reflected in the testing and regulation of residential stoves.

Keywords: air pollution; wood burning stoves; particulate matter; particle number concentration; indoor air quality; low-cost sensors; internet of things

---

1. Introduction

As a component of air pollution, particulate matter with an aerodynamic diameter that is equal to 2.5 μm or less ($PM_{2.5}$) has long been linked to adverse health effects. In terms of mortality, it causes seven-million deaths per year [1]. In terms of health effects, it causes inflammation and oxidative stress, which compromises pulmonary immunity and increases the susceptibility to infection [2]. As these particulates can move into every organ in the body, the illnesses that are associated with their presence range from lung cancer, bronchitis, and other respiratory infections, through to strokes, dementia, and Parkinson's disease [3]. Effects such as these are particularly pronounced for children, pregnancies, and the elderly [4]. While much research focuses on particulate emissions that are generated by industry and vehicles, in the United Kingdom (UK) the primary source for $PM_{2.5}$ is the domestic burning of wood and coal for heating [5]. Government estimates suggest that one in twelve UK homes is using residential stoves [6] and, in doing so, causing 38% of the nation's $PM_{2.5}$ emissions [5]. Growing in popularity, UK industry data suggest that stove sales are running between 150,000 and 200,000 units per year, with over one million being sold between 2010 and 2015 [7]. Several reasons have been posited for this, including perceived lower fuel costs where wood or biomass is recovered locally, particularly where this intersects with fuel poverty, with residential stoves becoming a lifestyle choice for those who already have a primary source of heating in their home [8], and the perception that wood burning stoves are low-carbon, because they can use renewable fuels [9]. Much of the existing literature on these residential stoves focuses on their efficiency [10,11] and outdoor emissions [12–14], with many also deploying monitoring equipment in order to establish the indoor PM emissions that originate from their use. Early work by Traynor et al. [15] measured indoor emissions from four wood burning stoves, finding that all of the stoves emitted particles indoor at some point during use. Canha et al. [16] found that wood burning used to heat one school classroom in rural Portugal contributed high levels of $PM_{2.5}$ to the indoor environment. Semmens et al. [17] examined 98 stoves over 48 h, finding average indoor $PM_{2.5}$ concentrations to exceed World Health Organisation ambient air quality guidelines and approach the United States Environment Protection Agency (U.S. EPA) 24-h standard equivalent. Piccardo et al. [18] tested indoor air emissions from nine stoves, finding indoor air pollution to be consistent with errors in self-installation and mismanagement. Wang et al. [19] tested one stove under lab conditions and four stoves in real-world settings. The number of tests conducted, or the real-world measurements taken, are unclear, but the study concludes that different emissions occur at different points during the burn cycle. Vicente et al. [20] tested one open fire and one wood stove under lab conditions, finding that the PM_{10} levels increased 12-fold for the former and 2-fold for the latter during operation. Allen et al. [21] upgraded stoves in 15 houses in order to understand the extent to which stove design can improve indoor air quality, finding that no consistent improvement occurs. Table 1 summarises this literature. While adding to understandings of indoor stove emissions, this body of scholarship also exhibits several limitations.

First, existing studies tend to judge indoor stove emissions against official average exposure guidelines [22]. This is a dominant approach in air quality research, but it serves to obfuscate emission 'peaks' by averaging them out of the results. For instance, while Semmens et al. [17] found that the 'reported number of times the wood stove was opened was not associated with $PM_{2.5}$ or any particle size fraction', this judgement was made in the context of a 48h average. This is problematic because epidemiologists are increasingly recognising that exposure to high intensities of PM over much shorter periods of time—hours rather than days—is linked to a range of health issues [23–26]. Indeed, Lin et al. [27] found a significant association between hourly peak $PM_{2.5}$ and mortality rates across six Chinese cities. Similarly, a systematic review of 196 articles found a positive relationship between short term PM exposure and cardiovascular, respiratory, and cerebrovascular mortality [28]. Several existing studies report stoves emitting peaks indoors, but these are either observed under controlled conditions [15,20,29,30] or have few real-world users or uses from which to derive data [19,21,22].

Second, the number of stove uses upon which conclusions are drawn is highly variable (see Table 1). This is less of an issue with lab-based testing, as the circumstances of use can be tightly controlled. However, low frequencies of use pose a challenge for studies into real-world emissions because one instance of stove management may not be identical to another. Relatedly, participants may actively change their behaviour if aware they are being observed. Known as 'participant reactivity', this can be produced by researchers through obvious and repeated intervention into a social setting. In order to minimise this influence and more accurately ascertain what indoor emissions are occurring through normal use, the sampling of a greater number of stove uses over a longer period of time, and without obvious researcher intervention in the social setting, is required.

Third, existing studies are not clear about the standard of stove being tested. The fuel accepted is outlined and the stove described, albeit inconsistently so (see Table 1), but the design regulations to which the stoves adhere, if at all, tend not to be detailed. This makes it difficult to generalise findings to categories of stove that share fundamental design features. Where stove standards are described, those chosen tend to have been approved by regulators outside the UK. For instance, [17,21,22,31] have focused on stoves that are approved by environmental regulators, but these are limited to the USA and Canadian contexts. Taken together, this relationship between indoor emissions and UK-specific regulations that govern stove design and testing requires investigation.

Fourth, few of the existing studies examine Ultra Fine Particles (UFP), which are defined as particles with a diameter of less than 100 nm, or Particle Number Concentration (PNC), which is defined as the total number of particles measured per cubic centimeter in a given sample. Measuring PNC along with the regular mass concentration measurements of $PM_{2.5}$ is important because PNC and $PM_{2.5}$ are not representative of each other [32], with Pearson's r lying between 0.09–0.64 and high levels of $PM_{2.5}$ not necessarily causing high levels of PNC or vice versa. Therefore, measures that are taken to reduce or regulate $PM_{2.5}$ may be different to those that are needed to tackle the problem of increasing PNC. Indeed, Penttinen et al. [33] found a stronger negative association between PNC and peak expiratory flow (PEF) than $PM_{2.5}$ amongst asthmatic children. Therefore, UFP may pose a substantial health risk since PNC exposure increases remarkably in the smallest size fractions.

When considering these limitations, this study has four aims. First, it seeks to determine real-world indoor PM exposure from the use of residential heating stoves over 30 days. This period was chosen to increase the number of uses from which data could be derived without instructing participants to use their stoves, minimise intrusion into the research setting, and more accurately capture 'real-world' use. Second, it detects and identifies the existence of peak indoor $PM_{2.5}$ and PM_1 levels as a result of stove use. Third, it seeks to clarify whether the level of indoor air pollution is originating from indoor or outdoor sources. Finally, it seeks to determine the extent to which these emissions are coming from a specific category of stoves; those that are certified as a 'Smoke Exempt Appliance' by the UK's Department for Environment, Farming, and Rural Affairs (DEFRA). These stoves are modified in order to restrict incoming air and limit smoke produced from combustion, differentiating them from the older equipment of focus in Semmens et al. [17]. If a stove passes the official testing process [34], they are certified to be exempt from the Smoke Control Area regulations covering most of the UK's towns and cities. However, this testing is limited to measuring outdoor air pollution via flue emissions and heat output; none of the applicable standards that are required by DEFRA are concerned with indoor PM emissions from stoves (see PD 6434: 1969; BS 3841: Part 1: 1994; BS 3841: Part 2: 1994). Even the latest 'EcoDesign' standards, which call up EN 16510:2018, do not introduce testing for indoor emissions. Indeed, when taken together, the DEFRA testing regime rests on a baseline assumption that stoves do not pollute indoors, or only do so when a fault is present. The results of this study test the validity of that foundational assumption. Taken together, this work makes three core contributions:

i It presents a framework in order to determine real-world indoor PM exposure from the use of residential heating stoves.

ii It can detect and identify the existence of peak indoor $PM_{2.5}$, PM_1, and PNC levels as a result of stove use.

iii It analyses the results in relation to the DEFRA regulations and determines the extent of these emissions from a specific category of stoves; those that are certified as a 'Smoke Exempt Appliance' by DEFRA.

In making these contributions, the study seeks to determine whether health risks are posed during normal operation and, in turn, whether DEFRA testing standards need modification in light of this reality.

The remainder of this paper is organised, as follows. Section 2 describes the experimental framework along with sensor calibration and evaluation in Section 2.2. Section 3 presents the findings and analysis, which is followed by the conclusion in Section 4.

Table 1. Overview of Existing Literature that has Monitoring Indoor Pollution from Residential Heating Stoves.

Study	Year-Study Site	No. of Sampled Stoves	Lab-Conditions or Real-World?	Heating Unit Type and Fuel Acceptance	No. of Uses Analysis Based on
Traynor et al. [15]	1987-USA	4	Lab/Real-world hybrid [1]	Wood stoves (3 'airtight', 1 'non-airtight Franklin model')	11
Allen et al. [21]	2009-Canada	15	Real-world (stove upgrade halfway through)	Wood stove (non-EPA-certified and EPA-certified)	Not provided (2 three-day samples taken over 6 days)
Noonan et al. [22]	2012-USA	21	Real-world (stove upgrade halfway through)	Wood stove (non-EPA-certified and EPA-certified)	Approx. 60 (1-4 samples taken from each home across 3 winters)
McNamara et al. [31]	2013-USA	50	Real-world	Wood stove (Non-EPA certified 'older model')	Not provided (4 separate 48h sampling visits over 2 winters)
Canha et al. [16]	2014 -Portugal	1	Real-world	Wood stove ('slow combustion stove')	1
Salthammer et al. [35]	2014-Germany	7	Real-world Wood stove ('closed')	6 Wood stove ('open')1	3 days for each stove
Piccardo et al. [18]	2014-Italy	9	Real-world	Wood stoves	183
Semmens et al. [17]	2015-USA	96	Real-world	Wood stoves ('older models' without 'modern control features focused on emission reduction')	192 (each stove used twice)
Vicente et al. [30]	2015-Portugal	1	Lab-conditions	Wood stove ('stainless steel with a cast iron grate')	Not provided
Mitchell et al. [29]	2016-UK and Ireland	1	Lab-conditions	Multi-fuel stove ('fixed grate stove with a single combustion chamber')	8
Wang et al. [19]	2020-China	5	Lab-conditions(1) Real-world(4)	Coal stoves (Real world—'steel stoves, cylindrical burning chamber, connected to a chimney')	Not provided
Vicente et al. [20]	2020-Portugal	2	Lab-conditions	Open fireplace and wood stove	7 (4 open fire, 3 wood stove)
Chakraborty et al.	2020-UK	20	Real-world	DEFRA-certified wood (14)- DEFRA-certified multi-fuel (5)- Defra-compliant open fire (1)	260 [2]

[1] The stoves were installed in a house but used under controlled conditions; [2] 280 uses in total but 20 removed due to incomplete data.

2. Materials and Methods

2.1. Sampling Area and Study Design

Sheffield (53°23′ N 1°28′ W), the chosen study site that is shown in Figure 1a, is a geographically diverse city that is located in the county of South Yorkshire, England. Built on several hills, it is situated at an elevation of 29 m–500 m above sea level, covers a total area of 367.9 km^2, and it has a growing population of 582,506 [36]. Sheffield has a temperate climate; July is considered to be the hottest month, with an average maximum temperature of 20.8 oC and January–February to be the coldest months. Air pollution in the city is primarily from road transport and industrial emissions and, to a lesser extent, fossil fuels run processes, such as energy supply and commercial or domestic heating systems [37].

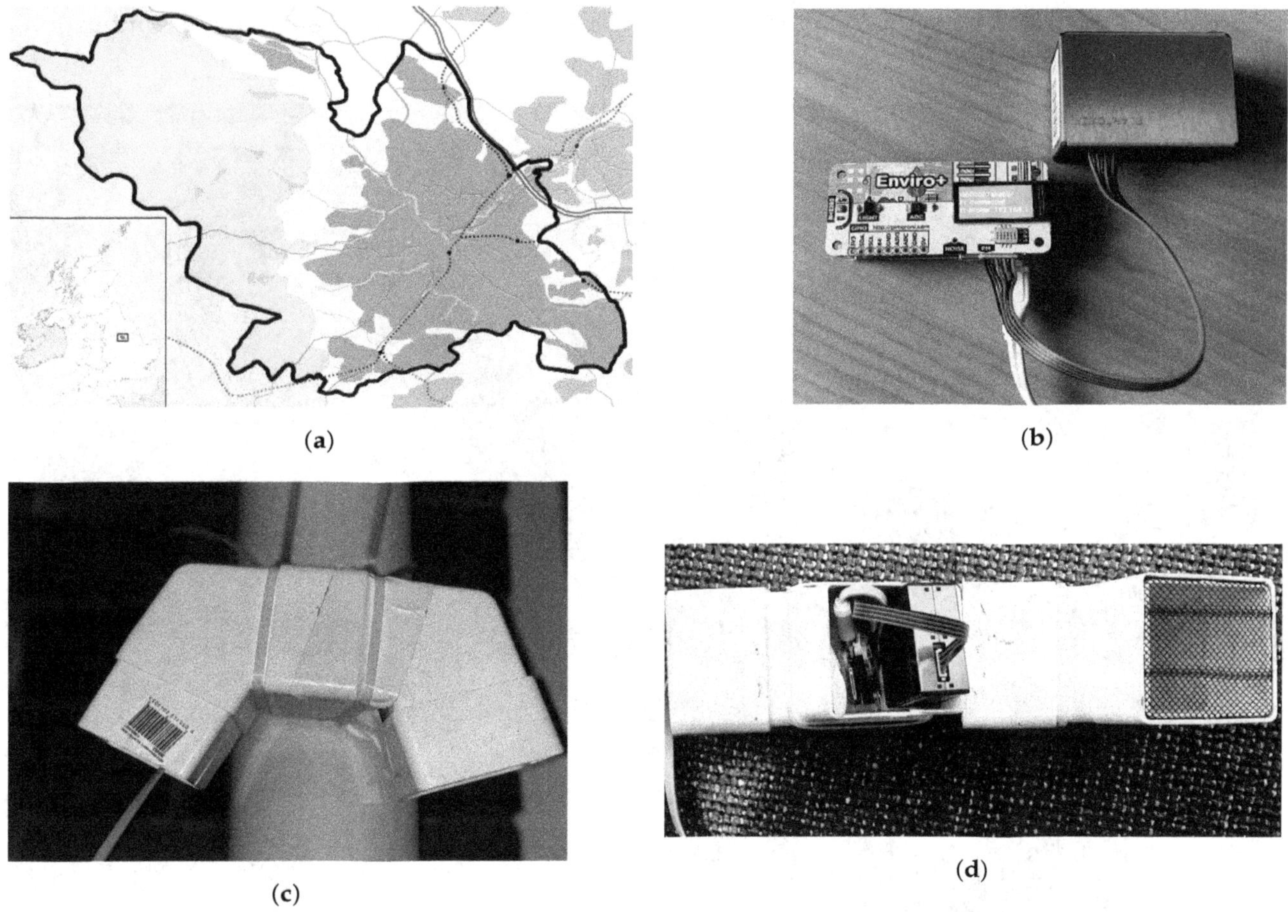

(a) (b) (c) (d)

Figure 1. Study region and the hardware setup. (**a**) Study Site: Sheffield Region, England; (**b**) Enviro+: A sample Indoor Air Quality Unit; (**c**) A sample Outdoor Air Quality Unit attached to a drainpipe outside a participant's house; and, (**d**) Enviro+ inside the casing for Outdoor Air Quality monitoring.

Twenty households with solid fuel stoves were recruited between January and April 2020. An indoor and outdoor low-cost air quality monitor was installed in each of the houses. The indoor sensor was placed at a minimum of 3 m distance from the wood burner for safety, but in the same room. The outdoor unit was put in weatherproof casing and then attached to a window or drain pipe outside of the house (see Figure 1c). Data from each household were collected over a total period of four weeks. Data were recorded on days when stoves were used and left unused, thus providing two groups of data. The control group contained 10 users who had stoves and, over a 30 day period, used them around 30% of the time. Control group data were taken from 20 days of non-usage. In total, 10 out of the 20 participants were identified as the control users for the study.

Pollutants that were measured real-time for both indoors and outdoors were PM_{10}, $PM_{2.5}$, PM_1, PNC (0.3 µm–1 µm), Nitrogen Dioxide (NO_2), Carbon Monoxide (CO), and Ammonia (NH_3). The meteorological parameters include temperature, Relative Humidity, and Atmospheric Pressure. The data were sampled every 145 s. Data for NO_2, CO, and NH_3 were omitted for research purposes and only visualised as trend levels due to the lack of calibration instruments. For indoor air pollution levels, the focus of our analysis was $PM_{2.5}$ and PM_1.

One participant from each household completed a survey prior to the measurement period and maintained a research diary throughout the study. Among other data, the research diary recorded stove usage timings, indicating when the stove was lit and when the last piece of fuel was added, type and total amount of fuel, and type and total amount of kindling used each time the stove was active. Any other activities carried out during stove use, such as cooking or lighting of candles, was also recorded. The air pollution level indoors was calculated between the time that the stove was lit until one hour after the last piece of fuel was added. This was done to allow for the complete combustion of the fuel that was fed to the stove.

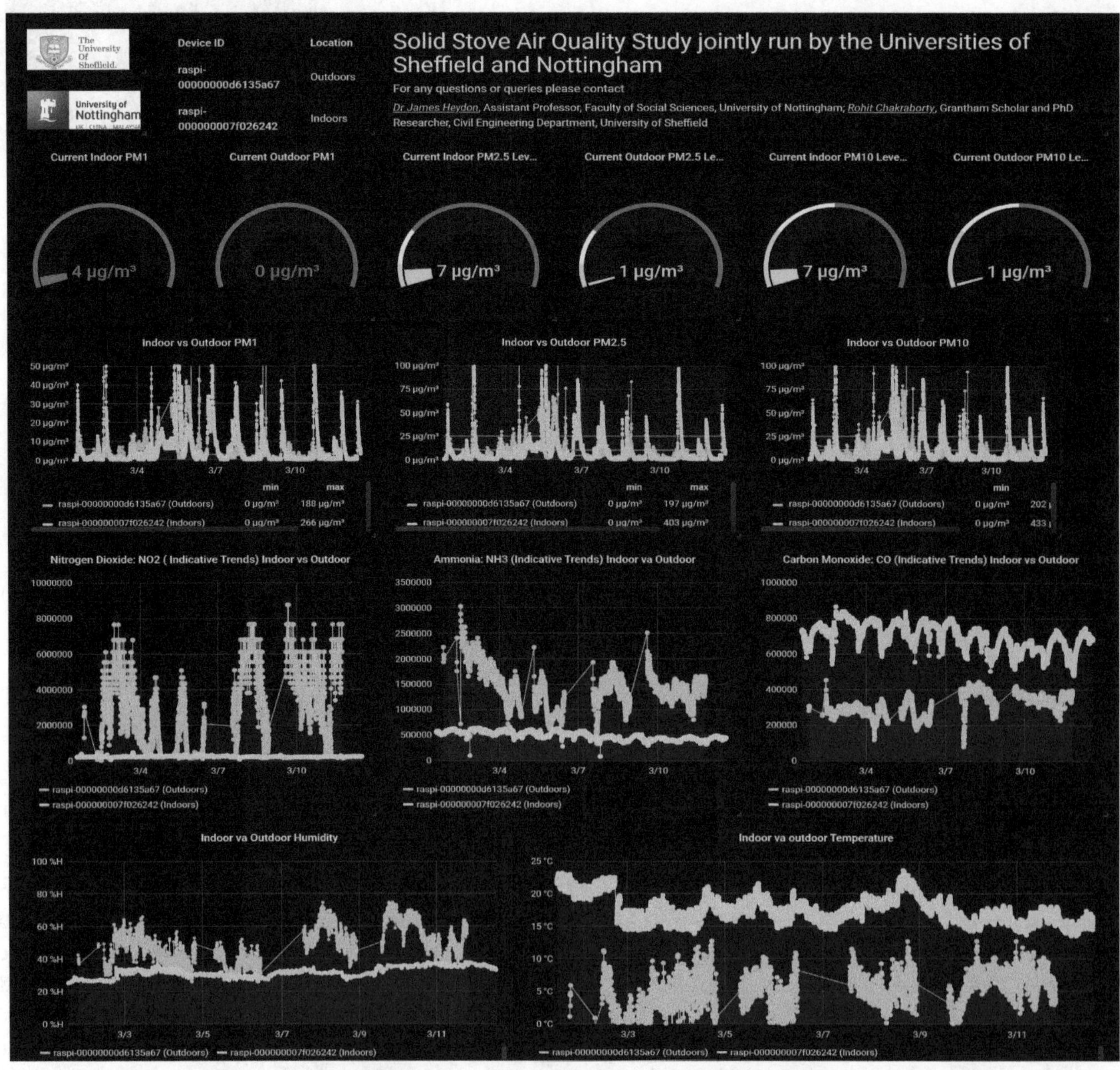

Figure 2. A sample dashboard displayed on the tablet.

Each participant was provided with a tablet computer. This displayed a dashboard containing real-time information on indoor and outdoor pollution levels that were collated from their sensors. A state-of-the-art cloud-based dashboard was built for each participant, as shown in Figure 2. The data from the monitoring units were sent to the cloud based server that was hosted by the University of Sheffield, which was then displayed on the dashboard. The information refreshed by default every minute. The graph panels plotted over a period of 30 days displayed daily average, minimum, and maximum values of each pollutant. Real-time sensor readings were also made available in the form of dynamic gauges.

2.2. Sensor Validation and Correction: Accuracy, Evaluation and Limitations

The Urban Flows Observatory [38] at the University of Sheffield have developed Enviro+ (Figure 1b), an air quality measurement device, in collaboration with Pimoroni, which is a local electronics company. Enviro+ is a pHat, which is an add-on board that sits on top of raspberry pi Zero and is suitable for both indoor and outdoor air quality measurement. Sensors onboard this pHat include a BME280, which is a weather sensor monitoring temperature, pressure. and relative humidity, an LTR-559 light and proximity sensor, a MICS6814 analog gas sensor monitoring NO_2, CO and NH_3, ADS1015 analog to digital converter (ADC), a MEMS microphone for noise measurement, and a 0.96″ colour LCD (160 × 80) for display. A connector for a particulate matter (PM) sensor is also available onboard, to which was connected the low-cost optical sensor PMS5003 (Plantower) Enviro+ with the connected PMS5003, which was used to conduct the particulate level measurement. Enviro+ with the connected PMS5003 was housed in a casing and installed outside the house for outdoor air pollution measurements (Figure 1c).

All of the units were collocated with Sheffield City Council's Reference Air Quality Monitoring station at Lowfield four weeks prior to the study. The high end Palas Fidas 200 instrument installed at Lowfield Station by Sheffield City Council was used as a reference in order to correct the PMS5003 sensors $PM_{2.5}$ measurement.

The procedure for correction of the collocated sensors is discussed below:

1. Raw data, including $PM_{2.5}$, PM_{10}, Temperature (T), and Relative Humidity (RH), were received every 160 s. This was converted to hourly averages in order to match the reference station data, because only hourly reference data are publicly available.
2. The hour average was excluded if less than 90% of the measurements were available in that hour average.
3. Humidity Correction: $PM_{2.5}$ concentrations can be relatively high from low-cost PM sensors at high RH levels. The hygroscopic growth of particles at high humidity, along with mist and fog particles, makes it detectable as particulates, as previously reported [39,40]. A Nephelometer, such as PMS5003, measures particulates based on light scattering principle. The particulates' refractive indices are dependent on relative humidity [41] and, thus, affects the sensor readings. While ambient temperature directly has a very limited role in sensors performance [40] (apart from extreme temperature), it affects the measurements indirectly. Jayaratne et al. [39] reports that, when the ambient temperature reaches the dew point temperature, the conditions become suitable for the formation of fog droplets in the air and fall within the detection size of such sensors. Figure 3 presents an example of the relationship between RH values and PM2.5 data from PMS5003 collocated. A Humidity-based bias correction approach was taken, as described here [42], while using the κ -Köhler theory [43]. The hygroscopic growth factor g (RH), as defined in Equation (1), where D_{dry} is the diameter of the dry particle and $D_{\text{wet}}(RH)$ is the diameter of the particle at a given RH value.

$$g(RH) = \frac{D_{\text{wet}}(RH)}{D_{\text{dry}}} \tag{1}$$

RH dependence [44] was established while using Equation (2), as follows:

$$g(RH) = \left(1 + \kappa \cdot \frac{RH}{100 - RH}\right)^{\frac{1}{3}} \quad (2)$$

where κ is a parameter that describes the degree of hygroscopicity of a particle and taken as 0.62, which is suitable for Sheffield [45]. Therefore, using Equations (1) and (2), hygroscopic growth factor g (RH) was calculated in order to obtain the humidity correction factor.

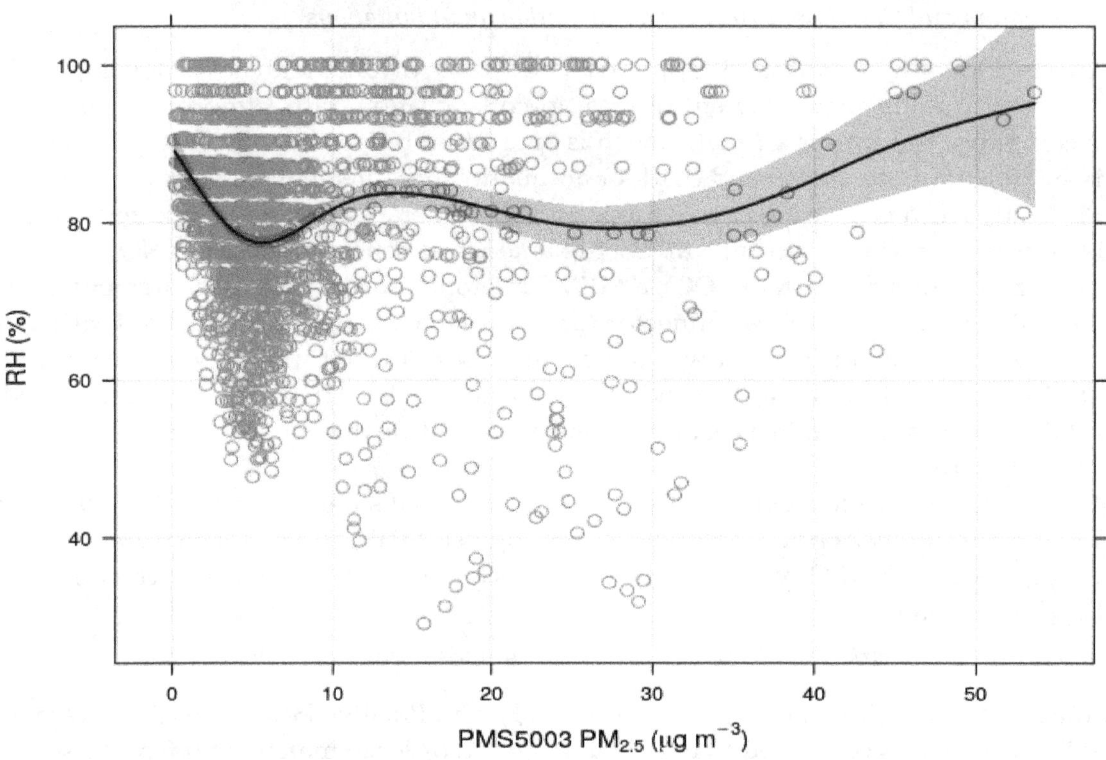

Figure 3. Distribution of $PM_{2.5}$ outputs on relative humidity (RH): LCS PMS5003.

Two additional PMS5003 have also been collocated at the same station permanently since 23rd April 2019 have been used to ensure correction factor accuracy. A conditional Quartile plot in Figure 4 below uses the corresponding values for both reference and low cost sensors, splitting the values into evenly spaced bins. For each low cost sensor value bin, the corresponding reference sensor values are identified and the median, 25/75t,h and 10/90 percentile (quantile) are calculated for that bin. The data are plotted in order to show how these values vary across all bins. The blue line shows the results for a perfect model i.e., zero error between low cost PMS5003 sensor and the reference Palas FIDAS 200 sensor. In the plot in Figure 4, the red line shows that the LCS tends to slightly over-report for $PM_{2.5}$ (NMB $\approx$ 0.2–0.3).

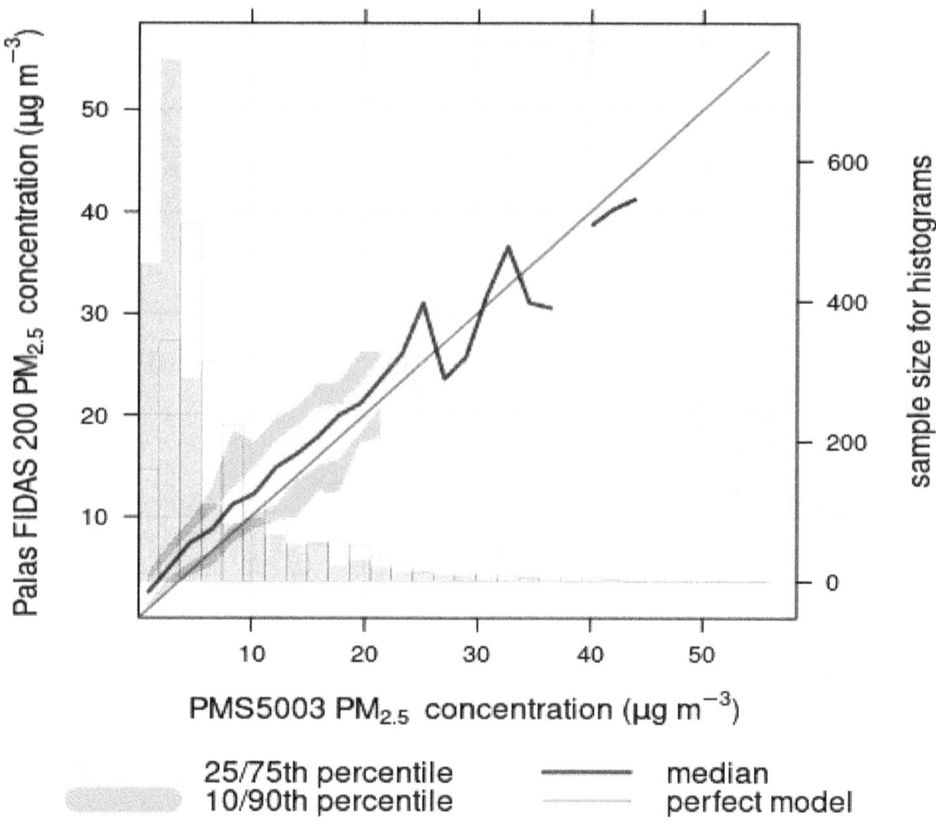

Figure 4. Conditional Quartile plot evaluating performance of low cost PMS5003 sensor/reference PALAS FIDAS sensor by showing how the corresponding sensor values vary together.

4. Concentration Range Correction: a correction was applied based on the relationship between pollutant concentration range and sensor performance. Multivariate Linear regression model were used in order to establish the relationship. Palas Fidas 200: $PM_{2.5ref}$ is used as the dependent variable and PMS5003 sensor data: $PM_{2.5lcs}$, T, and RH as predictors, as shown in Equation (3).

$$PM_{2.5ref} = \beta_0 + \beta_1 \times PM_{2.5lcs} + \beta_2 \times T + \beta_3 \times RH \quad (3)$$

β_0, β_1 and β_2 are calculated by training with the model generated. To note, β_3 is not used here, as it is obtained from the previous step.

5. Evaluation of LCS: PMS5003 corrected data are evaluated by comparing to the Palas Fidas 200 values in the holdout data set. From the field evaluation through collocation between January–April 2020, PMS5003 showed high linear correlation with reference instrument with R^2 value 0.81 for the hourly averaged data. This is an improvement in accuracy when compared to the findings from previous studies on evaluating Plantower sensors [10,46] with R^2 values lying between 0.71–0.77 for PMS5003 without applying any correction factors. The inter-sensor comparison showed a high correlation, with an R^2 value between 0.98–0.99. Figure 5, below, shows the scatter plot between the reference and corrected PMS5003 sensor.

Figure 6, below, also shows a consistently high linear correlation factor with an average R^2 value of 0.81 when analysed and split with relative humidity as the third variable.

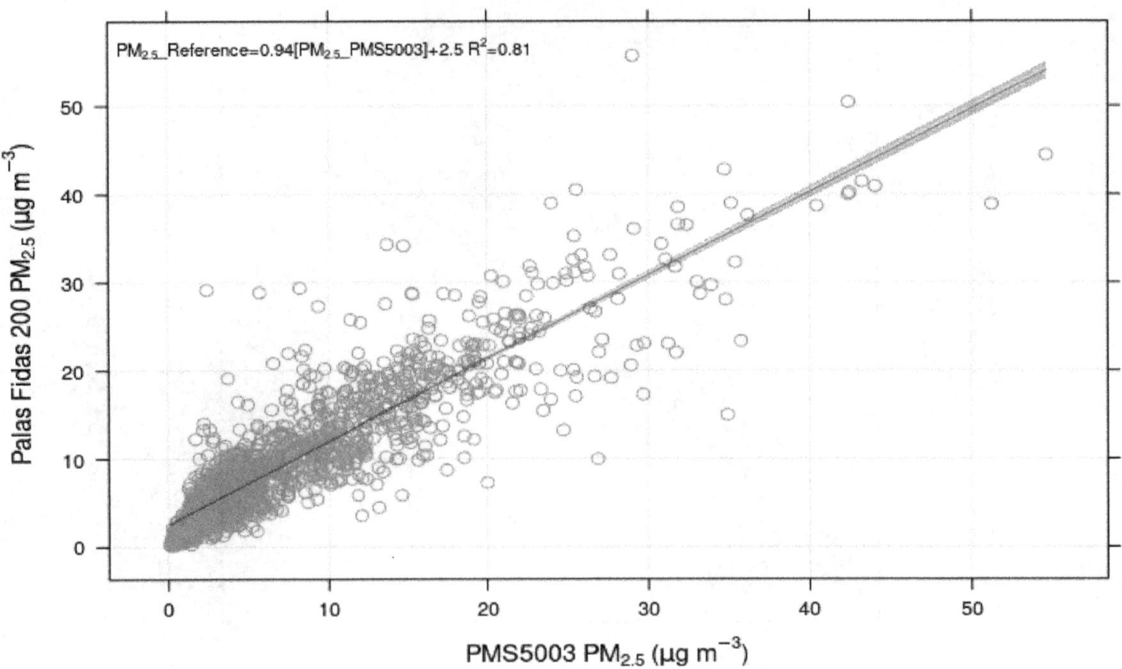

Figure 5. Scatter plot between PMS5003 (LCS) versus (vs.) Palas Fidas 200 (Reference Sensor) output: hourly averaged $PM_{2.5}$.

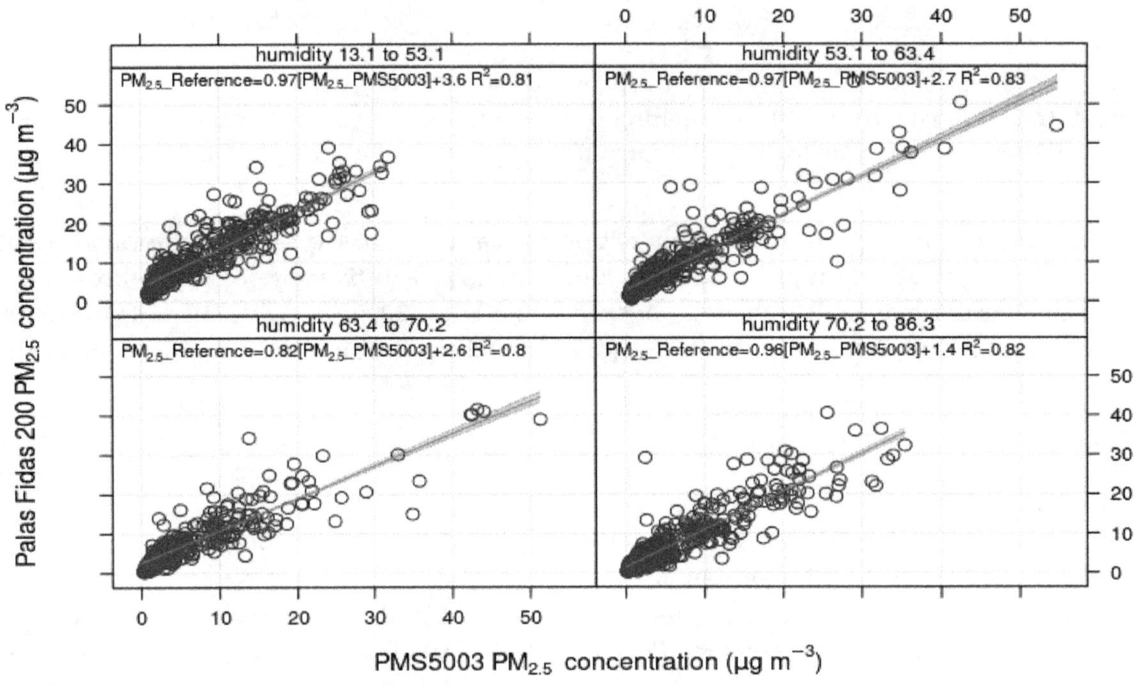

Figure 6. Scatter plot between PMS5003 (LCS) versus (vs.) Palas Fidas 200 (Reference Sensor) output: hourly averaged $PM_{2.5}$ with type humidity.

Table 2, below also shows the R^2 value for different concentrations of $PM_{2.5}$ and compared to the Daily Air Quality Index (DAQI) bands and breakpoints for $PM_{2.5}$, as set by DEFRA'S Air Quality Expert Group [47]. During the field evaluation, there was not enough data to evaluate the sensor for high and very high conditions (DAQI = 8–10).

Table 2. Concentration band analysis showing averaged coefficients of determination (R^2) for hourly averages of $PM_{2.5}$ from PMS5003 sensors against Reference Sensor Palas FIDAS 200 and compared to the Daily Air Quality Index (DAQI) bands.

DAQI	**1**	**2**	**3**	**4**	**5**	**6**	**7**	**8**	**9**	**10**
Band μgm^{-3}	Low 0–11	Low 12–23	Low 24–34	Moderate 35–41	Moderate 42–46	Moderate 47–52	High 53–58	High 59–64	High 65–69	Very High 70 or more
R^2	0.82	0.79	0.81	0.83	0.81	0.82	0.79	N/A	N/A	N/A

Sensor Limitations

The reference station does not provide PM_1 and PNC data and, therefore, this data cannot be subjected to this correction. Further research is underway in order to evaluate sensor performance and evaluation in this specific regard. Finally, the study has not been able to account for UFP due to these sensors being unable to detect or measure particles below 300 nm. As such, the measured PNC has been limited to a size of 0.3 µm–1 µm.

2.3. Monitoring Outdoor Air Quality and Adjusting for Weather: A Generalized Boosted Regression Model

While the data that were collected from the indoor unit (Figure 1b) were used to analyse the pollution emissions from the stoves, the first purpose of the outdoor unit (Figure 1c) was twofold. First, it was used for the general monitoring of outdoor PM levels. This allowed for the detection of any unusual levels of outdoor pollution that could impact the air quality indoors. Second, the sensors could indicate whether the outdoor air quality was also being influenced during stove use. While the outdoor sensors served the first purpose, achieving the second was complicated by multiple covariates, such as meteorological factors, local garden waste burning, neighbours using wood stoves, and traffic.

Figure 7 plots the average weekly variation of outdoor $PM_{2.5}$ and PM_{10} levels of the participants houses over the three-month period.

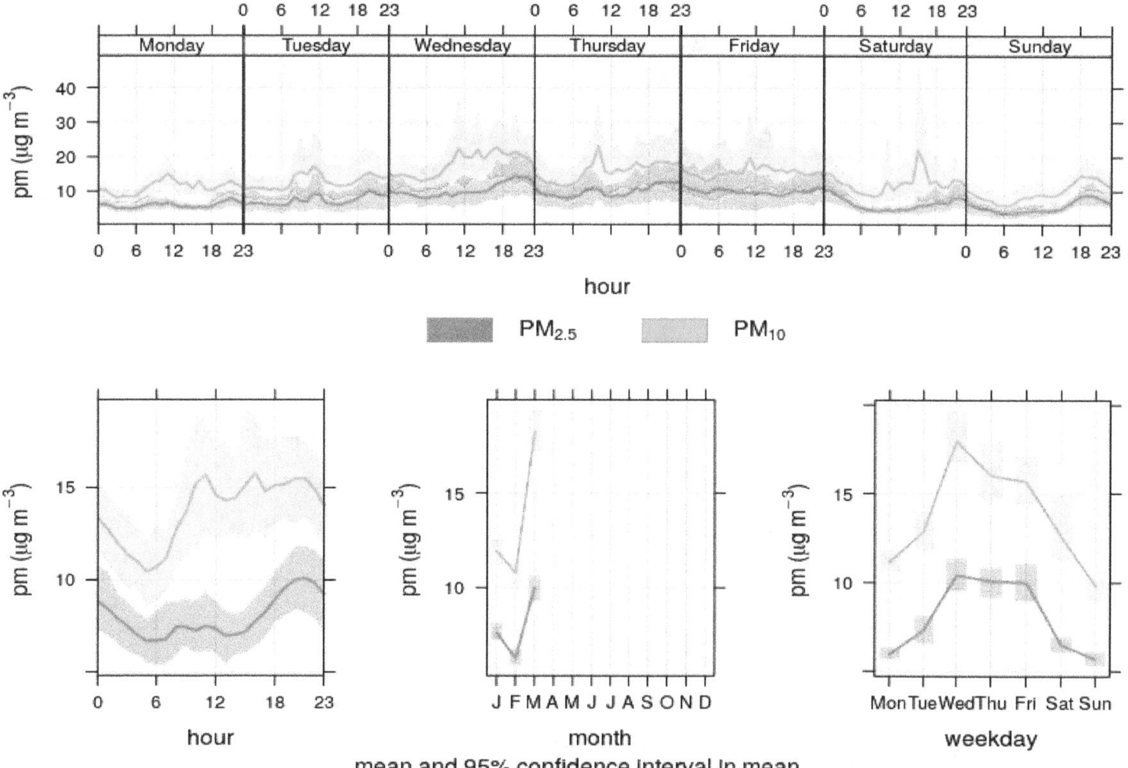

Figure 7. Outdoor Particulate Matter Variation plot.

Meteorology plays a crucial role in the estimation of levels of particulate matter. Therefore, when trying to understand the trends of outdoor pollution levels, it can be very challenging to determine whether a pollution episode is caused by local emissions or meteorology. Therefore, a Machine Learning (ML) based algorithm based on Generalised Boosted Regression Model [48] was used in order to explore and adjust for the non-linear relationships between the meteorological covariates and particulate matter $PM_{2.5}$ levels. The partial dependencies in Figure 8 show the relationship between $PM_{2.5}$ and the covariates that were used in the model while holding the value of other covariates at their mean levels. As can be seen, wind speed (16.1%) and wind direction (12.7%) play a crucial role in determining $PM_{2.5}$ levels; hence, its impact should be accounted for in order to better understand the air quality around the participating households.

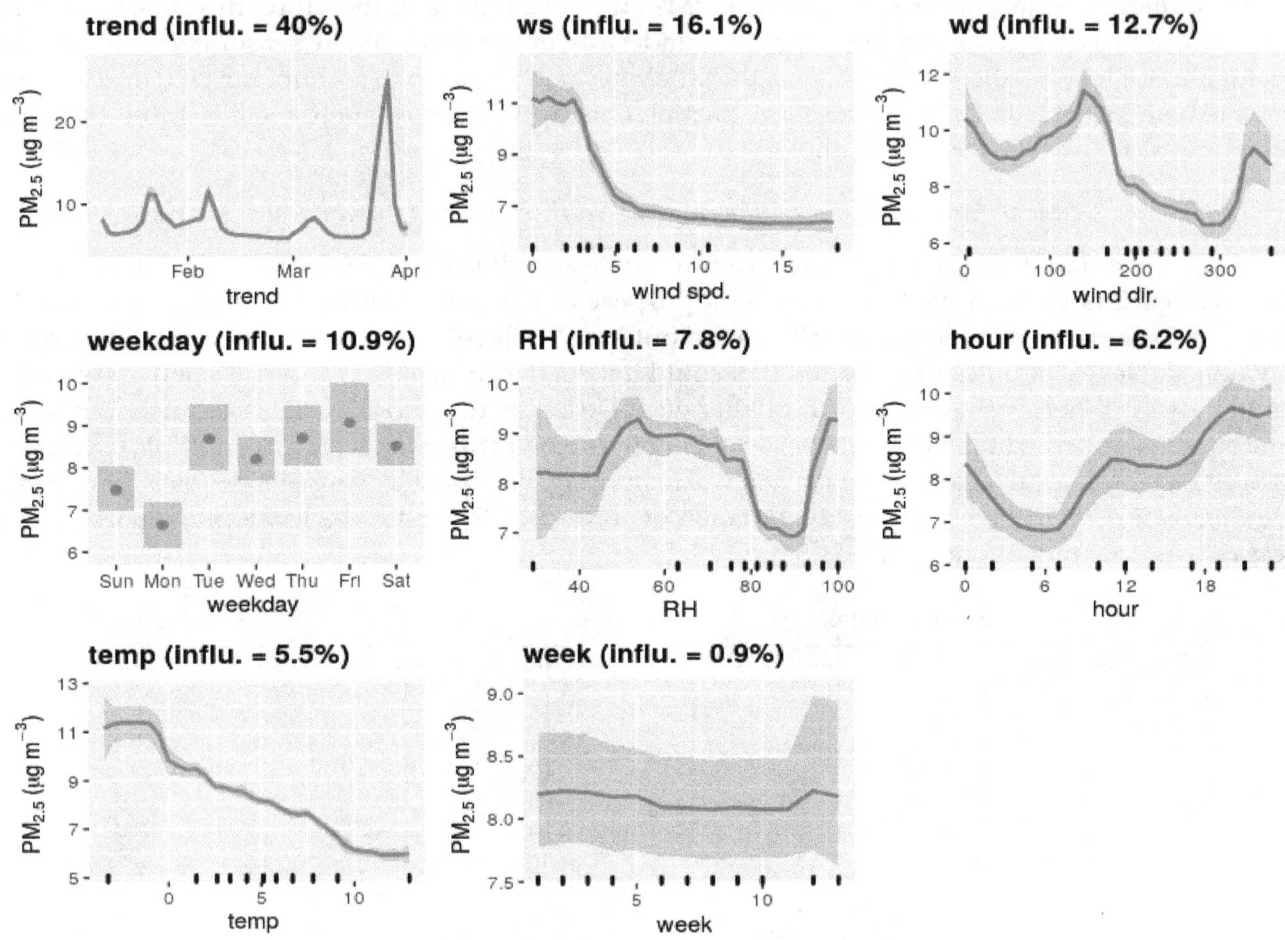

Figure 8. Influence of different covariates on outdoor $PM_{2.5}$ levels.

The popular R deweather and openair package [49] was used in creating the prediction model and plotting. The model is formed, as shown in Equation (4).

$$[PM_{2.5}] = RH + \bar{u} + \phi + T\theta + t_{hour} + t_{weekday} + tJD \tag{4}$$

where $\bar{u}$ is the mean hourly wind speed, ϕ is the mean hourly wind direction (degrees, clockwise from the north), and $T\theta$ is the mean hourly temperature (°C). Variables representing hour of the day, t_{hour}, day of the week, $t_{weekday}$, and day of the year, tJD were also considered for the model development.

From Figures 7 and 8, it is evident that, during weekdays, the outdoor levels of $PM_{2.5}$ and PM_{10} are higher than during the weekend. It can also be seen that the levels are considerably higher outside during the evening, which corresponds to the usage pattern of stoves by participants. This indicates that even DEFRA-certified solid fuel stoves could affect the local air quality outdoors. A more sophisticated source apportionment study is required in order to further investigate this. The high level of ($PM_{2.5}$ and PM_{10} gradually decreases throughout the night, with the lowest levels being attained at around 5:30 am–6:00 am GMT (see Figure 7 hourly plot). Ten-fold cross validation [50] was used for evaluating the model performance and the model fitting results are shown in Figure 9.

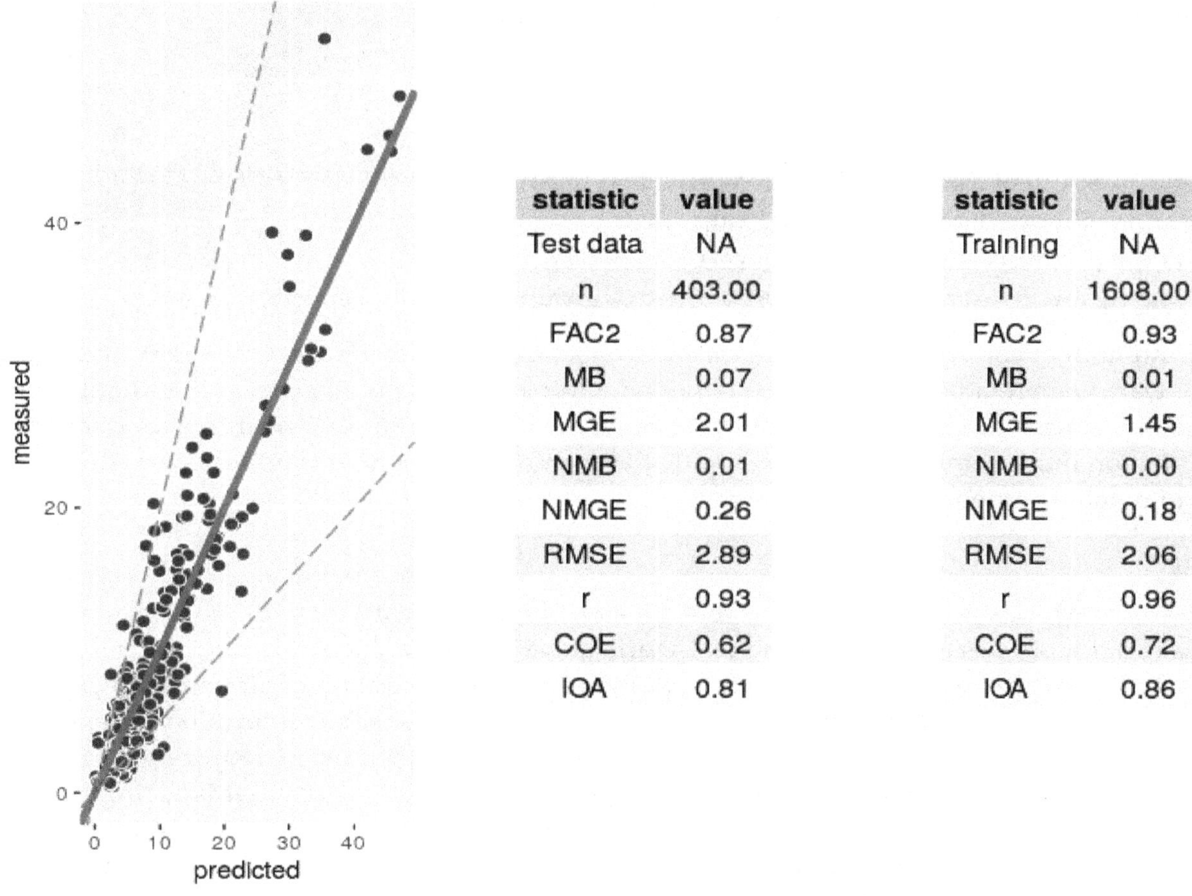

statistic	value
Test data	NA
n	403.00
FAC2	0.87
MB	0.07
MGE	2.01
NMB	0.01
NMGE	0.26
RMSE	2.89
r	0.93
COE	0.62
IOA	0.81

statistic	value
Training	NA
n	1608.00
FAC2	0.93
MB	0.01
MGE	1.45
NMB	0.00
NMGE	0.18
RMSE	2.06
r	0.96
COE	0.72
IOA	0.86

Figure 9. Generalised Boosted Regression Model to explore and remove weather impact on outdoor pollution level: Model Evaluation.

2.4. Data Processing and Storage

The real-time sensor data are pushed to a cloud-based database over WiFi while using a Python script running on the Raspberry Pi Zero. The data are stored in a database designed and installed in a virtual server hosted by the University of Sheffield. These data are then made available to be accessed through an Application Programming Interface (API), called Enviro-API, developed for data retrieval and displayed on the dashboard. The API developed along with the database had two goals:

- The ability to ingest a high volume of time series data with dynamic data from the sensors.
- The ability to return this time series data with basic querying parameters such as sensor ID and timestamps.

This allowed for us to create an end-to-end secure infrastructure for real-time sensor data collection, storage, and retrieval system for our study.

2.5. Data Analysis

Missing data have been treated. The usage days were only included if 90% of the hourly data were available. Data analyses were performed while using Excel, R, and Python programming languages.

The statistical significance of the results was calculated based on Welch's t-test (Moser & Stevens, 1992) whlie using the standard equations:

$$t = \frac{m_A - m_B}{\sqrt{\frac{S_A^2}{n_A} + \frac{S_B^2}{n_B}}}, \tag{5}$$

and the degree of freedom of Welch t-test is calculated, as follows:

$$df = (\frac{S_A^2}{n_A} + \frac{S_B^2}{n_B})^2 / (\frac{S_A^4}{n_A^2(n_A - 1)} + \frac{S_B^4}{n_B^2(n_B - 1)}) \tag{6}$$

- A and B represent the Control and Experimental group.
- m_A and m_B represent the means of groups of samples A and B, respectively.
- n_A and n_B represent the sizes of group A and B, respectively.
- S_A and S_B are the standard deviation of the two groups A and B, respectively.

The strength of correlations was classified as weak (±0.1–0.3), moderate (±0.3–0.5), and strong (±0.5–1). Data were removed during such periods while stoves were lit in order to avoid data being influenced by emissions from cooking, burning candles, or incense sticks. The influence of outdoor air pollution on indoor emissions data was anticipated, but adjustment was unnecessary, due to the absence of notable outdoor pollution levels.

2.6. Study Limitations

The study exhibits several limitations that are associated with variability in the research setting due to its exploratory design and focus on real-world stove use. First, the study does not account for the impact of room size, seal, ventilation, and dwelling age on the duration of air pollution exposure witnessed. Nor does it relate the levels of air pollution to specific stages of the combustion cycle. Further study is needed in order to understand these aspects of indoor air pollution, requiring a sampling frame that is determined by more than the stove type and a research design that is appropriate for lab conditions. Second, despite using outdoor sensors to illustrate that the indoor air pollution is not coming from outside sources (see Section 3.2), further details on air pollution at the indoor-outdoor interface were beyond the design of this study. This is a characteristic of air pollution research more broadly, as reflected in the UK government's recent multi-million-pound call for research that is able to develop solutions to air pollution problems at the indoor/outdoor interface [51]. Relatedly, windspeed could influence the infiltration rate of outdoor air indoors, but, again, this was beyond the remit here. As such, further research into this relationship is recommended. Finally, the influence of sensor data on participant stove management practice has not been explored in detail. This will be drawn out more fully in a separate paper.

3. Results and Discussion

Table 3 summarises the daily $PM_{2.5}$ and PM_1 mean, and hourly peak $PM_{2.5}$ and PM_1 mean from 20 households and 260 stove usages, along with the statistical analysis and distribution. Data on the average pieces of fuel per use (FP) and kindling per use (KP), along with the average duration of use, have also been presented. The hourly indoor mean $PM_{2.5}$ and PM_1 concentrations that were observed during stove usage ranged from 2.27 μg/m^3 and 1.11 μg/m^3 to 47.60 μg/m^3 and 36.15 μg/m^3, respectively, with a high coefficient of variation 0.9 for $PM_{2.5}$ and 0.94 for PM_1. The hourly PNC average that was observed indoors in the particle size range (0.3–1 μm diameter) was 2607 particles/0.1 litre (L) of air when each stove was used, but the hourly peak PNC average observed was 4345 particles/0.1 L with an hourly maximum of 9978 particles/0.1 L. The average number of fuel pieces (9.58 wooden logs) and kindling (8.37 pieces) used varied significantly between the households, with a coefficient of variation 0.69 and 0.67, respectively. The average duration of use was approximately 4 h, with most households using their stove between 6 pm and 10 pm.

3.1. Increase in Indoor Pollution Levels during Stove Use

The findings indicate that average indoor $PM_{2.5}$ (mean = 12.21 μg/m^3 SD = 10.36, 95%CL: 8.16, 12.68) and PM_1 (mean = 8.34 μg/m^3 SD= 7.64, 95%CL: 5.29, 9.42) are higher when the stoves are lit when compared to the period in which they are not in use with $PM_{2.5}$ levels (mean = 4.12 μg/m^3 SD= 3.61, 95% CL: 2.82, 4.82), and PM_1 levels (mean = 2.54 μg/m^3 SD= 2.61, 95%CL: 1.59, 3.04). Statistical analysis estimates that the difference in concentrations between these two groups is significantly different for both $PM_{2.5}$ (Welch's t(57.0448) = −5.0531, $p < 0.0001$) and PM_1 (Welch's t(56.6291) = −4.9197, $p < 0.0001$).

The analysis in the three quartiles—(i) <25 percentile, (ii) >25 <75 percentile, and (iii) >75 percentile, representing low, medium, and peak concentrations, showed an increase for $PM_{2.5}$(223.92%, 241.23%, 127.84%) and PM_1(254.38%, 238.02%, 209.32%). The overall average concentrations were higher for $PM_{2.5}$ by 196.23% and PM_1 by 227.80% when used.

Figure 10 density rug plots show the distribution and levels of PM_{25} and PM_1 for users. Figure 11 compare the control group's indoor pollution levels with the experimental group. For reasons of visualisation, scaling the x-axis in the graph (see Figure 10) is limited to 60 μg/m^3.

Figure 10 reveals that the levels of PM that people are exposed to can vary, with a maximum peak average of 47.60 μg/m^3 for $PM_{2.5}$ and 36.15 μg/m^3 for PM_1. While calculating the averages smooths the graph, these findings demonstrate that some users are exposed to maximum values of up to 160 μg/m^3 $PM_{2.5}$. Control users experience much lower indoor particulate levels when their stoves are not lit when compared to users that do, as indicated by Figure 11.

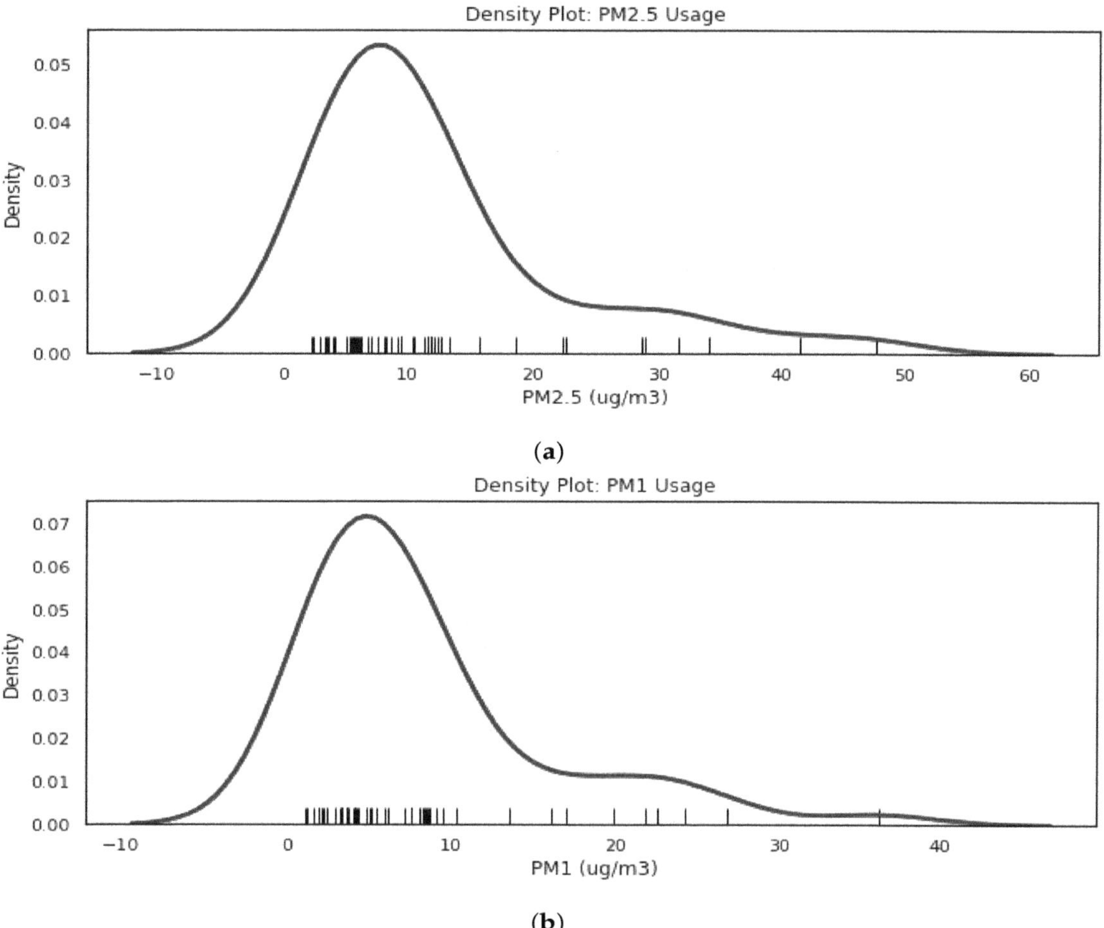

Figure 10. Conditional distribution density plot shows the overall indoor concentration levels during the usage of wood burners. (**a**) $PM_{2.5}$ distribution; (**b**) PM_1 distribution. Note. While the analysis includes the full range of data, for display purposes only the x-axis is truncated to 60 μg/m^3.

Table 3. Statistical summary and distribution of hourly mean and peak particulate matter (PM) (μg/m^3), daily usage fuel pieces and kindling pieces, and Pearson's r value.

	Duration (Hours)	FP	Peak $PM_{2.5}$	Mean $PM_{2.5}$	Peak PM_1	Mean PM_1	KP
mean	4.06	9.07	27.34	12.21	19.44	8.34	10.37
std	1.63	6.32	31.26	10.36	22.37	7.64	7.04
min	1.1	1	0.23	2.27	0	1.11	0
25%	2.95	4	9	5.66	5.79	3.51	6
50%	3.95	8	16.87	9.26	11.375	5.5	9
75%	4.94	11	34	12.50	22.77	8.87	15
max	9.2	32	195.83	47.60	121	36.15	39
Coefficient of Variation	0.39	0.69	1.14	0.90	1.14	0.94	0.67
Pearson's r							
Duration (hours)	1						
FP	0.55	1					
Peak $PM_{2.5}$	0.4	0.44	1				
Mean $PM_{2.5}$	0.017	0.17	0.75	1			
Peak PM_1	0.38	0.43	0.97	0.75	1		
Mean PM_1	0.021	0.15	0.73	0.98	0.76	1	
KP	−0.007	−0.004	−0.04	0.019	−0.034	0.039	1

Wood burner usage data from 20 households collected between January and April 2020.

In Figure 10, comparing the concentration levels between usage and non-usage days for the control group also illustrates an increase for $PM_{2.5}$(139.52%, 327.48%, 320.66%) and PM_1(132.16%, 413.11%, 366.56%) when stoves are used. The overall average concentrations were higher for $PM_{2.5}$ by 432.91% and PM_1 by 281.22%.

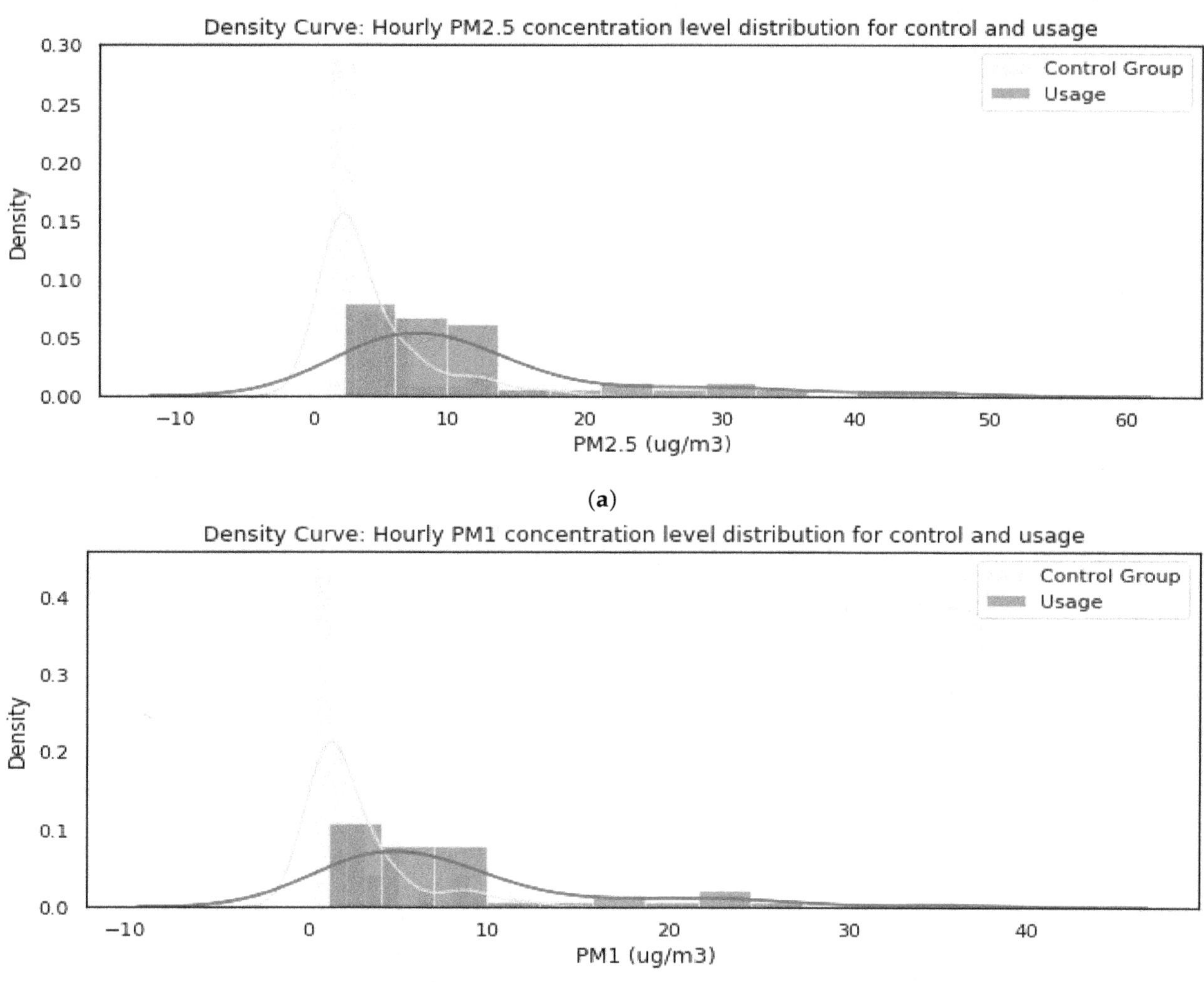

(**a**)

(**b**)

Figure 11. Control group compared to usage shows higher indoor concentration levels during the usage of wood burners with larger variation. (**a**) $PM_{2.5}$ distribution comparison; (**b**) PM_1 distribution. Note. While the analysis includes the full range of data, for display purposes only the x-axis is truncated to 60 μg/m^3.

3.2. *Indoor Outdoor Interface: Average Indoor* $PM_{2.5}$ *Levels Are Higher and Weakly Correlated with Outdoor Average* $PM_{2.5}$ *Levels*

The average indoor $PM_{2.5}$ levels are higher (mean = 12.21 μg/m^3 SD = 10.36, 95%CL: 8.16, 12.68) than the outdoor $PM_{2.5}$ levels (mean = 7.99 μg/m^3 SD= 5.51, 95%CL: 3.60, 8.93) during stove usage. From Figure 12, below, it is clear that indoor and outdoor values vary significantly between 10–45 μg/m^3 concentration levels. This variation is because the mean and hourly peak indoor PM lies within this range and, thus, the indoor levels are much higher than the corresponding outdoor levels. Further analysis of average indoor and outdoor $PM_{2.5}$ levels indicated a weak correlation (R^2 = 0.19) between them, which suggests that outdoor air quality is not a driving factor behind the high indoor pollution levels that were seen during stove usage.

While we acknowledge that indoor $PM_{2.5}$ levels can impact outdoor air quality, no measurements were taken from the chimney/flue. The air quality sensor outside the house indicates immediate outdoor air pollution levels and, thus, it is difficult to measure any leakage at the interface. Future research studies should focus on indoor air pollution and its influence on outdoor air quality in order to address this limitation of our study.

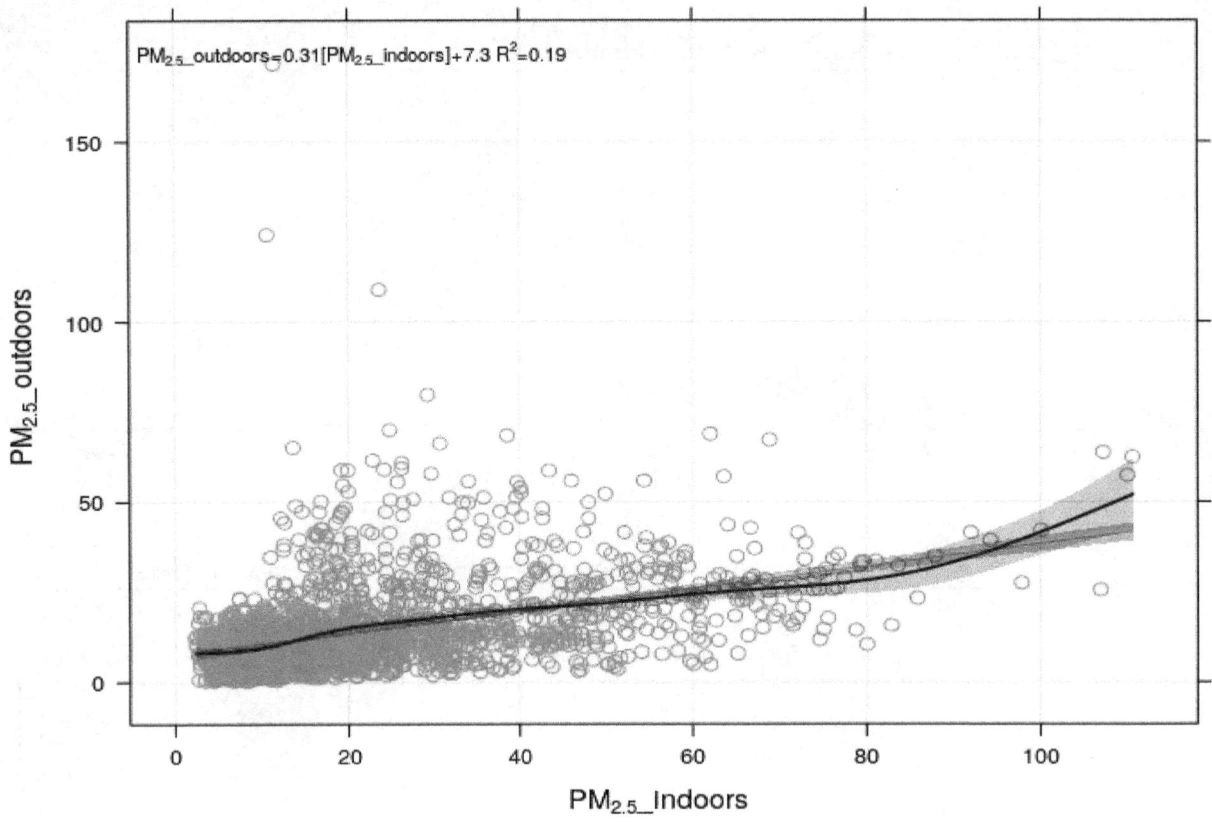

Figure 12. Indoor $PM_{2.5}$ vs. Outdoor $PM_{2.5}$.

3.3. Hourly Peak PM Average Higher than Daily PM Average

The analysis of Table 3 shows hourly peak $PM_{2.5}$ and PM_1 is strongly correlated with daily mean $PM_{2.5}$ and PM_1 (r = 0.75). Statistical analysis shows that the hourly peak mean $PM_{2.5}$ (27.34 μg/m^3, 95% CL:18.38, 37.77) and PM_1 (19.44 μg/m^3, 95% CL:12.04, 28.30) are significantly higher than the daily mean $PM_{2.5}$ (12.21 μg/m^3, 95% CL: 8.16, 13.68) and PM_1 (8.34 μg/m^3, 95% CL: 5.29, 9.43) by 123.91% and 133.09%, respectively. Hourly $PM_{2.5}$ and PM_1 peak mean and the daily mean concentrations varied between households with the minimum and maximum, being 19.2 μg/m^3–86.83 μg/m^3 and 17.79 μg/m^3–84.47 μg/m^3, respectively.

There exists high variation in exposure concentrations, concerning both short peaks and daily levels. This characteristic is related to the "real-world" nature of the study. The research diary tool provided data on not only the amount of fuel and kindling pieces used, but also their type. On average, participants used 9.58 pieces of solid fuel and 8.32 pieces of kindling per use. The number of fuel pieces used varied between a minimum of seven to a maximum of 40, while kindling varied between a minimum of one and a maximum of 32. All participants used dried and seasoned logs, but the sizes varied. There was also a diversity of kindling used, taking the form of firelighters, newspapers, balls of paper, twigs, sawdust, packing cardboard, greeting cards, and even empty egg boxes. Echoing the findings of existing studies [20,35,52]. This means that the same wood burner may emit different levels of indoor air pollution depending on the quantity and type of fuel and kindling used. While suggesting a link between indoor air pollution and fuel quantity, and type of fuel and kindling, following other studies in the next section, demonstrates that this is actually linked with the stove door being opened.

Epidemiology studies and policymaking are focused around hourly average concentration monitoring by regulatory air quality stations. This leads to the omission of short-term high exposure through the "flooding" of indoor spaces with $PM_{2.5}$ and PM_1. Very few studies have reflected on short term peak concentration exposure. Lin et al.'s study [27] associated increased risk factors with hourly peak concentrations of $PM_{2.5}$. Similarly, Delfino et al. [53] associated peak PM levels with Asthma attacks in children, but in outdoor environments. Therefore, the present study encourages future researchers to study the occurrences and effects of relatively short-term peak PM exposure on human health.

3.3.1. *Hourly Peak Average PM Has a Moderate Correlation to the Pieces of Fuel Used*

While Table 3 indicates a weak correlation between fuel pieces and mean $PM_{2.5}$($r = 0.17$), and with PM_1 ($r = 0.15$), comparing the hourly peak concentration of $PM_{2.5}$ ($r= 0.44$) and PM_1 ($r = 0.43$) exhibits a moderate correlation with fuel pieces. The scatter plots in Figures 13 and 14 chart the relation between peak hourly levels to the possible co-factors of fuel amount and duration of usage. In Figure 13a,b, higher concentration peak levels are clustered towards the left of the x-axis. This indicates a non-linear relationship with fuel pieces.

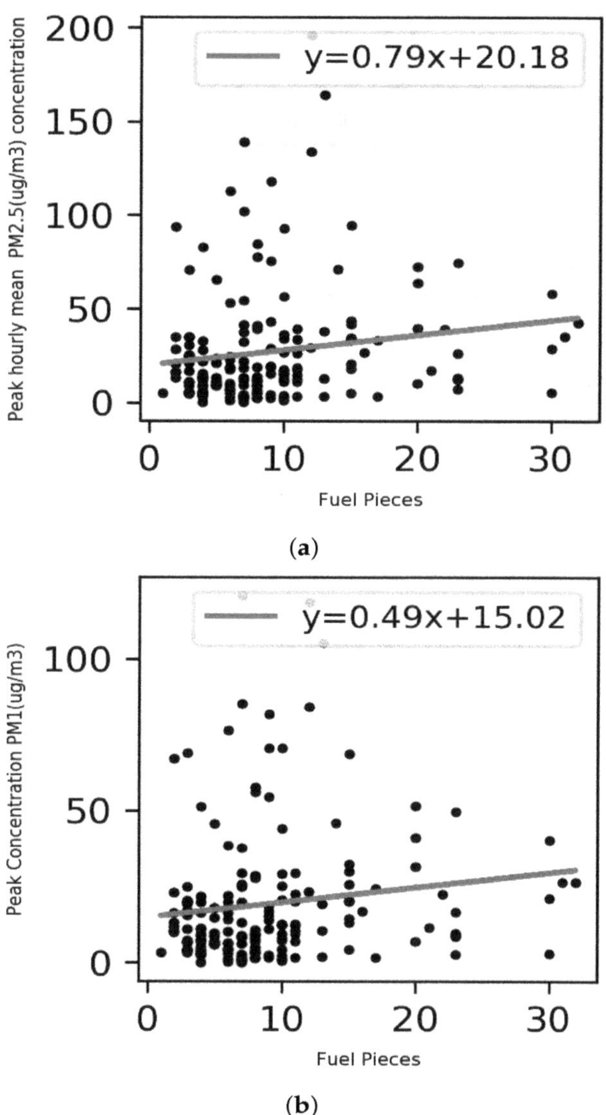

Figure 13. Scatter plot between PM concentration and Fuel Pieces. (**a**) $PM_{2.5}$ vs. fuel pieces; (**b**) PM_1 vs. fuel pieces;

While correlation between fuel pieces and hourly mean concentration is weak, it is stronger when compared to the hourly peak concentration. Therefore, the findings suggest that the peak hourly concentrations are often higher by a minimum of 250% and a maximum of 400% when participants have refuelled their stove more than once during a usage compared to one refuel or none at all. As such, the findings indicate that the 'flooding' of indoor space occurs as a result of the stove door being opened for refuelling. This accords with several existing real-world [21,22,35], and lab-based [20] studies into stoves outside the UK. While the findings point to the opening of the stove door as the origin for indoor PM emissions, further lab-based research is required into how this might relate to duration, timings, and the point in the burn cycle at which the opening occurs.

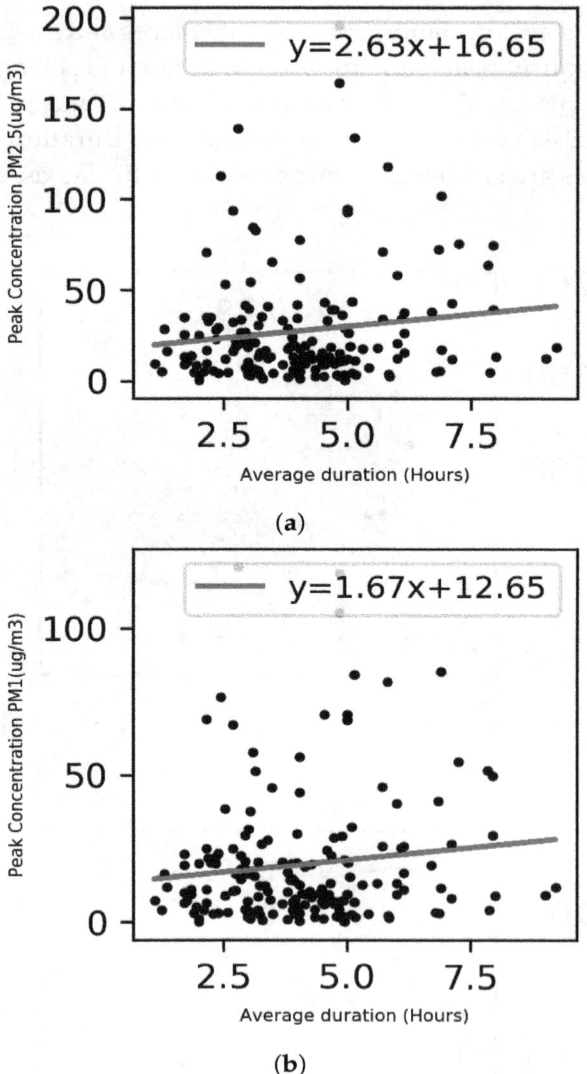

Figure 14. Scatter plot between PM concentration and duration.(**a**) $PM_{2.5}$ vs. duration; (**b**) PM_1 vs. duration.

The hourly peak concentrations explain the shape of the rug plots, as seen in Figure 10. The shape of the curves exhibit a distinct broad frequency distribution in the lower PM concentration. This indicates that most of the sensor readings are lower during stove use, but there are also smaller spikes towards the right of x-axis, indicating sensor readings that correspond to higher levels of PM pollution. A 'leakage' would result in a more uniform shape, and, thus, the presence of the smaller spikes cannot be explained. This echoes Salthammer et al.'s findings [35] and provides further support for the theory of opening doors being the cause of the indoor air pollution seen rather than a leakage, which appears to be more common to open fires than 'closed' stoves (see [54]). The PM fraction gets dispersed quickly throughout the room due to its smaller size, reverting to lower hourly average concentrations.

3.3.2. *Hourly Peak Averages Illustrate a Moderate Correlation with Duration of Use*

Table 3 also illustrates a non-linear relationship between the duration of use and mean $PM_{2.5}$ (r = 0.017). This is similar to PM_1 (r = 0.021), although, again, comparing the hourly peak concentrations of $PM_{2.5}$ (r= 0.4) and PM_1 (r = 0.38), it exhibits a moderate correlation with the duration of use. The scatter plots in Figure 14a,b also reflect this, with higher levels of peak values being continuously registered during the stove use.

Longer usage is associated with greater numbers of fuel pieces used. This result supports the explanation for the 'flooding' phenomenon observed, with higher short-term peak concentrations being seen during longer periods of use, because these periods are sustained by more refueling actions. This accords with [20], who also found the lighting and refueling aspects of stove management to form the main pollutant-generating phases of operation.

4. Conclusions

The present study aimed to understand the extent to which PM was emitted indoors and under real-world conditions by DEFRA-certified residential stoves. The findings indicate that real-world indoor PM exposure from these stoves is higher when lit as compared to the period in which they are not in use. When compared to periods of non-use, the overall average concentrations were higher for $PM_{2.5}$ by 196.23% and PM_1 by 227.80%. Peak hourly concentrations of PM were often found to be higher by 250–400% when the participants had refueled their stove more than once during a single usage. The findings also provide information on PNC, with an average hourly peak of 9978 particles/0.1 L emitted during a single usage. These 'flooding' events correlated with the opening of the stove door, which indicated that such incidents occurred as fuel was added. Data from outdoor sensors clarified that this was not originating from outdoors. On the basis of these results, it is recommended that DEFRA testing standards be modified in order to account for these normative health risks. The PM that is released into the home is not an aberration from normal use, but results directly from it. This is because real-world operation cannot occur without opening the stove door. It may be that with regulatory encouragement stove designs can be modified in a way that limits such instances. In the meantime, or in the event that appropriate modification cannot be achieved, it is also recommended that new residential stoves be accompanied by a health warning at the point of sale in order to indicate the normative health risks posed to users.

Author Contributions: Conceptualization, R.C. and J.H.; methodology, R.C. and J.H.; formal analysis, R.C.; investigation, R.C. and J.H.; data curation, R.C.; writing—original draft preparation, R.C.; writing—review and editing, R.C., J.H., L.M. and M.M.; visualization, R.C.; supervision, L.M. and M.M.; project administration, R.C. and J.H.; funding acquisition, R.C., L.M., M.M. and J.H. All authors have read and agreed to the published version of the manuscript.

Acknowledgments: We are grateful to Urban Flows Observatory, The University of Sheffield for funding the sensors and The University of Nottingham for funding tablet computers. We are grateful to The Grantham Centre for Sustainable Futures for funding the research and APC charges. We would like to thank the Associate Editor and Reviewers for constructive suggestions and valuable comments, helping us to improve this work.

Abbreviations

The following abbreviations are used in this manuscript:

n the number of complete pairs of data.

FAC2 fraction of predictions within a factor of two
MB mean bias.
MGE mean gross error.
NMB normalised mean bias.
NMGE normalised mean gross error.
RMSE root mean square error.
r Pearson correlation coefficient.
COE the Coefficient of Efficiency
IOA the Index of Agreement

References

1. World Health Organisation. 2014. Available online: https://www.who.int/mediacentre/news/releases/2014/air-pollution/en// (accessed on 14 June 2020).
2. Martins, N.R.; da Graça, C.G. Impact of PM2.5 in indoor urban environments: A review. *Sustain. Cities Soc.* **2018**, *42*, 259–275. [CrossRef]
3. Fu, P.; Guo, X.; Cheung, F.M.H.; Yung, K.K.L. The association between PM 2.5 exposure and neurological disorders: A systematic review and meta-analysis. *Sci. Total Environ.* **2019**, *655*, 1240–1248. [CrossRef] [PubMed]
4. Schraufnagel, D.E.; Balmes, J.R.; Cowl, C.T.; De Matteis, S.; Jung, S.H.; Mortimer, K.; Perez-Padilla, R.; Rice, M.B.; Riojas-Rodriguez, H.; Sood, A.; et al. Air Pollution and Noncommunicable Diseases: A Review by the Forum of International Respiratory Societies' Environmental Committee, Part 2: Air Pollution and Organ Systems. *Chest* **2019**, *155*, 417–426. [CrossRef]
5. DEFRA. Clean Air Strategy **2019**, *1*, 9. Available online: https://assets.publishing.service.gov.uk/government/uploads/system/uploads/attachment_data/file/770715/clean-air-strategy-2019.pdf (accessed on 9 July 2020).
6. Department for Business, Energy & Industrial Strategy. *Summary Results of the Domestic Wood Use Survey*; Department for Energy and Climate Change: London, UK, 2016; pp. 67–80. Available online: https://assets.publishing.service.gov.uk/government/uploads/system/uploads/attachment_data/file/833061/Summary_results_of_the_domestic_wood_use_survey_.pdf (accessed on 09 July 2020).
7. Font, A.; Fuller, G. *Report: Airborne Particles from Wood Burning in UK Cities*; Environmental Research Group -King's College London, UK, 2017; pp. 1–52. Available online: https://uk-air.defra.gov.uk/library/reports?report_id=953 (accessed on 14 July 2020).
8. Air Quality Expert Group. *The Potential Air Quality Impacts from Biomass Combustion*; Air Quality Expert Group: Defra, UK, 2017; p. 80. Available online: https://uk-air.defra.gov.uk/assets/documents/reports/cat11/1708081027_170807_AQEG_Biomass_report.pdf (accessed on 14 July 2020).
9. Eisentraut, A.; Adam, B.; International Energy Agency, Paris. Heating without global warming. *Featured Insight* **2014**. Available online: https://www.iea.org/reports/heating-without-global-warming (accessed on 15 July 2020).
10. Bertrand, A.; Stefenelli, G.; Bruns, E.A.; Pieber, S.M.; Temime-Roussel, B.; Slowik, J.G.; Prévôt, A.S.; Wortham, H.; El Haddad, I.; Marchand, N. Primary emissions and secondary aerosol production potential from woodstoves for residential heating: Influence of the stove technology and combustion efficiency. *Atmos. Environ.* **2017**, *169*, 65–79. [CrossRef]
11. Winijkul, E.; Bond, T.C. Emissions from residential combustion considering end-uses and spatial constraints: Part II, emission reduction scenarios. *Atmos. Environ.* **2016**, *124*, 1–11. [CrossRef]
12. Nyström, R.; Lindgren, R.; Avagyan, R.; Westerholm, R.; Lundstedt, S.; Boman, C. Influence of Wood Species and Burning Conditions on Particle Emission Characteristics in a Residential Wood Stove. *Energy Fuels* **2017**, *31*, 5514–5524. [CrossRef]
13. Avagyan, R.; Nyström, R.; Lindgren, R.; Boman, C.; Westerholm, R. Particulate hydroxy-PAH emissions from a residential wood log stove using different fuels and burning conditions. *Atmos. Environ.* **2016**, *140*, 1–9. [CrossRef]

14. Win, K.M.; Persson, T. Emissions from residential wood pellet boilers and stove characterized into start-up, steady operation, and stop emissions. *Energy Fuels* **2014**, *28*, 2496–2505. [CrossRef]
15. Traynor, G.W.; Apte, M.G.; Carruthers, A.R.; Dillworth, J.F.; Grimsrud, D.T.; Gundel, L.A. Indoor air pollution due to emissions from wood-burning stoves. *Environ. Sci. Technol.* **1987**, *21*, 691–697. [CrossRef]
16. Canha, N.; Almeida, S.M.; Freitas, M.d.C.; Wolterbeek, H.T.; Cardoso, J.; Pio, C.; Caseiro, A. Impact of wood burning on indoor PM2.5 in a primary school in rural Portugal. *Atmos. Environ.* **2014**, *94*, 663–670. [CrossRef]
17. Semmens, E.O.; Noonan, C.W.; Allen, R.W.; Weiler, E.C.; Ward, T.J. Indoor particulate matter in rural, wood stove heated homes. *Environ. Res.* **2015**, *138*, 93–100. [CrossRef] [PubMed]
18. Piccardo, M.T.; Cipolla, M.; Stella, A.; Ceppi, M.; Bruzzone, M.; Izzotti, A.; Valerio, F. Indoor pollution and burning practices in wood stove management. *J. Air Waste Manag. Assoc.* **2014**, *64*, 1309–1316. [CrossRef] [PubMed]
19. Wang, D.; Li, Q.; Shen, G.; Deng, J.; Zhou, W.; Hao, J.; Jiang, J. Significant ultrafine particle emissions from residential solid fuel combustion. *Sci. Total Environ.* **2020**, *715*, 1–7. [CrossRef] [PubMed]
20. Vicente, E.D.; Vicente, A.M.; Evtyugina, M.; Oduber, F.I.; Amato, F.; Querol, X.; Alves, C. Impact of wood combustion on indoor air quality. *Sci. Total Environ.* **2020**, *705*, 135769. [CrossRef] [PubMed]
21. Allen, R.W.; Leckie, S.; Millar, G.; Brauer, M. The impact of wood stove technology upgrades on indoor residential air quality. *Atmos. Environ.* **2009**, *43*, 5908–5915. [CrossRef]
22. Noonan, C.W.; Navidi, W.; Sheppard, L.; Palmer, C.P.; Bergauff, M.; Hooper, K.; Ward, T.J. Residential indoor PM 2.5 in wood stove homes: Follow-up of the Libby changeout program. *Indoor Air* **2012**, *22*, 492–500. [CrossRef]
23. Chen, C.; Li, H.; Niu, Y.; Liu, C.; Lin, Z.; Cai, J.; Li, W.; Ge, W.; Chen, R.; Kan, H. Impact of short-term exposure to fine particulate matter air pollution on urinary metabolome: A randomized, double-blind, crossover trial. *Environ. Int.* **2019**, *130*, 104878. [CrossRef]
24. Madsen, C.; Rosland, P.; Hoff, D.A.; Nystad, W.; Nafstad, P.; Næss, Ø.E. The short-term effect of 24-h average and peak air pollution on mortality in Oslo Norway. *Eur. J. Epidemiol.* **2012**, *27*, 717–727. [CrossRef]
25. Darrow, L.A.; Klein, M.; Sarnat, J.A.; Mulholland, J.A.; Strickland, M.J.; Sarnat, S.E.; Russell, A.G.; Tolbert, P.E. The use of alternative pollutant metrics in time-series studies of ambient air pollution and respiratory emergency department visits. *J. Expo. Sci. Environ. Epidemiol.* **2011**, *21*, 10–19. [CrossRef]
26. Bhaskaran, K.; Hajat, S.; Armstrong, B.; Haines, A.; Herrett, E.; Wilkinson, P.; Smeeth, L. The effects of hourly differences in air pollution on the risk of myocardial infarction: Case crossover analysis of the MINAP database. *BMJ* **2011**, *343*, 1–11. [CrossRef]
27. Lin, H.; Ratnapradipa, K.; Wang, X.; Zhang, Y.; Xu, Y.; Yao, Z.; Dong, G.; Liu, T.; Clark, J.; Dick, R.; et al. Hourly peak concentration measuring the PM2.5-mortality association: Results from six cities in the Pearl River Delta study. *Atmos. Environ.* **2017**, *161*, 27–33. [CrossRef]
28. Orellano, P.; Reynoso, J.; Quaranta, N.; Bardach, A.; Ciapponi, A. Short-term exposure to particulate matter (PM10 and PM2.5), nitrogen dioxide (NO_2), and ozone (O_3) and all-cause and cause-specific mortality: Systematic review and meta-analysis. *Environ. Int.* **2020**, *142*, 105876. [CrossRef]
29. Mitchell, E.J.; Lea-Langton, A.R.; Jones, J.M.; Williams, A.; Layden, P.; Johnson, R. The impact of fuel properties on the emissions from the combustion of biomass and other solid fuels in a fixed bed domestic stove. *Fuel Process. Technol.* **2016**, *142*, 115–123. [CrossRef]
30. Vicente, E.D.; Duarte, M.A.; Calvo, A.I.; Nunes, T.F.; Tarelho, L.; Alves, C.A. Emission of carbon monoxide, total hydrocarbons and particulate matter during wood combustion in a stove operating under distinct conditions. *Fuel Process. Technol.* **2015**, *131*, 182–192. [CrossRef]
31. Mcnamara, M.; Thornburg, J.; Semmens, E.; Ward, T.; Noonan, C. Coarse particulate matter and airborne endotoxin within wood stove homes. *Indoor Air* **2013**, *23*, 498–505. [CrossRef] [PubMed]
32. de Jesus, A.L.; Rahman, M.M.; Mazaheri, M.; Thompson, H.; Knibbs, L.D.; Jeong, C.; Evans, G.; Nei, W.; Ding, A.; Qiao, L.; et al. Ultrafine particles and PM2.5 in the air of cities around the world: Are they representative of each other? *Environ. Int.* **2019**, *129*, 118–135. [CrossRef]
33. Penttinen, P.; Timonen, K.L.; Tiittanen, P.; Mirme, A.; Ruuskanen, J.; Pekkanen, J. Ultrafine particles in urban air and respiratory health among adult asthmatics. *Eur. Respir. J.* **2001**, *17*, 428–435. [CrossRef]
34. Testing of Solid Fuel Stoves. 2013. Available online: https://www.bsria.com/uk/news/article/testing-of-solid-fuel-stoves (accessed on 15 July 2020).

35. Salthammer, T.; Schripp, T.; Wientzek, S.; Wensing, M. Impact of operating wood-burning fireplace ovens on indoor air quality. *Chemosphere* **2014**, *103*, 205–211. [CrossRef]
36. Sheffield City Council. *2011 Census First Results: Population Estimates*; Sheffield City Council: Sheffield, UK, 2011; pp. 1–6. Available online: https://www.sheffield.gov.uk/content/dam/sheffield/docs/your-city-council/facts-figures/2011%20Census%20July%20Release%20-%20Population%20Estimates.pdf (accessed on 15 July 2020)
37. Sheffield City Council. *Air Quality: Action Plan*; Sheffield City Council: Sheffield, UK, 2015; p. 76. Available online: https://www.sheffield.gov.uk/content/dam/sheffield/docs/pollution-and-nuisance/air-pollution/air-quality-management/Air%20Quality%20Action%20Plan%202015.pdf (accessed on 15 July 2020)
38. Urban Flows Observatory. 2018. Available online: https://urbanflows.ac.uk/ (accessed on 17 July 2020).
39. Jayaratne, R.; Liu, X.; Thai, P.; Dunbabin, M.; Morawska, L. The influence of humidity on the performance of a low-cost air particle mass sensor and the effect of atmospheric fog. *Atmos. Meas. Tech.* **2018**, *11*, 4883–4890. [CrossRef]
40. Bulot, F.M.; Johnston, S.J.; Basford, P.J.; Easton, N.H.; Apetroaie-Cristea, M.; Foster, G.L.; Morris, A.K.; Cox, S.J.; Loxham, M. Long-term field comparison of multiple low-cost particulate matter sensors in an outdoor urban environment. *Sci. Rep.* **2019**, *9*, 1–13. [CrossRef] [PubMed]
41. Hänel, G. The properties of atmospheric aerosol particles as functions of the relative humidity at thermodynamic equilibrium with the surrounding moist air. *Adv. Geophys.* **1976**, *19*, 73–188. [CrossRef]
42. Streibl, N. Influence of Humidity on the Accuracy of Low-Cost Particulate Matter Sensors. Technical Report. 2017. Available online: https://www.researchgate.net/profile/Norbert_Streibl/publication/320474792_Influence_of_Humidity_on_the_Accuracy_of_Low-Cost_Particulate_Matter_Sensors/links/59e7ad15aca272bc423d0b97/Influence-of-Humidity-on-the-Accuracy-of-Low-Cost-Particulate-Matter-Sensors.pdf (accessed on 17 July 2020).
43. Köhler, H. The nucleus in and the growth of hygroscopic droplets. *Trans. Faraday Soc.* **1936**, *32*, 1152–1161. [CrossRef]
44. Petters, M.D.; Kreidenweis, S.M. A single parameter representation of hygroscopic growth and cloud condensation nucleus activity-Part 3: Including surfactant partitioning. *Atmos. Chem. Phys.* **2013**, *13*, 1081–1091. [CrossRef]
45. Svenningsson, B.; Rissler, J.; Swietlicki, E.; Mircea, M.; Bilde, M.; Facchini, M.C.; Decesari, S.; Fuzzi, S.; Zhou, J.; Mønster, J.; et al. Hygroscopic growth and critical supersaturations for mixed aerosol particles of inorganic and organic compounds of atmospheric relevance. *Atmos. Chem. Phys.* **2006**, *6*, 1937–1952. [CrossRef]
46. South Coast Air Quality Management District (SCAQMD). Field Evaluation of SainSmart. 2017. Available online: http://www.aqmd.gov/docs/default-source/aq-spec/field-evaluations/sainsmart---field-evaluation.pdf (accessed on 20 August 2020).
47. Air Quality Expert Group. Fine Particulate Matter (PM 2.5) in the United Kingdom. 2012. Available online: https://uk-air.defra.gov.uk/assets/documents/reports/cat11/1212141150_AQEG_Fine_Particulate_Matter_in_the_UK.pdf (accessed on 20 August 2020)
48. Ridgeway, G. Generalized Boosted Models: A guide to the gbm package. *Compute* **2007**, *1*, 1–12. [CrossRef]
49. Carslaw, D.C.; Taylor, P.J. Analysis of air pollution data at a mixed source location using boosted regression trees. *Atmos. Environ.* **2009**, *43*, 3563–3570. [CrossRef]
50. Tong, J.C. Cross-Validation. *Encycl. Syst. Biol.* **2013**, 508–508._941. [CrossRef]
51. The Natural Environment Research Council. 2020. Available online: https://bit.ly/3kjWfn8 (accessed on 19 September 2020)
52. Ezzati, M.; Kammen, D.M. The health impacts of exposure to indoor air pollution from solid fuels in developing countries: Knowledge, gaps, and data needs. *Environ. Health Perspect.* **2002**, *110*, 1057–1068. [CrossRef]
53. Delfino, R.J.; Zeiger, R.S.; Seltzer, J.M.; Street, D.H.; McLaren, C.E. Association of asthma symptoms with peak particulate air pollution and effect modification by anti-inflammatory medication use. *Environ. Health Perspect.* **2002**, *110*, 607–617. [CrossRef]

54. Castro, A.; Calvo, A.I.; Blanco-Alegre, C.; Oduber, F.; Alves, C.; Coz, E.; Amato, F.; Querol, X.; Fraile, R. Impact of the wood combustion in an open fireplace on the air quality of a living room: Estimation of the respirable fraction. *Sci. Total Environ.* **2018**, *628–629*, 169–176. [CrossRef]

5

Pollution Characteristics of Particulate Matter ($PM_{2.5}$ and PM_{10}) and Constituent Carbonaceous Aerosols in a South Asian Future Megacity

Afifa Aslam [1], Muhammad Ibrahim [1], Imran Shahid [2], Abid Mahmood [1], Muhammad Kashif Irshad [1], Muhammad Yamin [3], Ghazala [1], Muhammad Tariq [1] and Redmond R. Shamshiri [4,*]

[1] Department of Environmental Sciences & Engineering, Government College University Faisalabad, Faisalabad 38000, Pakistan; afifaaslam22@gcuf.edu.pk (A.A.); mibrahim@gcuf.edu.pk (M.I.); drabid@gcuf.edu.pk (A.M.); kashifirshad@gcuf.edu.pk (M.K.I.); ghazala@gcuf.edu.pk (G.); tariqnazir159@gmail.com (M.T.)

[2] Environmental Science Centre, Qatar University, Doha P.O. Box 2713, Qatar; ishahid@qu.edu.qa

[3] Department of Farm Machinery & Power, Faculty of Agricultural Engineering & Technology, University of Agriculture, Faisalabad 38040, Pakistan; yamin529@uaf.edu.pk

[4] Leibniz Institute for Agricultural Engineering and Bioeconomy, Max-Eyth-Allee 100, 14469 Potsdam-Bornim, Germany

* Correspondence: rshamshiri@atb-potsdam.de

Abstract: The future megacity of Faisalabad is of prime interest when considering environmental health because of its bulky population and abundant industrial and anthropogenic sources of coarse particles (PM_{10}) and fine airborne particulate matter ($PM_{2.5}$). The current study was aimed to investigate the concentration level of $PM_{2.5}$ and PM_{10}, also the characterization of carbonaceous aerosols including organic carbon (OC), elemental carbon (EC) and total carbon (TC) in $PM_{2.5}$ and PM_{10} samples collected from five different sectors (residential, health, commercial, industrial, and vehicular zone). The data presented here are the first of their kind in this sprawling city having industries and agricultural activities side by side. Results of the study revealed that the mass concentration of $PM_{2.5}$ and PM_{10} is at an elevated level throughout Faisalabad, with ambient $PM_{2.5}$ and PM_{10} points that constantly exceeded the 24-h standards of US-EPA, and National Environment Quality Standards (NEQS) which poses harmful effects on the quality of air and health. The total carbon concentration varied between 21.33 and 206.84 $\mu g/m^3$, and 26.08 and 211.15 $\mu g/m^3$ with an average of 119.16 ± 64.91 $\mu g/m^3$ and 124.71 ± 64.38 $\mu g/m^3$ for $PM_{2.5}$ in summer and winter seasons, respectively. For PM_{10}, the concentration of TC varied from 34.52 to 289.21 $\mu g/m^3$ with an average of 181.50 ± 87.38 $\mu g/m^3$ (for summer season) and it ranged between 44.04 and 300.02 $\mu g/m^3$ with an average of 191.04 ± 87.98 $\mu g/m^3$ (winter season), respectively. No significant difference between particulate concentration and weather parameters was observed. Similarly, results of air quality index (AQI) and pollution index (PI) stated that the air quality of Faisalabad ranges from poor to severely pollute. In terms of AQI, moderate pollution was recorded on sampling sites in the following order; Ittehad Welfare Dispensary > Saleemi Chowk > Kashmir Road > Pepsi Factory, while at Nazria Pakistan Square and Allied Hospital, higher AQI values were recorded. The analysis and results presented in this study can be used by policy-makers to apply rigorous strategies that decrease air pollution and the associated health effects in Faisalabad.

Keywords: particulate matter; aerosols; vehicular exhaust; industrial activity; WHO; NEQS; US-EPA

1. Introduction

In many developing countries, increasing industrialization and overpopulation becomes the reason for escalating air pollution [1]. According to various researches conducted in high-income countries situated in the Asian region, the level of many air pollutants is normally beyond the ambient air quality standards and WHO guidelines. In many developing countries, the use of non-renewable fuel like biomass and diesel is associated with the increasing level of air pollution at the regional level. Airborne particulate matter is abundant in the atmosphere and is the foremost indicator of the quality of air in a specified area. Chemical composition, concentration and size of particulate matter varied widely and are delimited universally under acceptable standards built on size elements ranging from $PM_{2.5}$ to PM_{10} to Total Suspended Particles (TSP), while PM_4 was also identified as the respirable size fraction [2]. Particulate matter instigates from a diversity of anthropogenic (e.g., rapid industrialization, agricultural activities, refineries, waste incineration, biomass burning, motor vehicles, utilities, brick kiln, industrial emissions power plants, factories, large population and heavy traffic) are responsible for bad air quality in the cities due to elevated levels of gaseous and particulate pollutants [3] and natural (e.g., dust storm and sea spray) sources, besides secondary formation processes. However, also mineral dust transport from deserted areas is considered a significant source for regional pollution in Asia [4]. Henceforth, for the air quality management and epidemiological studies, the assessment of the concentration of atmospheric particulate matter (PM) and its associated toxic constituents is a prerequisite [5]. It was consistently confirmed by epidemiological studies that there is a strong association between ambient particulate matter comprising toxic components and cardiovascular- and respiratory-related upsurges in mortality and morbidity, particularly in urban areas [6]. This connection has been revealed to be stronger for $PM_{2.5}$ rather than for PM_{10} or total suspended particles since $PM_{2.5}$ can infiltrate deep into the alveolar areas of the human lungs [6]. The transport and distribution of particulate matter in the atmosphere are distinctly allied with meteorological parameters such as air temperature, relative humidity, atmospheric pressure, wind direction, speed, and rainfall [7]. In various parts of the world, different monitoring programs on atmospheric PM have been directed which exposed varied instabilities and disproportions among the trace element constituents and particulate matter [8].

Components of carbonaceous aerosol, elemental carbon (EC) and organic carbon (OC), account for a large element of atmospheric particulate matter and, on average, subsidize 20–35% of coarse particulate and 20–45% of fine particulate [9]. Carbonaceous aerosols have a chief role in the interactions of light-particles within the atmosphere and are one of the significant components of fine and coarse particulate matter; they are therefore associated with the negative climatic and environmental impacts and the worsening in public health and air quality [10]. Elemental carbon is often used as a substitute for black carbon (BC) and is discharged into the atmosphere mostly through the processes of combustion [11]. Elemental carbon is primarily accountable for the absorption of light in the atmosphere, which sturdily influence the radiative balance of the earth [12]. The six main sources of elemental carbon have been recognized using organic tracers as coal combustion, biomass burning, vehicle exhaust, cigarette smoke, cooking and vegetative detritus [13]. Carbonaceous aerosols were found dominant in $PM_{2.5}$ (which is attained from agricultural waste and wood-fuel burning) and have a strong effect on the decline in visibility and air-quality and also stimulates radiative forcing on a regional scale [14].

In Pakistan, control of air pollution has not yet become a democratic issue because of a lack of suitable information for policy and decision-makers, though some infrequent reports that identify airborne particulate matter as a great health and environmental concern in urban regions of Pakistan are present [15]. Generally, the concentration of particulate matter is many folds higher than the acceptable limits documented by the World Health Organization (WHO), National Environmental Quality Standards (NEQS) and the United States Environmental Protection Agency (US-EPA). According to the World Bank [16], the annual burden of health because of particulate matter was 1% of the GDP and is accountable for 700 deaths among children and 22,000 premature deaths amongst adults in Pakistan. However, due to the absence of air quality management competencies, the country is suffering from the deterioration of air quality. Evidence from many international bodies and governmental organizations has indicated that air pollution is a momentous risk to the health of residents, environment and quality of life [17]. According to a study directed by the World Health Organization, Bombay, Calcutta and Tehran were found to be the most contaminated cities in Asia [18]. Similarly, Faisalabad (the textile city of Pakistan) is also highlighted to be an extremely polluted city in this study. Due to increased industrialization and construction of commercial zones and rapid urbanization, the atmosphere of the city is getting worst day by day [19]. This state of concern stimulated us to conduct a comprehensive study on the status of air pollution in Faisalabad. As a result of the burning issue of air pollution and associated health impacts, a study was planned to examine the quality of air in Faisalabad city for which 12 different sites were selected and categorized as residential, commercial, industrial, and health centers.

Keeping in view the facts discussed above, the present study was conducted with the following objectives: (a) to measure the quality of air with its allied consequences within varying activity zones of Faisalabad city; (b) to compare the ambient air quality of Faisalabad with air pollution indexes such as NEQS -Pakistan, National Ambient Air Quality Standards (NAAQS)-US-EPA and WHO; and (c) to provide an opportunity to conduct additional studies on source identification, impact assessment, and trend analysis for this zone. It is expected that the current study will be supportive for designing and establishing emission regulations and abatement strategies in the future.

2. Methodology

2.1. Study Area and Sampling Sites

Faisalabad is the third-largest city of Pakistan and a major industrial hub (dominated by textile and chemical industries); consequently, the air quality of the city is a major environmental problem. It covers an area of 1230 km^2 and is occupied by more than four million people. The summer season is very hot with a humid climate while a cold winter (falls to 0 °C some days) is experienced by the Faisalabad city. The climate of the city touches extreme hotness and humidity during summer and cold during winter. The sampling sites were banquets around Faisalabad and its vicinities. Twelve sampling sites were nominated based on current anthropogenic activities accountable for atmospheric pollution, and the dominant direction of wind for pollutant dispersion and distribution in the area. The locations were selected based on the zones in the city. The selected locations are comprised of medical units, residential areas, commercial areas, industrial areas, and automobiles rich areas. The average wind speed of 3–6 km in winter and 6–13 km in summer was observed. The map showing the locations on the Faisalabad (Figure 1) represents the coordinates of the location within Faisalabad geography.

2.2. Data Collection

Data of meteorological parameters were obtained from Agromet. The PM concentrations were determined by the first author herself. We took the samples from all the locations and then measurements were made in the Lab. We took sample readings sector-wise and readings for all the residential sites were taken at the same time. A similar trend was followed for commercial, industrial, health, and automobile sites. At Provincial and Federal EPAs, Data Logging systems retrieve the data about the quality of ambient air from air monitoring stations with the help of data processing software. The seasonal average was intended to find out the difference in the mass concentration of $PM_{2.5}$ in summer and winter seasons. For this study, 12 discrete sampling sites under five diverse sectors (residential, health, commercial, industrial, and automobile vehicles) were selected for the evaluation of $PM_{2.5}$ and PM_{10} with the help of high air volume sampler. The interpretations were taken at three diverse times (morning, noon, and evening) daily from November 1 to December 31 for winter and from May 1 to June 30 for summer. It should be noted that wind speed and direction influence the rate of diffusion of pollution. The temperature inversion is also directly linked to solar radiation making the air softer, hence the air converts into fog because pollutants and dust are no longer raised from the surface. This can become a problem in metropolises where numerous pollutants exist. The data of temperature and radiation used in this study was collected from the Agricultural Metrological Cell Agromet Bulletin.

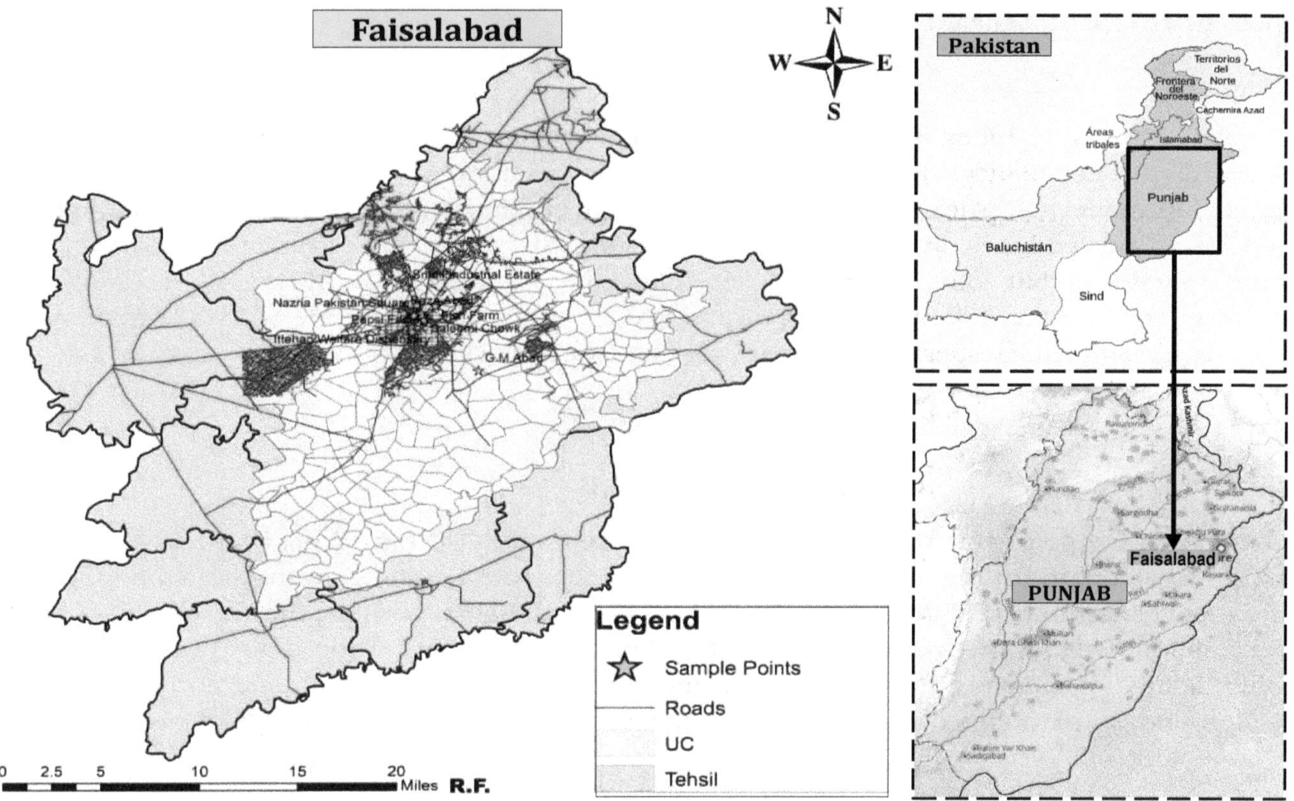

Figure 1. Study sites within the Faisalabad city and location of the city within the political map of Pakistan.

2.3. Sample Analysis (Chemical, Gravimetric, and Carbonaceous Aerosols)

After sample collection, the filter papers were kept in exact environmental conditions at a relative humidity of 30–40% and temperature of 20–23 °C for 24 h as per the US-EPA standard. Before mass analysis, the filter paper with fine and coarse particulates samples was equilibrated for 24 h in silica gel desiccators to abolish the effect of humidity and to attain accurate particulate matter measurements. The $PM_{2.5}$ and PM_{10} masses of each sample were determined gravimetrically by deducting the initial average mass of the blank filter from the final average mass of the sampled filter. Gravimetric analysis is the determination of particulate concentration based on weight difference. Individual filters (Teflon®, 46.2 mm) were weighed on an electronic micro-balance pre and post field sampling. Particulate matter <2.5 μm was collected from ambient air on the filters throughout the sample duration of 24 h. The net variances between pre- and post-sampling filter weights were used to estimate the mass concentration in the ambient air of the city. After post weighing, filters can be stored for a minimum of one year. Using the post-sample and pre-sample filter weights, the total filter mass gain ($PM_{2.5}$) and the concentration of PM_{10} were respectively calculated from Equation (1) and Equation (2):

$$PM_{2.5} = (M_{Post} - M_{Pre})(10^3)/M_{Pre} \quad (1)$$

$$PM10 = (Wf - Wi)(106)/V \quad (2)$$

Here $PM_{2.5}$ is the total mass gain in μg, M_{Post} is the post sample filter weight in mg, M_{Pre} is the pre-sample filter weight in μg, W_f is the filter paper weight, W_i is the initial mass of filter paper, and V is the total air sampled in m^3. In the current study, elemental carbon was determined by a two-step combustion method described by [20]. Filters were heated for 2 h at 340 °C in an oxygen atmosphere to remove organic carbon (OC). The calibration procedure was done using tartaric acid dyed in aluminum foil. While, total carbon was determined by a combustion method, where all material on the filter is combusted in pure oxygen at 1000 °C and the resulting CO_2 is measured by non-dispersive IR photometry (NDIR, Maihak) [21].

2.4. Air Quality and Pollution Index

An Air Quality Index is defined as a complete scheme that converts the weighed values of parameters related to individual air pollution (e.g., the concentration of pollutant) into a sole number or set of numbers [22]. Air Quality Index (AQI) is a tool to detect the present scenario of air quality. AQI was calculated based on the arithmetic mean of the ratio of the concentration of pollutants to the standard value of that pollutant such as PM_{10}, $PM_{2.5}$, NO_2, and SO_2. The average is then multiplied by 100 to arrive at the AQI index. The pollutant AQI and the pollution index (PI) of the potentially noxious element were respectively derived from Equations (3) and (4):

$$\mathrm{AQI} = (\mathrm{W} * \mathrm{C}/\mathrm{Cs}) \tag{3}$$

$$\mathrm{PI} = \mathrm{Cn}/\mathrm{Bn} \tag{4}$$

where W is the pollutant weighted, C is the observed value ($PM_{2.5}$, PM_{10}, SO_2 and NO_2), Cs is the CPCB standard for the residential area [23], B_n is the background concentration, and C_n is the measured concentration of the element. It should be noted that in Pakistan, the National Air Quality Index is followed. Moreover, it should be highlighted that the pollution index of the potentially contaminated elements is the ratio between the concentration of toxic elements and the reference background concentration of the consistent elements obtained from a previous published study [24].

3. Results

3.1. Mass Concentration of $PM_{2.5}$ and PM_{10} in Winter Season

The results of the present investigations in Faisalabad city for which 12 different sites were selected and categorized as residential, commercial, industrial and health centers are presented (Figure 2). The concentration of $PM_{2.5}$ had the following decreasing order in the air samples collected near medical centers of Mian Trust Hospital (38.50 ± 0.30 μg/m^3), Ittehad Welfare Dispensary (37.35 + 0.45 μg/m^3), Allied Hospital (36.65 ± 0.27 μg/m^3). While, in residential areas, the highest concentration of fine particulate was found in Ghulam Muhammad Abad (39.1 ± 0.50 μg/m^3) followed by Raza Abad (35.2 ± 0.23 μg/m^3), Saleemi Chowk (33.83 ± 0.74 μg/m^3), Kashmir Road (32.25 ± 0.14 μg/m^3). In the selected commercial areas, Nazria Pakistan (43.63 ± 0.59 μg/m^3) was the most polluted site of the city. The average concentration of $PM_{2.5}$ in small industrial estate was recorded as (37.93 ± 0.19 μg/m^3) followed by Pepsi factory area (37.17 ± 0.62 μg/m^3); while in and near transport station (35.08 ± 0.61 μg/m^3) was analyzed. In contrast, the highest concentration level of PM_{10} was found in the samples collected from Nazria Pakistan (800.85 ± 0.93 μg/m^3) followed by Mian Trust Hospital (586.6 ± 3.88 μg/m^3), Allied Hospital (584.62 ± 3.41 μg/m^3), Small Industrial Estate (469.1 ± 0.57 μg/m^3) and Ghulam Muhammad Abad (440.2 ± 0.10 μg/m^3); while the lowest concentration was analyzed in the ambient air of Saleemi Chowk (280.18 ± 0.12 μg/m^3). The average particulate concentration can be compared with guideline values proposed by WHO, US-EPA and NEQS. During the present study, the concentration of $PM_{2.5}$ was within the guideline value of the US-EPA and NEQS (35 μg/m^3) but still higher than the WHO guideline value of 25μg/m^3 [25] for 24 h average, while coarse particles were drastically exceeding the safe limits of all the quality standards. The composition of particulate matter is strongly reliant with its sources, i.e., anthropogenic or natural (Figure 2).

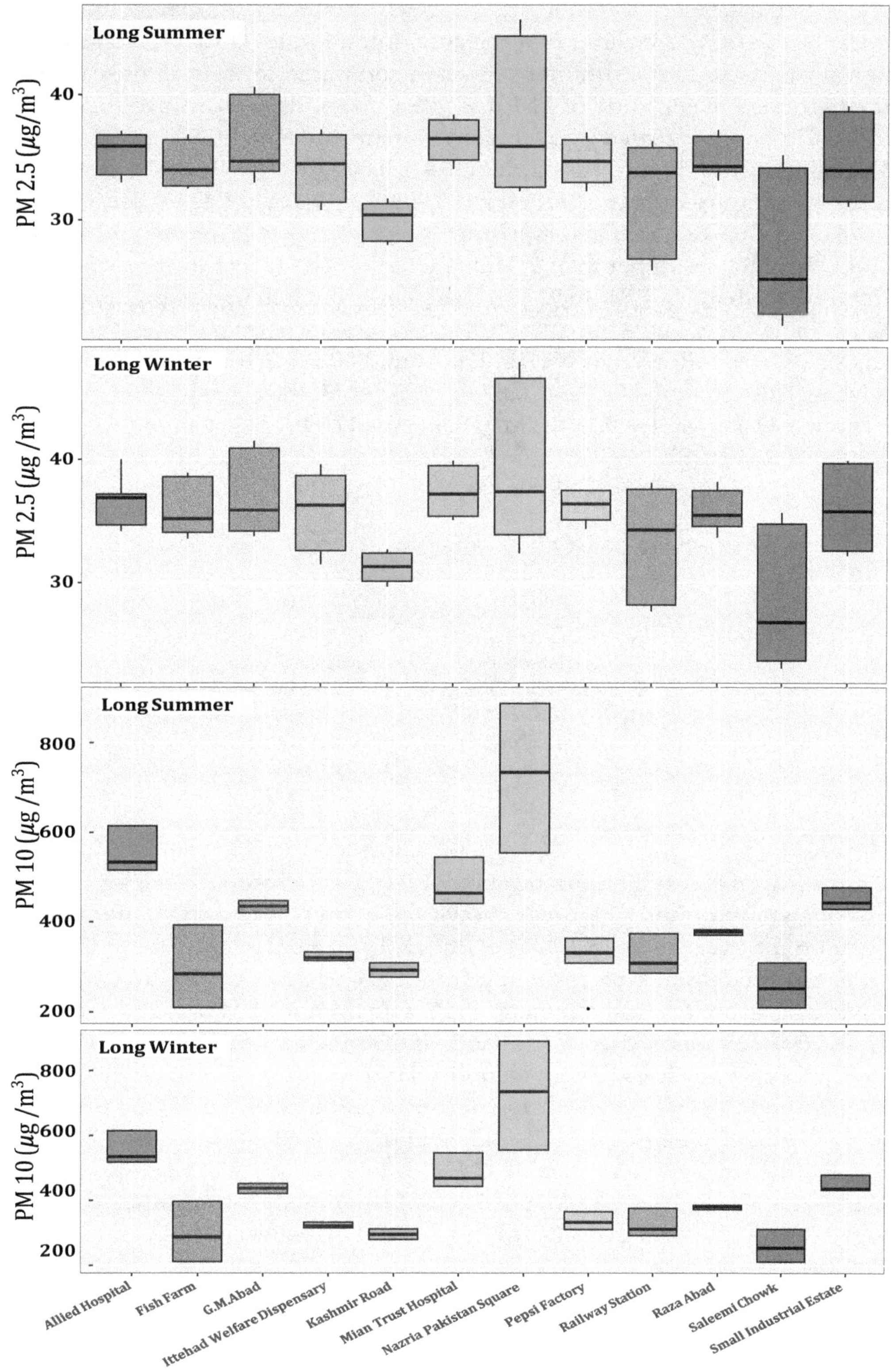

Figure 2. Average mass concentrations of $PM_{2.5}$ and PM_{10} in long summer and long winter in Faisalabad city.

3.2. Mass Concentration of $PM_{2.5}$ and PM_{10} in Summer Season

Figure 2 shows the mass concentration of fine ($PM_{2.5}$) and coarse (PM_{10}) particles calculated from the samples collected during the winter (Dec 2016 to Jan 2017) and summer (May to June 2017) seasons from different selected areas of Faisalabad. It was clear from the results that the concentration levels of both particulates were lower in the summer season as compared to the winter season. According to the results, it was cleared that the contamination level that the concentration level of both particulates was lower in the summer season as compared to the winter season. According to the results, the highest concentration of $PM_{2.5}$ was found in the air samples collected from Nazria Pakistan (42.5 ± 0.57 μg/m^3) followed by a Ghulam Muhammad Abad (38.4 ± 0.34 μg/m^3), Mian Trust Hospital and Small Industrial Estate air samples (36.9 ± 0.34 μg/m^3). While, the mass volume of $PM_{2.5}$ was lower in the air samples of Station Chowk (33.5 ± 0.35 μg/m^3), Raza Abad (34.2 ± 0.28 μg/m^3), Saleemi Chowk (31.8 ± 0.94 μg/m^3) and Kashmir Road (31.2 ± 0.21 μg/m^3) when compared with US-EPA and NEQS guidelines rather than WHO safe limits. The rest of the areas were slightly higher in $PM_{2.5}$ concentrations than US-EPA and NEQS safe limits but still highly polluted if compared with WHO guidelines. Table 1 represents the concentration values of coarse particles obtained after analysis. The decreasing order was followed as Nazria Pakistan (800.6 ± 2.16 μg/m^3), Allied Hospital (477.4 ± 3.7 μg/m^3), Mian Trust Hospital (477.3 ± 0.82 μg/m^3), Small Industrial Estate (465.4 ± 1.16 μg/m^3), Ghulam Muhammad Abad (440.1 ± 0.13 μg/m^3), Raza Abad (380.1 ± 0.08 μg/m^3) with the lowest value obtained at Kashmir Road (297.3 ± 0.50 μg/m^3). According to the results, the mass concentration of PM_{10} is exceeding the safe guidelines of all the selected air quality standards throughout the study area.

Table 1. (National Air Quality Index, CPCB, October 2014).

Category	Range
Good	0–50
Satisfactory	51–100
Moderately Polluted	101–200
Poor	201–300
Very Poor	301–400
Severe	401–500

3.3. Seasonal Impact on $PM_{2.5}$ and PM_{10} Concentration

Figure 2 represents the average concentration of $PM_{2.5}$ and PM_{10} in the long summer and winter seasons during the study period. Figure 2 shows the highest fine particulates concentration in a commercial area on the average 39.18 ± 4.70 μg/m^3 and 40.73 ± 2.9 μg/m^3 with the lowest obtained concentration in the residential sector ranging from 33.91 ± 3.27 μg/m^3 and 35.1 ± 2.93 μg/m^3 for summer and winter season, respectively. The WHO safe limit for $PM_{2.5}$ is 25 μg/m^3 and for PM_{10} is 50 μg/m^3 (WHO, 2005). Similarly, the US-EPA and NEQS safe limit for $PM_{2.5}$ is 35 μg/m^3 and PM_{10} is 150 μg/m^3 [26,27]. Coarse particles (PM_{10}) were similar in trend as shown by $PM_{2.5}$ with decreasing trend as commercial areas, hospital areas, industrial areas, automobile station and residential areas were in the range of 575.19 ± 66.26 μg/m^3, 499.81 ± 148.62 μg/m^3, 409.63 ± 59.44 μg/m^3, 379.63 ± 0.81 μg/m^3 and 350.98 ± 74.29 μg/m^3 in winter and 573.14 ± 321.64 μg/m^3, 427.52 ± 86.30 μg/m^3, 405.63 ± 84.49 μg/m^3, 349.41 ± 74.48 μg/m^3 and 349.46 ± 74.49 μg/m^3 in the summer season, respectively. Correspondingly, Figure 2 showed a strong positive correlation between $PM_{2.5}$ and PM_{10} in both winter and summer seasons on average.

3.4. Analysis of Carbonaceous Aerosols in fine ($PM_{2.5}$) and Coarse Particulate (PM_{10}) Samples

The concentration level of carbonaceous aerosols is presented in Figures 3–5 for summer and winter seasons. It was clear that concentration of EC and OC was higher in winter (Figures 3 and 4) which was quite similar with previous studies. According to the results, EC was found in low concentration on average in the samples of fine particles collected from Kashmir Road (8.56 ± 1.86 μg/m^3) in summer season while highest EC contamination was found in the ambient air of Nazria Pakistan (89.67 ± 1.52 μg/m^3). A similar trend was found for OC with the lowest concentration in the $PM_{2.5}$ samples collected from Kashmir Road (19.93 ± 0.42 μg/m^3) categorized as one of the residential sites, while the highest values were obtained in the ambient air of Nazria Pakistan (178.4 ± 3.51 μg/m^3) nominated as the busiest commercial zone of Faisalabad with a variety of businesses (Figure 4).

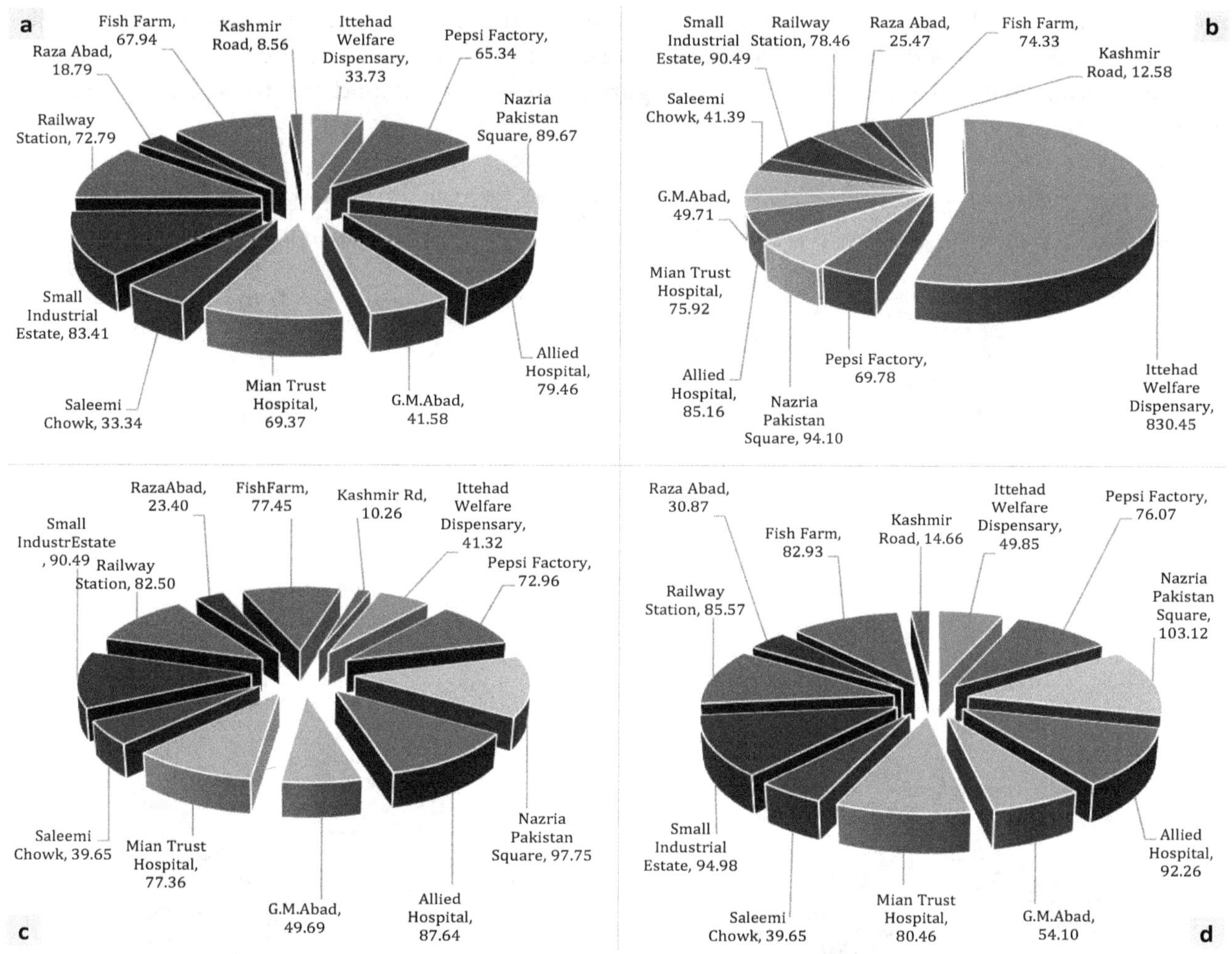

Figure 3. Concentration of carbonaceous species—elemental carbon (EC) in $PM_{2.5}$ in summer (**a**); $PM_{2.5}$ in winter (**b**); PM_{10} in summer (**c**); PM_{10} in winter (**d**). All units are expressed in μg/m^3.

When considering TC for the investigated sites (Figure 5), it was observed that Nazria Pakistan was heavily contaminated (268.08 ± 5.03 μg/m^3) followed by Small Industrial Estate (248.23 ± 5.79 μg/m^3), Fish Farm (235.68 ± 5.02 μg/m^3), Allied Hospital (224.34 ± 3.62 μg/m^3), Vehicular station (223.21 ± 3.82 μg/m^3), Mian Trust Hospital (202.77 ± 8.42 μg/m^3) and Pepsi Factory (184.84 ± 4.27 μg/m^3) with positive OC/EC correlation which indicates the common source of emission of TC in these zones. While, residential areas (G.M Abad, Saleemi Chowk, Raza Abad and Kashmir Road) were less contaminated (114.71 ± 2.76 μg/m^3; 64.97 ± 3.58 μg/m^3; 51.16 ± 3.14 μg/m^3; and 28.49 ± 2.27 μg/m^3, respectively), as compared to the other sites indicating negative OC/EC correlation. While samples of fine particulates collected in the winter season from the same investigating sites were analyzed for carbonaceous aerosol concentration. The levels of EC and OC were higher in winter as compared to the summer season. This may be due to the more wood and fossil fuel burning to warm up the surroundings as well as extra consumption of diesel and petrol by vehicles to warm up the engines in sizzling cold weather. The concentration of TC was much higher in the ambient air of Nazria Pakistan (277.5 ± 4.9 μg/m^3) followed by Small Industrial Estate (262.01 ± 3.68 μg/m^3), Fish Farm (248.07 ± 5.28 μg/m^3), Allied Hospital (237.13 ± 2.89 μg/m^3), Vehicular Station (234.4 ± 4.08 μg/m^3), Mian Trust Hospital (215.97 ± 8.82 μg/m^3), Pepsi Factory (194.27 ± 4.30 μg/m^3) and Ittehad Welfare Dispensary (142.58 ± 7.69 μg/m^3). On the other side, mix community of domestic zones showed less concentration of both EC and OC but still falls in contamination categories that are not safe for human health. Kashmir Road was detected with the least concentration of TC (38 ± 3.43 μg/m^3) while Ghulam Muhammad Abad was higher in TC (130.04 ± 2.6 μg/m^3) concentration in the residential zone. A positive OC/EC correlation was observed in the maximum of the investigating sites in the winter season. Samples of coarse particulate were also analyzed to evaluate the concentration level of carbonaceous aerosols collected from the investigating sites of Faisalabad city for summer and winter seasons, respectively.

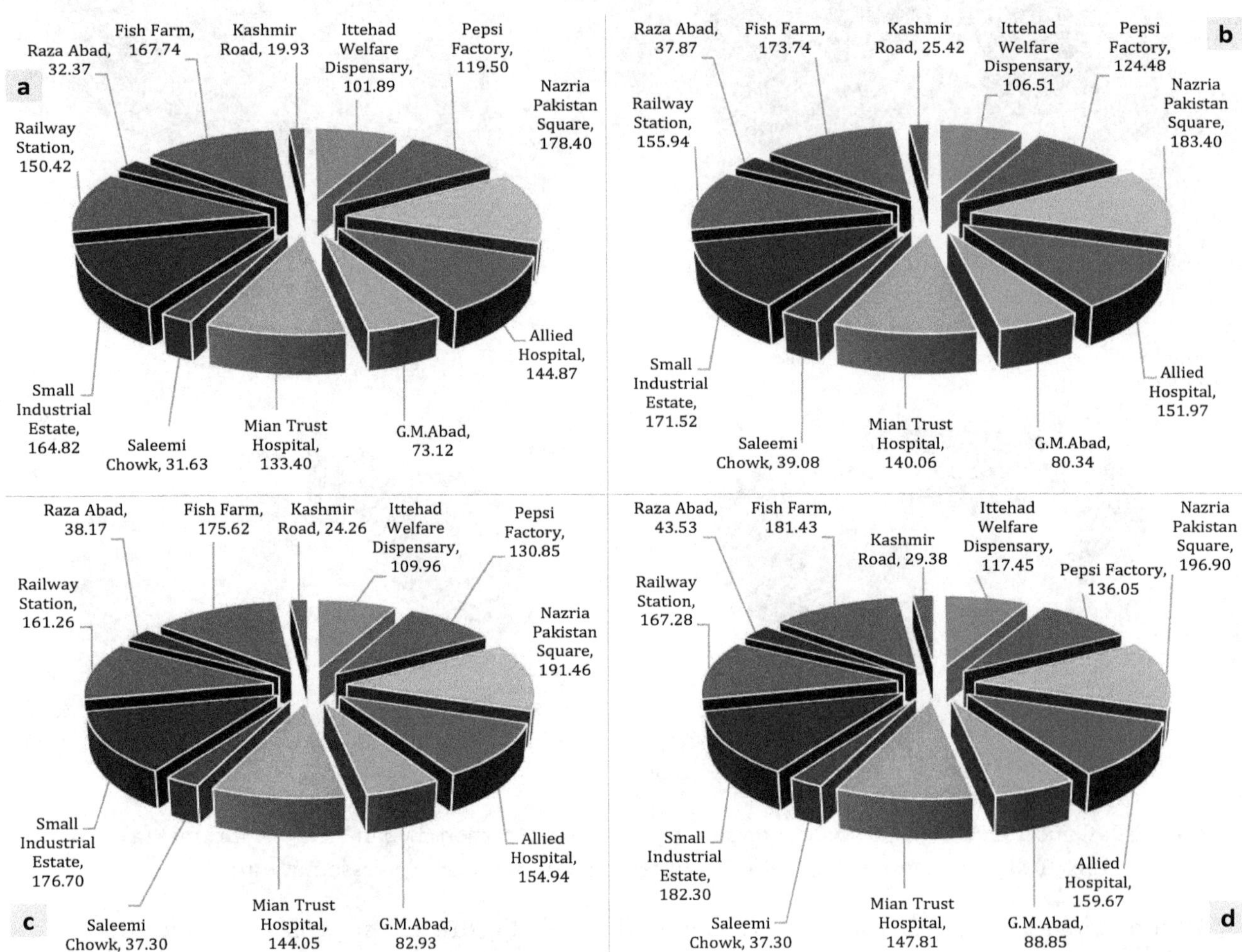

Figure 4. Concentration of carbonaceous species—organic carbon (OC) in $PM_{2.5}$ in summer (**a**); $PM_{2.5}$ in winter (**b**); PM_{10} in summer (**c**); PM_{10} in winter (**d**). All units are expressed in μg/m^3.

We found similar trends of concentration level as experienced with aerosols available in fine particulates but higher in concentration than observed in $PM_{2.5}$ samples. Commercial areas of Faisalabad were enriched with TC (289.21 ± 2.75 μg/m^3 and 300.02 ± 3.25 μg/m^3 for Nazria Pakistan and 253.06 ± 5.59 μg/m^3 and 264.36 ± 4.16 μg/m^3 for Fish Farm) at an elevated level among all the sites. Coarse particulate samples collected from the Small Industrial Estate were also found to be extremely high (267.19 ± 4.28 μg/m^3 and 277.28 ± 3.16 μg/m^3) after Nazria Pakistan followed by Vehicular Station (243.75 ± 3.66 μg/m^3 and 252.85 ± 2.9 μg/m^3), Allied Hospital (242.58 ± 5.24 μg/m^3 and 251.93 ± 4.44 μg/m^3), Mian Trust Hospital (221.4 ± 6.51 μg/m^3 and 228.27 ± 5.53 μg/m^3) and Ittehad Welfare Dispensary (151.28 ± 12.09 μg/m^3 and 167.31 ± 4.57 μg/m^3) for summer and winter seasons, respectively. Concentration level of carbonaceous aerosol was higher in Ghulam Muhammad Abad (132.62 ± 3.4 μg/m^3 and 142.95 ± 3.65 μg/m^3) while considering the residential zone of the city followed by Saleemi Chowk (76.95 ± 4.87 μg/m^3 and 76.95 ± 3.9 μg/m^3), Raza Abad (61.57 ± 3 μg/m^3

and 74.4 ± 2.32 μg/m^3) and Kashmir Road (34.52 ± 2.02 μg/m^3 and 44.04 ± 1.97 μg/m^3) for both the seasons accordingly.

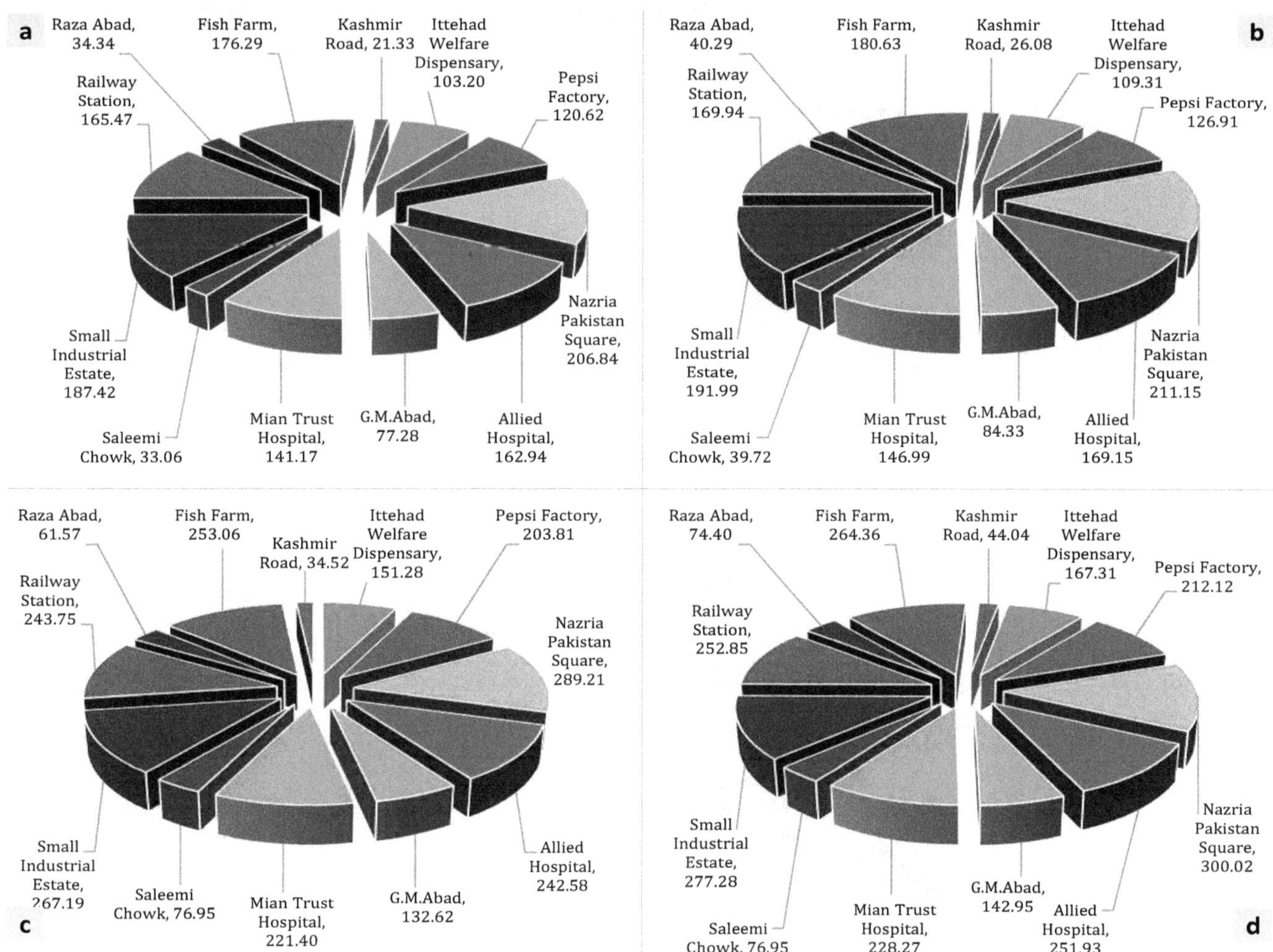

Figure 5. Concentration of carbonaceous species—total carbon (TC) in $PM_{2.5}$ in summer (**a**); $PM_{2.5}$ in winter (**b**); PM_{10} in summer (**c**); PM_{10} in winter (**d**). All units are expressed in μg/m^3.

3.5. Air Quality and Pollution Index

The air temperature of the study area fluctuated between 18–25 °C in winter which is considered a typical range while the trend in May–June 2017, as shown in Figure 6, showed an increase to 37.2 °C on average. Figure 6 also demonstrates the trend of relative humidity with a mean value of that varies from 39.45% in summer to 60.4% in winter. Table 1 presented the categories of air quality according to the AQI while Figure 7 illustrates the AQI index of selected sites of Faisalabad city with detrimental outcomes. It was found that the ambient air of Faisalabad city ranges from moderately polluted with the sequence of Ittehad Welfare Dispensary > Saleemi Chowk > Kashmir Road > Pepsi Factory, while severely polluted air was found in the vicinity of Nazria Pakistan Square, followed by Allied Hospital. The air quality of Mian Trust Hospital and the Small Industrial Estate was categorized as very poor besides GM Abad, Raza Abad and Station Chowk where the air quality was poor. The overall air quality of Faisalabad city was not good for health and other activities that require urgent attention from Government institutes and ministries involved in making and implementing policies to safeguard the environment.

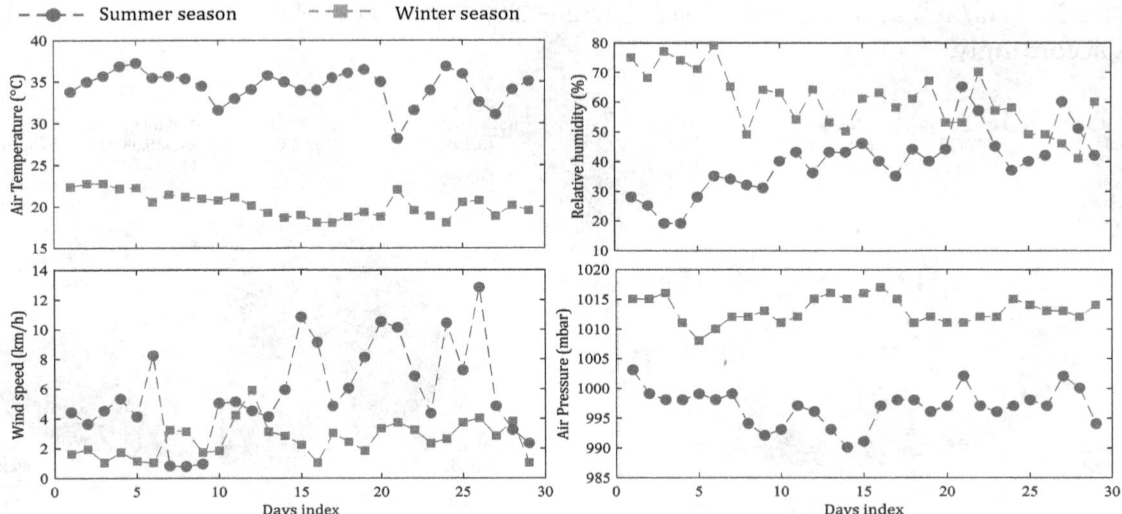

Figure 6. Relation of weather parameters in summer and winter seasons.

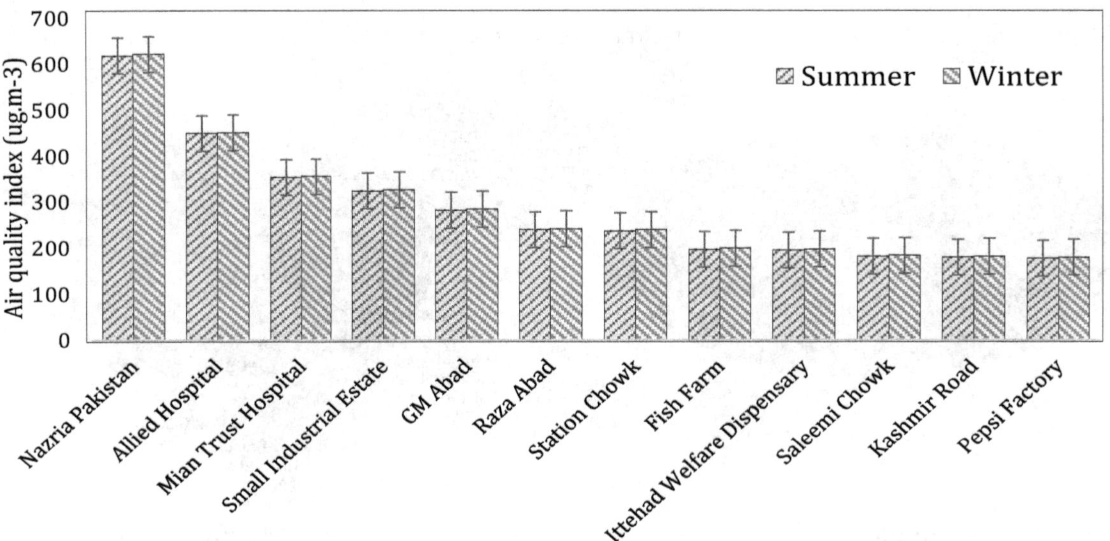

Figure 7. Air Quality Index (AQI) of Faisalabad city after analysis.

The average values of pollution index (PI) for each potential toxic element at selected sites of Faisalabad city for both seasons have been shown in Figure 8. In some residential areas, PI of $PM_{2.5}$ was found in the average level of pollution. While the PI of PM_{10} was estimated for the same areas showed a middle level of pollution $1 < PI \leq 2$, and sample site which is located near Nazria Pakistan Square suggested a high level of environmental pollution $PI > 4$. The $PM_{2.5}$ concentrations for almost all the sampling sites also showed a low level of environmental pollution of $PI \geq 1$. At Saleemi Chowk, Fish Farm, and Kashmir Road, the PI of PM_{10} showed a low level of environmental pollution $PI \leq 1$, while samples collected in the vicinities of Ittehad Welfare Dispensary, Pepsi Factory, Station Chowk, Raza Abad, GM Abad, and Mian Trust Hospital showed the middle level of pollution ($2 < PI \leq 3$) in the environment. It should be noted that the samples of Allied Hospital and Nazria Pakistan showed the highest environmental pollution level of $3 < PI < 4$. This can be due to the toxic elements in urban dust which accumulate and originate mainly from traffic, paint, and many other nonspecific urban sources in the megacity.

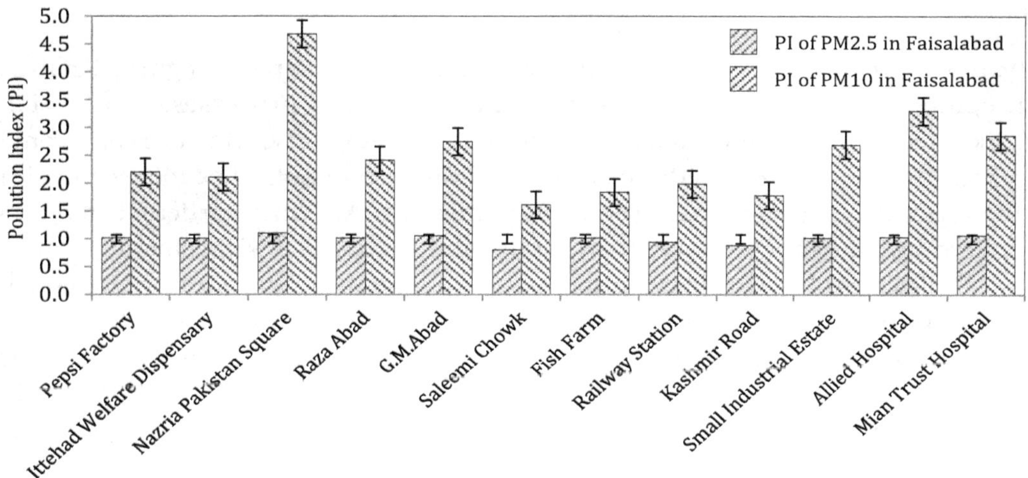

Figure 8. Pollution index flow diagram of $PM_{2.5}$ and PM_{10} in Faisalabad.

In order to show the difference between the mass concentrations of $PM_{2.5}$ and PM_{10} were statistically significant between the 12 locations, four sets of null hypotheses with H_0: $\mu(Location1) = \mu(Location2) = \ldots = \mu(Location12)$ were tested against the alternative hypotheses that the means of mass concentrations in the 12 locations were not equal. Based on the very small p-value that resulted from the one-way analysis of variance tests ($p < 0.0001$), all null hypotheses were rejected at any significant level and we concluded that the difference between locations was statistically significant. This result has been also shown in Figure 9 by means of the four scatter plots that demonstrate a visual comparison between the mean values of $PM_{2.5}$ and PM_{10} data in the 12 locations with respect to the daytime (morning, afternoon, evening) and season (summer or winter). Taking plot labeled (c) of Figure 9 as an example (mean of PM_{10} data in summer), the difference between morning concentrations of $PM_{2.5}$ in the 12 locations is clearly visible. It should be noted that for 61 days of winter the total number of measurements was 549. That is 61 days multiplied by 3 daytime (morning, afternoon, and evening) multiplied by 3 replications for each time.

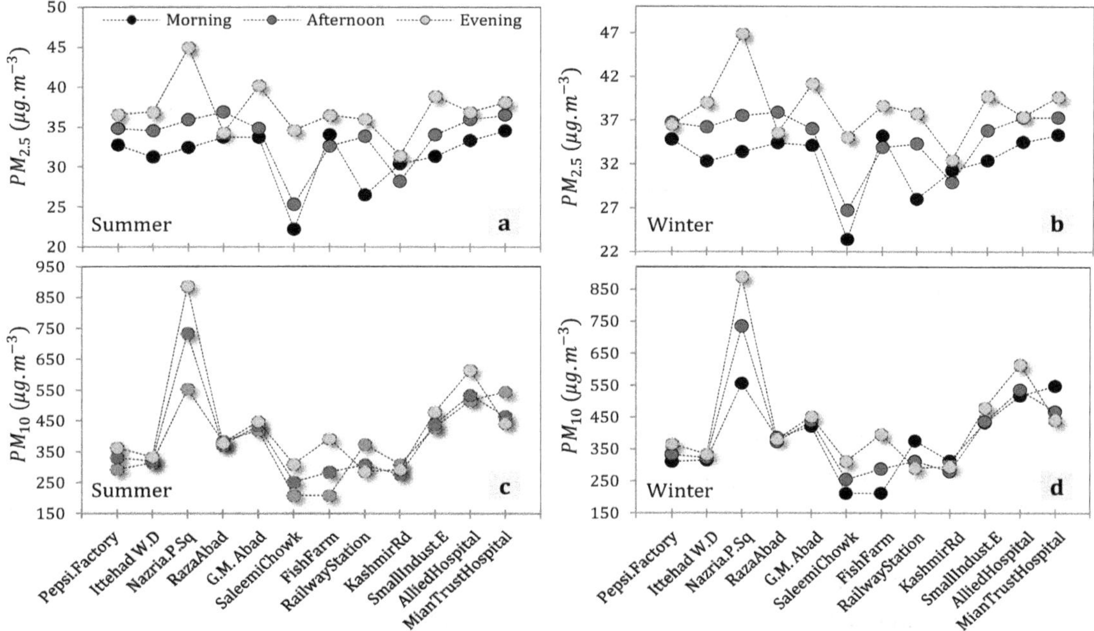

Figure 9. A comparison between mean values of $PM_{2.5}$ and PM_{10} data with respect to the daytime (morning, afternoon, evening) and season (summer or winter) in the 12 locations for (**a**): $PM_{2.5}$ in summer, (**b**): $PM_{2.5}$ in winter, (**c**) PM_{10} in summer, (**d**) PM_{10} in winter.

4. Discussion

Usually, the sources of gaseous pollutants are measured in three categories: natural emission mechanism, combustion sources and industrial manufacturing processes. Industrial sources of particulates, like steel, heavy traffic loads, indiscriminate burning of solid wastes and cement factories are the main sources of PM [2,28] besides gaseous and noise pollutants. The increasing air pollution is, after losses of properties, crops and increased health care costs. Airborne particulate matter is abundant in the atmosphere and varies extensively (temporally and spatially) in size, chemical composition and concentration. Emissions of particulate matter towards air are the focal environmental challenges for the transport and industrial sectors [4]. According to the WHO report, South Asia has developed as one of the most polluted zones in the globe because of its increasing population and rapid industrialization. Acquaintance to PM leads to more appointments to the emergency room or doctor. Health effects include premature deaths with existing lung and heart diseases, lung damage, coughing, aggravated asthma, wheezing and shortness of breath. Specifically, in the dry and cold season in major urban areas of Pakistan, people of almost all ages suffer from throat infections as reported [2,4,17]. In Pakistan, like the other developing countries, the emissions from vehicles have been conquered by emissions from poorly maintained and old vehicles that subsidize to heightened mass concentrations of carbon monoxide and fine particulates [17,28].

The current study focused on particulate volume has reported the highest concentration of $PM_{2.5}$ and PM_{10} in a commercial area on the average 39.18 ± 4.70 μg/m^3, 573.14 ± 321.64 μg/m^3 and 40.73 ± 2.9 μg/m^3, 575.19 ± 225.66 μg/m^3 for summer and winter, respectively (Figure 2). While, the lowest concentration of $PM_{2.5}$ and PM_{10} was obtained in the residential sector ranging (33.91 ± 3.27 μg/m^3), (35.1 ± 2.93 μg/m^3) and (349.46.75 ± 74.49 μg/m^3), (350.98 ± 74.29 μg/m^3) for summer and winter season correspondingly with a strong positive correlation between $PM_{2.5}$ and PM_{10} in both seasons on average. According to the guidelines, the reference value for $PM_{2.5}$ and PM_{10} are WHO (25 μg/m^3 and 50 μg/m^3) [17,25], NEQS and US-EPA (35 μg/m^3, 150 μg/m^3) [26,27] and most samples examined in the present study had values higher than the reference values. Elemental carbon is discharged from a variety of ignition procedures, categorized as a short-lived climate forcer that put up to atmospheric warming and also allied with human mortality and morbidity [28]. Common sources of atmospheric primary and secondary organic carbon antecedents are biomass burning, vehicular exhaust, biogenic emission and industrial emissions [29]. During the winter season, a higher level of pollutants especially the mass concentration of $PM_{2.5}$ persists in the ambient air of Faisalabad, owing to reduced atmospheric dispersion due to high relative humidity. Similarly, it was observed that $PM_{2.5}$ and PM_{10} sources were frequently localized as depicted by high concentrations at low wind speeds, mostly by the emissions from road vehicles [2,29]. This demonstrates the fact that $PM_{2.5}$ and PM_{10} concentrations were lower in summer than in winter (Figure 2) due to an increase in wind speed and temperature.

Prior studies conducted in the carbonaceous aerosols were assessed to account for about 50–60% of the total mass of $PM_{2.5}$ in metropolises in Jordan, Israel and Palestine [30]. Not unexpectedly, since production and processing of oil was prevalent transversely in the Middle East, substantial oil burning was valued to contribute 18% to total mass of PM_{10} and 69% to the total mass of $PM_{2.5}$ in Jeddah, Saudi Arabia [31]. Likewise, in Faisalabad, Pakistan, the quality of air not only reflects the impact of regional and local dust but also momentous local sources which include numerous industries and a heavy traffic weight. In municipal areas, the higher concentrations of $PM_{2.5}$ and PM_{10} are symbolic of the higher density of traffic as presented in the current study (Figures 2 and 4). Additionally, the burden of particulates is higher in the daytime than nighttime one, demonstrating more urban activities throughout day time. In Faisalabad, the textile industry, the topographical configuration and the geographical location make the problem of air pollution so perilous that it is very crucial to study it (Figure 5). The current study aimed at finding out whether or not the situation of air pollution in Faisalabad was previously seriously abundant to warrant the establishment of a regular air quality management system through which intercession measures can be premeditated and executed.

The analysis result of ambient air samples of selected sites of Faisalabad city displays that the level of particulate matter in most of the areas of the city is above the indorsed levels of the WHO, NEQS and US-EPA. Most of the city's commercial and residential areas are within the sensitive zone with the maximum concentrations of PM, which is constant with their proximity to the city's industrial areas.

When compared with the other studies conducted in other cities, it was found that the $PM_{2.5}$ level at Industrial Estate I-10 and IJP Road has reached the critical level (>35 μg/m^3) whereas at Industrial Estate I-9 it was moderate to the high level (31.9 μg/m^3 to 41.1 μg/m^3) [27]. While, the mean concentration of $PM_{2.5}$ and PM_{10} for Peshawar city during the study period has been calculated to be respectively 172 μg/m^3 and 480 μg/m^3 [32]. A similar high mass concentration of particulate matter was observed by [33] at Lahore, Pakistan and documented that the average $PM_{2.5}$ mass was 190 μg/m^3, and ranged from 89 μg/m^3 to 476 μg/m^3, far over US-EPA standards. Much higher PM_{10} mass concentration was experienced in Faisalabad when compared with other megacities [34], In addition, the PM_{10} concentrations were quite higher than the annual mean PM_{10} concentrations in Eastern Mediterranean and Africa [35,36] (WHO. Ambient (outdoor) 2014), Malaysia [34] and Bogota, Egypt, Los Angeles and Mexico [34]. It was also identified that PM_{10} is the dominant pollutant in the index value [37]. While, according to the results obtained after the analysis of particulate matters samples, the highest concentration of elemental carbon was 103.12 ± 1.46 μg/m^3 and the highest concentration of organic carbon was 196.9 ± 1.79 μg/m^3. While, 300.02 ± 3.25 μg/m^3 was the highest TC concentration found in the samples of coarse particulate matter collected in the vicinity of Nazria Pakistan (Figure 3). When compared, it was found that these concentrations are comparatively higher than in other metropolises in the areas like Punjab, India (116 μg/m^3), Hangzhou, China (119 μg/m^3), Kolkatta, India (197 μg/m^3), New Delhi, India (219 μg/m^3) and Lahore, Pakistan (233 μg/m^3) [38–40]. For elemental carbon, a large number of sources are identified, e.g., biomass and coal-fired power plant, two-stroke vehicles, fossil fuel burning, diesel engines and low burning efficiency. Elemental carbon is also utilized as a tracer for vehicular emission [17,40]. It was stated by [41] that diesel and gasoline motor vehicles and traffic exhaust are key sources of elemental carbon, followed by biomass burning. Organic carbon can be released straight from sources identified as primary carbon as a result of biomass and fossil combustion or can be formed as a result of a chemical reaction recognized as secondary organic carbon [42]. Temperature means are also under the normal limit but the increasing trend shows the alarming state of affairs and the same case is with radiations. Relative humidity has a value that is normal and considered healthy but a decreasing trend precedes the deterioration of ambient air quality. AQI and PI indicated that the ambient air quality of Faisalabad city falls from poor to severely polluted categories which are not safe to breathe and perform our daily activities.

5. Conclusions

Studying particle matters with aerodynamic diameters below 10 μm and 2.5 μm have received research attention for atmospheric pollution characteristics due to their severe effects on the human health issue. In this paper, we studied PM_{10} and $PM_{2.5}$ and highlighted that atmospheric pollution has become a significant issue as a result of growing industries in the megacity of Faisalabad, leading to the increased risk factors for chronic respiratory diseases in elderly and accelerated loss of lung function in newborns. To determine the pollution characteristics of particular matter, as well as the source and factors affecting them, we concentrated our study on 12 different sites that were selected and categorized as residential, commercial, industrial and health centers. Results of our study showed that the PM concentrations measured during current study periods (Dec 2016–Jan 2017) at various zones of Faisalabad were surprisingly higher than summer (May–June 2017). The enormous difference between fine ($PM_{2.5}$) and coarse (PM_{10}) particulate specifies that Faisalabad is inclined by a high loading of "coarse" particulate dust. Commercial areas are heavily polluted with fine and coarse particulate pollution. The average levels of pollution for fine and particulate matter were recorded as 39.18 ± 4.70, 573.14 ± 321.64 and 40.73 ± 2.9, 575.19 ± 225.66 during summer and winter, respectively (values in μg/m^3). The average $PM_{2.5}$ and PM_{10} concentrations were higher as compared to other major cities like

Islamabad, Lahore, and Peshawar. The quality of ambient air of Faisalabad has deteriorated beyond the safe limits set by WHO, US-EPA and NEQS. We also concluded that carbonaceous aerosols are in higher concentration in the air of the study sites. The air quality of Faisalabad city ranges from poor to severely polluted category which is highly unsafe for human health. These demands for an effort to introduce appropriate pollution control and management plans such as plantation and green belts for the betterment of civic life. A sustainable solution to improve air quality in Faisalabad would be to reduce emission by replacing high-energy consuming industries with renewable and clean energy sources, besides other strategies that reduce the use of fossil energy. Future studies may involve the use of wavelet analysis to explore the temporal characteristics of $PM_{2.5}$ and PM_{10}, or to investigate the relationship between meteorological factors and PM_{10}.

Author Contributions: A.A., M.I., I.S. and A.M. conceived and designed the experiments; A.A., M.T. and A.M. performed the experiments. M.K.I., G., M.Y., M.T. used software; R.R.S. worked on the final review and editing, figures and analysis; A.A., M.T., R.R.S. performed formal analysis and writing the draft; M.I., A.M. and G. did project funding and finding acquisition; M.I., I.S., A.M. and G. performed supervision of the experiments. All authors have read and agreed to the published version of the manuscript.

Acknowledgments: The authors acknowledge the financial support by the Open Access Publication Fund of the Leibniz Association, Germany, the partial research funding and editorial supports from Adaptive AgroTech Consultancy International and the administrative supports from Benjamin Mahns at the Leibniz Institute for Agricultural Engineering and Bioeconomy in Potsdam, Germany.

References

1. Hamid, A.; Akhtar, S.; Atique, S.A.; Huma, Z.; Mohay Uddin, S.G.; Asghar, S. Ambient air quality & noise level monitoring of different areas of Lahore (Pakistan) and its health impacts. *Pol. J. Environ. Stu.* **2019**, *28*, 623–629.
2. Javed, W.; Wexler, A.S.; Murtaza, G.; Iqbal, M.M.; Zhao, Y.; Naz, T. Chemical characterization and source apportionment of atmospheric particles across multiple sampling locations in Faisalabad, Pakistan. *Clean Soil Air Water* **2016**, *44*, 753–765. [CrossRef]
3. Gurjar, B.R.; Butler, T.M.; Lawrence, M.G.; Lelieveld, J. Evaluation of emissions and air quality in megacities. *Atmos. Environ.* **2008**, *42*, 1593–1606. [CrossRef]
4. Shahid, I.; Kistler, M.; Mukhtar, A.; Ghauri, B.M.; Cruz, C.R.S.; Bauer, H.; Puxbaum, H. Chemical characterization and mass closure of PM10 and PM2.5 at an urban site in Karachi, Pakistan. *Atmos. Environ.* **2016**, *128*, 114–123. [CrossRef]
5. Cheng, F.J.; Lee, K.H.; Lee, C.W.; Hsu, P.C. Association between particulate matter air pollution and hospital emergency room visits for pneumonia with septicemia: A retrospective analysis. *Aerosol Air Qual. Res.* **2019**, *19*, 345–354. [CrossRef]
6. Correia, A.W.; Pope III, C.A.; Dockery, D.W.; Wang, Y.; Ezzati, M.; Dominici, F. The effect of air pollution control on life expectancy in the United States: An analysis of 545 US counties for the period 2000 to 2007. *Epidemiology* **2013**, *24*, 23. [CrossRef]
7. Pakbin, P.; Hudda, N.; Cheung, K.L.; Moore, K.F.; Sioutas, C. Spatial and temporal variability of coarse (PM10−2.5) particulate matter concentrations in the Los Angeles area. *Aerosol. Sci. Technol.* **2010**, *44*, 514–525. [CrossRef]
8. Leghari, S.K.; Zaidi, M.A.; Ahmed, M.; Sarangzai, A.M. Assessment of suspended particulate matters level and role of vegetation in ambient air of North-East Balochistan, Pakistan. *Fuuast J. Biol.* **2013**, *3*, 37–43.
9. Dinoi, A.; Cesari, D.; Marinoni, A.; Bonasoni, P.; Riccio, A.; Chianese, E.; Tirimberio, G.; Naccarato, A.; Sprovieri, F.; Andreoli, V.; et al. Inter-comparison of carbon content in $PM_{2.5}$ and PM_{10} collected at five measurement sites in southern Italy. *Atmosphere* **2017**, *8*, 243. [CrossRef]

10. Choomanee, P.; Bualert, S.; Thongyen, T.; Salao, S.; Szymanski, W.W.; Rungratanaubon, T. Vertical variation of carbonaceous aerosols with in the $PM_{2.5}$ fraction in Bangkok, Thailand. *Aerosol. Air Qual. Res.* **2020**, *20*, 43–52. [CrossRef]
11. Ji, D.; Zhang, J.; He, J.; Wang, X.; Pang, B.; Liu, Z.; Wang, L.; Wang, Y. Characteristics of atmospheric organic and elemental carbon aerosols in urban Beijing, China. *Atmos. Environ.* **2016**, *125*, 293–306. [CrossRef]
12. Li, C.; Chen, P.; Kang, S.; Yan, F.; Hu, Z.; Qu, B.; Sillanpää, M. Concentrations and light absorption characteristics of carbonaceous aerosol in $PM_{2.5}$ and PM_{10} of Lhasa city, the Tibetan Plateau. *Atmos. Environ.* **2016**, *127*, 340–346. [CrossRef]
13. Li, Y.C.; Yu, J.Z.; Ho, S.S.; Schauer, J.J.; Yuan, Z.; Lau, A.K.; Louie, P.K. Chemical characteristics and source apportionment of fine particulate organic carbon in Hong Kong during high particulate matter episodes in winter 2003. *Atmos. Res.* **2013**, *120*, 88–98. [CrossRef]
14. Bisht, D.S.; Tiwari, S.; Dumka, U.C.; Srivastava, A.K.; Safai, P.D.; Ghude, S.D.; Chate, D.M.; Rao, P.S.; Ali, K.; Prabhakaran, T.; et al. Tethered balloon-born and ground-based measurements of black carbon and particulate profiles within the lower troposphere during the foggy period in Delhi, India. *Sci. Total Environ.* **2016**, *573*, 894–905. [CrossRef]
15. Colbeck, I.; Nasir, Z.A.; Ali, Z. The state of ambient air quality in Pakistan-a review. *Environ. Sci. Poll. Res.* **2010**, *17*, 49–63. [CrossRef]
16. World Bank and Institute for Health Metrics and Evaluation. *The Cost of Air Pollution: Strengthening the Economic Case for Action*; World Bank Group: Washington, DC, USA, 2016.
17. Niaz, Y.; Zhou, J.; Iqbal, M.; Nasir, A.; Dong, B. Ambient air quality evaluation: A comparative study in China and Pakistan. *Pol. J. Environ. Stu.* **2015**, *24*, 1723–1732. [CrossRef]
18. Gunawardana, C.; Goonetilleke, A.; Egodawatta, P.; Dawes, L.; Kokot, S. Source characterization of road dust based on chemical and mineralogical composition. *Chemosphere* **2012**, *87*, 163–170. [CrossRef]
19. Shahid, M.A.K.; Hussain, K.; Awan, M.S. Characterization of Solid Aerosols related to Faisalabad Environment and their probable sources. *Coden Jnsmac.* **2012**, *52*, 09–29.
20. Cachier, H.; Bremond, M.P.; Buat Ménard, P. Determination of atmospheric soot carbon with a simple thermal method. *Tellus B* **1989**, *41*, 379–390. [CrossRef]
21. Shahid, I.; Kistler, M.; Shahid, M.Z.; Puxbaum, H. Aerosol Chemical Characterization and Contribution of Biomass Burning to Particulate Matter at a Residential Site in Islamabad, Pakistan. *Aeros. Air Qual. Res.* **2019**, *19*, 148–162. [CrossRef]
22. Ott, W.R. *Environmental Indices: Theory and Practices*; Ann Arbor Science Publishers Inc.: Ann Arbor, MI, USA, 1978.
23. Central Pollution Control Board (CPCB). *Guidelines for National Ambient Air Quality Monitoring*; Series: NAAQM/25/2003- 04; Central Pollution Control Board: Delhi, India, 2009.
24. Chen, T.B.; Zheng, Y.M.; Lei, M.; Huang, Z.C.; Wu, H.T.; Chen, H.; Fan, K.K.; Yu, K.; Wu, X.; Tian, Q.Z. Assessment of heavy metal pollution in surface soils of urban parks in Beijing, China. *Chemosphere* **2005**, *60*, 542–551. [CrossRef]
25. World Health Organization. *Mental Health: Facing the Challenges, Building Solutions: Report from the WHO European Ministerial Conference*; WHO Regional Office Europe: Geneva, Switzerland, 2005.
26. Abbas, M.; Tahira, A.; Jamil, S. Air quality monitoring of particulate matter (PM2.5 & PM10) at Niazi and Daewoo bus station, Lahore. *FUUAST J. Biol.* **2017**, *7*, 13–18.
27. Hassan, M.; Malik, A.H.; Waseem, A.; Abbas, M. Air pollution monitoring in urban areas due to heavy transportation and industries: A case study of Rawalpindi and Islamabad. *J. Chem. Soc. Pak.* **2013**, *35*, 1623.
28. Huang, X.H.; Bian, Q.J.; Louie, P.K.K.; Yu, J.Z. Contributions of vehicular carbonaceous aerosols to $PM_{2.5}$ in a roadside environment in Hong Kong. *Atmos. Chem. Physics.* **2014**, *14*, 9279–9293. [CrossRef]
29. Weinhold, B. Global bang for the buck: Cutting black carbon and methane benefits both health and climate. *Environ. Health Perspect.* **2012**, *120*, A245. [CrossRef]
30. Abdeen, Z.; Qasrawi, R.; Heo, J.; Wu, B.; Shpund, J.; Vanger, A.; Sharf, G.; Moise, T.; Brenner, S.; Nassar, K.; et al. Spatial and temporal variation in fine particulate matter mass and chemical composition: The Middle East consortium for aerosol research study. *Sci. World J.* **2014**, 878704. [CrossRef]
31. Khodeir, M.; Shamy, M.; Alghamdi, M.; Zhong, M.; Sun, H.; Costa, M.; Chen, L.-C.; Maciejczyk, P. Source apportionment and elemental composition of $PM_{2.5}$ and PM_{10} in Jeddah City, Saudi Arabia. *Atmos. Pollut. Res.* **2012**, *3*, 331–340. [CrossRef]

32. Alam, K.; Rahman, N.; Khan, H.U.; Haq, B.S.; Rahman, S. Particulate matter and its source apportionment in Peshawar, Northern Pakistan. *Aerosol Air Qual. Res.* **2015**, *15*, 634–647. [CrossRef]
33. Husain, L.; Dutkiewicz, V.A.; Khan, A.J.; Ghauri, B.M. Characterization of carbonaceous aerosols in urban air. *Atmos. Environ.* **2007**, *41*, 6872–6883. [CrossRef]
34. Safar, Z.S.; Labib, M.W. Assessment of particulate matter and lead levels in the Greater Cairo area for the period 1998–2007. *J. Advan. Res.* **2010**, *1*, 53–63. [CrossRef]
35. World Health Organization. *Ambient (Outdoor) Air Quality and Health*; Fact sheet No. 313; 2014; Available online: https://www.who.int/news-room/fact-sheets/detail/ambient-(outdoor)-air-quality-and-health (accessed on 25 April 2020).
36. Ul-Saufie, A.; Yahya, A.; Ramli, N.; Hamid, H. Future PM_{10} concentration prediction using quantile regression models. In *International Conference on Environmental and Agriculture Engineering*; IACSIT Press: Singapore, 2012; Volume 37.
37. Nigam, S.; Rao, B.P.S.; Kumar, N.; Mhaisalkar, V.A. Air quality index-A comparative study for assessing the status of air quality. *Res. J. Eng. Technol.* **2015**, *6*, 267–274. [CrossRef]
38. Cao, J.; Shen, Z.; Chow, J.C.; Qi, G.; Watson, J.G. Seasonal variations and sources of mass and chemical composition for PM10 aerosol in Hangzhou, China. *Particuology* **2009**, *7*, 161–168. [CrossRef]
39. Awasthi, A.; Agarwal, R.; Mittal, S.K.; Singh, N.; Singh, K.; Gupta, P.K. Study of size and mass distribution of particulate matter due to crop residue burning with seasonal variation in rural area of Punjab, India. *J. Environ. Monit.* **2011**, *13*, 1073–1081. [CrossRef] [PubMed]
40. Alam, K.; Mukhtar, A.; Shahid, I.; Blaschke, T.; Majid, H.; Rahman, S.; Khan, R.; Rahman, N. Source apportionment and characterization of particulate matter (PM10) in urban environment of Lahore. *Aerosol. Air Qual. Res.* **2014**, *14*, 1851–1861. [CrossRef]
41. Ghauri, B.; Lodhi, A.; Mansha, M. Development of baseline (air quality) data in Pakistan. *Environ. Monit. Assess.* **2007**, *127*, 237–252. [CrossRef]
42. Seinfeld, J.H.; Pandis, S.N. *Atmospheric Chemistry and Physics: From Air Pollution to Climate Change*; John Wiley & Sons: Hoboken, NJ, USA, 2016.

6

Wintertime Residential Biomass Burning in Las Vegas, Nevada: Marker Components and Apportionment Methods

Steven G. Brown [1,2,*], **Taehyoung Lee** [3], **Paul T. Roberts** [1] **and Jeffrey L. Collett Jr.** [2]

[1] Sonoma Technology Inc., 1455 N. McDowell Blvd., Suite D, Petaluma, CA 94954, USA; paul@sonomatech.com

[2] Department of Atmospheric Science, Colorado State University, Fort Collins, CO 80523, USA; collett@atmos.colostate.edu

[3] Department of Environmental Science, Hankuk University of Foreign Studies, Yongin 427-010, Korea; thlee@hufs.ac.kr

* Correspondence: sbrown@sonomatech.com

Abstract: We characterized residential biomass burning contributions to fine particle concentrations via multiple methods at Fyfe Elementary School in Las Vegas, Nevada, during January 2008: with levoglucosan on quartz fiber filters; with water soluble potassium (K^+) measured using a particle-into-liquid system with ion chromatography (PILS-IC); and with the fragment $C_2H_4O_2{}^+$ from an Aerodyne High Resolution Aerosol Mass Spectrometer (HR-AMS). A Magee Scientific Aethalometer was also used to determine aerosol absorption at the UV (370 nm) and black carbon (BC, 880 nm) channels, where UV-BC difference is indicative of biomass burning (BB). Levoglucosan and AMS $C_2H_4O_2{}^+$ measurements were strongly correlated (r^2 = 0.92); K^+ correlated well with $C_2H_4O_2{}^+$ (r^2 = 0.86) during the evening but not during other times. While K^+ may be an indicator of BB, it is not necessarily a unique tracer, as non-BB sources appear to contribute significantly to K^+ and can change from day to day. Low correlation was seen between UV-BC difference and other indicators, possibly because of an overwhelming influence of freeway emissions on BC concentrations. Given the sampling location—next to a twelve-lane freeway—urban-scale biomass burning was found to be a surprisingly large source of aerosol: overnight BB organic aerosol contributed between 26% and 33% of the organic aerosol mass.

Keywords: biomass burning; organic aerosol; black carbon; levoglucosan; Las Vegas; source apportionment; aerosol mass spectrometer; elementary school

---

1. Introduction

1.1. Residential Wintertime Biomass Burning and Its Fine Particle Tracers

Biomass burning includes both residential biomass burning for home heating during the wintertime, and smoke transported from wildfires or prescribed burns. In the winter, wildfires and prescribed burns in the Las Vegas area are minimal, so the main biomass burning influence is from residential burning. Understanding the impact of residential biomass burning on aerosol concentrations in urban areas is of particular interest, since emissions are potentially controllable through burn-prevention or fireplace change-out programs [1,2] and because residential biomass burning can lead to high concentrations during the evening and overnight, when emissions are trapped in a shallow boundary layer [3,4]. These short, high-concentration events can have acute health impacts [5,6], and specific health effects have also been associated with inhaling biomass burning aerosol [7–10]. Biomass burning emissions include not just black carbon (BC) and organic matter (OM), but also carcinogens such as benzene and polycyclic aromatic hydrocarbons (PAHs) [11].

Biomass burning is typically apportioned using: (1) the organic molecule levoglucosan, either via chemical analysis of filters or semi-continuously via instruments such as the Aerodyne High Resolution Aerosol Mass Spectrometer (HR-AMS); (2) potassium; and (3) multi-channel Aethalometer data. Levoglucosan is an anhydrous sugar produced in the combustion of cellulose [12–16]. It is typically quantified by extracting aerosol collected on quartz fiber filters and analyzing the aerosol by gas chromatograph-mass spectrometer (GC-MS) or other analytical techniques. While levoglucosan is a good tracer for biomass burning, Sullivan *et al.* and others have found that the relationship of levoglucosan to organic aerosol in biomass burning emissions can vary widely by fuel type and burning conditions [16]. Levoglucosan may not be fully conserved during transport due to atmospheric oxidative processes [17–21], so using levoglucosan observations may not capture the complete impact of primary biomass burning smoke emissions at a receptor. Hennigan *et al.* [17] in a laboratory study, found that under typical summertime OH concentrations, levoglucosan is stable for 0.7–2.2 days. Since our study occurred during wintertime, and the main source of levoglucosan is local biomass burning with little transport time or distance, levoglucosan is likely stable enough here to be used as a robust tracer for primary biomass burning emissions. In addition to being quantified by filter collection, levoglucosan and related compounds also emitted by combustion of cellulose or hemi-cellulose can be quantified on a semi-continuous basis by the HR-AMS, where the ion $C_2H_4O_2^+$ at mass-to-charge ratio (m/z) 60 is commonly used to indicate biomass burning; $C_2H_4O_2^+$ is proportional to the amount of levoglucosan in the sampled aerosol [22–25]. Levoglucosan is not the only source of this ion, since other organic species such as other anhydrosugars (e.g., mannosan and galactosan) and organic acids also contribute to its mass, but levoglucosan and structurally related molecules in biomass burning smoke typically are the dominant source of $C_2H_4O_2^+$ ion [17,25].

Mohr *et al.* and Takegawa *et al.* have found that the additional signal at m/z 60 is likely from long chain alkanoic acids or other acid compounds [26,27]. Cubison *et al.* further demonstrated that without biomass burning influence, ambient aerosol includes a m/z 60 background level of ~0.3% of OM, likely due to acids and other compounds [28]. Lee *et al.* suggest that increased/decreased levoglucosan yield in biomass burning smoke may be offset to some extent by corresponding decreases/increases in other molecules that also yield $C_2H_4O_2^+$ ions, resulting in a fairly stable $OA/C_2H_4O_2^+$ ratio across fuel and burn types [25]. In Spain, Minguillon *et al.* found that levoglucosan-apportioned biomass burning was less than AMS-apportioned BB, possibly because alkanoic acids contributed to the m/z 60 signal. Thus, a combination of filter-based levoglucosan plus higher-time-resolution AMS $C_2H_4O_2^+$ measurements should effectively bound the contribution of biomass burning to OA [29].

Potassium is also produced from the combustion of wood lignin. Elemental potassium (K) and soluble potassium (K^+) are commonly used as tracers when using data from X-ray fluorescence (XRF) and IC analysis of filter samples [30–33]. Other prevalent sources of potassium, such as dust, sea salt, or cooking aerosol, can confound use of this tracer [34–36]. In experiments of different biomass fuels, Sullivan *et al.* found poor correlation between emissions of K^+ and levoglucosan among the fuel types, whereas Lee *et al.* showed that emissions of K^+ were higher under flaming conditions compared to smoldering conditions; AMS $C_2H_4O_2^+$ emissions were comparable between conditions (both K^+ and AMS $C_2H_4O_2^+$ were reported in terms of ratio to total PM) [16,25]. These results are consistent with other studies suggesting that K^+ may have a modest correlation at best with organic tracers of biomass burning. Zhang *et al.* found an $r^2 = 0.59$ using 24-h filter data during wintertime in the southeastern U.S., but a much lower correlation in summer; K had lower spatial variability than levoglucosan did [35]. In Mexico City, Aiken *et al.* found that levoglucosan had a modest correlation with $PM_{2.5}$ K ($r^2 = 0.67$), and that non-biomass burning sources typically accounted for two-thirds of K concentrations [34]. In source profiles, the ratio between K and levoglucosan can be quite variable, ranging between 0.03 and 0.16 [11,25,37–39]. In part because of this variability and confounding alternative potassium sources, Minguillon *et al.* suggested that, based on comparisons of K from 24-h filter measurements to K from AMS, levoglucosan, and other measurements, K was an unreliable tracer for their sites because of the influence of other sources [29]. While K is nonvolatile and not subject to chemical destruction during plume aging, results from the studies referenced above and others suggest that apportionment using K can have high uncertainties.

Multi-channel Aethalometers (e.g., Magee Scientific AE22 used here) provide measurements of absorption from sampled aerosol at multiple wavelengths at 880 nm and at 370 nm. The absorption measurement at 880 nm defines the concentration of black carbon (BC), while the 370 nm measurement is the absorption of the aerosol in the UV [40–42]. Aerosols are sampled continuously and impacted on a filter tape, where the absorption measurement is taken. With the Aethalometer, the absorption measurement is then converted to a black carbon concentration using an assumed mass extinction coefficient of 16.6 m^2/g [40,41]. If measuring only true black carbon, the calculated mass from either channel is the same; when PAHs or other "brown carbon" material are present, the response in the UV channel is different than in the BC channel, where this difference in response is defined as UV-BC. The UV-BC difference has been attributed to the presence of wood smoke, meaning that higher UV-BC difference values are indicative of increased wood smoke. Studies in the northeastern U.S. report that there is good agreement between UV-BC and levoglucosan [3,4], and multiple studies have exploited this difference to apportion traffic and wood smoke aerosol [41,43,44]. While there is evidence that multi-channel data can be used to indicate or apportion wood smoke, Harrison *et al.* caution that this method is very dependent on the choice of Angstrom exponent in the calculations, and that apportioning wood smoke via this method in an urban environment is challenging [42]. Here, we report UV-BC difference, and compare trends in BC and UV-BC difference with other wood smoke measures.

1.2. Study Area: Las Vegas

Las Vegas, Nevada, in a shallow bowl area with mountains to the west and north, is a relatively isolated, large urban area with a population exceeding 1.9 million in the greater metropolitan area (as of 2010). Unlike areas in the northern and northeastern United States, Las Vegas is not widely recognized as having a tradition of home heating from residential wood combustion; rather, most homes are heated by natural gas or electricity. However, the few studies that have been conducted on Las Vegas aerosol have suggested biomass burning as a moderate source of wintertime aerosol.

A key study, Green *et al.*, focused on approximately 50 24-h filter samples collected in 2000–2001 [45]. The major components of $PM_{2.5}$ were BC, OM and crustal elements, with carbonaceous material contributing over 50% of the total mass at an urban site, East Charleston. Ammonium sulfate and nitrate concentrations were generally quite low, about 12% of the total $PM_{2.5}$ mass. Though no formal apportionment was completed, extensive data analysis led the authors to surmise that, although gasoline and diesel vehicle emissions are likely an important source, other sources such as residential biomass burning may also be a significant contributor. Another study, the Southern Nevada Air Quality Study (SNAQS), used 10–12 24-h $PM_{2.5}$ filter samples at four sites in January 2003 to apportion $PM_{2.5}$ [46]. 80% of the mass was from carbonaceous aerosol, and 38%–49% of the $PM_{2.5}$ was attributed to mobile sources. Biomass burning contributed 11%–21% of the mass. Dust, ammonium sulfate, and ammonium nitrate comprised the remainder of the mass. These apportionments were based on a standard suite of filter analyses, including OC and EC by thermal optical reflectance (TOR), sulfate and nitrate by IC, and metals by XRF. No continuous data were used, nor were specific tracers for biomass burning available other than K, which has additional, significant non-BB sources. Without more specific tracers or higher-time-resolution data, the apportionment of OC has a high uncertainty. In addition, the temporal pattern of OC could not be examined, since only 24-h filters were collected on a small number of days.

2. Methodology

2.1. Monitoring Location

Measurements were made next to a classroom and playground in Las Vegas, Nevada, during January 2008 at Fyfe Elementary School, directly adjacent to and 18 meters from the US Highway 95 highway soundwall (Supplementary Materials Figure S1); this monitor is 60 m from the middle of the first set of lanes, and 90 m from the middle of the farthest set of lanes. Additional details on monitoring location and the influence of traffic have been reported elsewhere [47,48].

2.2. Measurement Methods

Collection of black carbon (BC) and meteorological data are further described in Brown *et al.* [47,48]. Wind speed and direction were measured with an RM Young AQ 5305-L at 10 meters above ground level (AGL). BC data at 880 nm (BC channel) and 370 nm (UV channel) were collected using a Magee Scientific Aethalometer model AE-22 with a $PM_{2.5}$ inlet at 5 L/min. Time-stamp and filter tape spot saturation corrections were made using the Washington University Air Quality Lab AethDataMasher Version 6.0e (St. Louis, MO, USA). An Aerodyne HR-AMS was used to quantify OM and biomass burning organic aerosol (BBOA) tracers. The HR-AMS is a widely used instrument described in detail elsewhere [49–51]; specifics of its operation in this study are detailed in Brown *et al.* [47]. Ambient air is drawn through a $PM_{2.5}$ cyclone and is sampled through a critical orifice into an aerodynamic lens; a narrow particle beam with a 50% transmission efficiency of 900 nm diameter particles is thus created so that, essentially, PM_1 is measured [52,53]. Particles are sampled through a $PM_{2.5}$ cyclone, and then accelerated via supersonic expansion of gas molecules into a vacuum at the end of the aerodynamic lens. Particles are collected by inertial impaction onto a heated surface (600 °C), and non-refractory species such as nitrate, sulfate, ammonium, and OM are thermally vaporized. Vaporized gases undergo electron impact ionization, and the charged fragments enter a time-of-flight mass spectrometer (ToF-MS) region, where they are separated by mass-to-charge ratio (m/z). After correction for interferences from ambient gases such as N_2 and O_2, mass spectra are analyzed for each 2-min averaged sample. AMS data were processed and analyzed using the standard AMS analysis software, Squirrel version 1.51, implemented with Wavemetric's Igor Pro (version 6.20). Detection limits for individual ions are provided elsewhere [54]; the focus of this work is on m/z 60, which has a detection limit of 0.001 μg/m^3.

Collection and chemical analysis of quartz fiber filters are detailed elsewhere [55]. Briefly, 8″ × 10″ filters in Tisch 231 $PM_{2.5}$ plates were used in hi-volume samplers (nominal flow rate 68 m^3/h) to collect aerosol at multiple times of day: 0500–0900 LST, 0900–1100 LST, 1100–1700 LST, and 1700–0500 LST. Filters were pre-baked, individually wrapped in aluminum foil, and kept in a freezer before and after sampling. Flow checks were done in the morning and evening (e.g., prior to 0900 and prior to 1700). Only a limited number of samples could be analyzed, so 12 overnight samples were selected, since this is the period of highest OM concentrations. Chemical analysis was done by GC-MS for levoglucosan and more than 20 PAHs, the latter reported in Olson *et al.* [55].

Semi-continuous measurements of $PM_{2.5}$, K^+, sulfate, nitrate, ammonium, and other major ions were made using a Particle Into Liquid Sampler (PILS) coupled to two ion chromatographs (IC). The detailed design and operation of the PILS is described elsewhere [56–59] and is briefly summarized here. The PILS nucleates aerosol particles to form water droplets by mixing a denuded aerosol stream with supersaturated steam. The nucleated droplets are collected into a flowing liquid stream by inertial impaction. The liquid stream, containing an internal LiBr standard to determine dilution by condensed water vapor, is split into two streams which are injected every 15 min to two ion chromatographs (Dionex, DX-500) for measurement of major inorganic ion (NO_3^-, SO_4^{2-}, NH_4^+, Cl^-, Na^+, K^+, Ca^{2+} and Mg^{2+}) concentrations. K^+, the focus of our analysis, has a detection limit of 0.06 μg/m^3 [60].

A $PM_{2.5}$ cyclone (16.7 LPM, URG-2000-30EH) and two URG annular denuders (URG-2000-30X242-3CSS) were used upstream of the PILS/IC. The first denuder was coated with Na_2CO_3 for removal of acidic gases, and the second denuder was coated with phosphorous acid to remove basic gases. Denuders were exchanged every 5–6 days after calibration and blank checks. Blanks were taken by sampling particle-free air, drawn through a High Efficiency Particulate-Free Air (HEPA) capsule filter (Pall Corporation, New York, NY, USA), through the PILS/IC system after a calibration check standard (NO_3^-, SO_4^{2-}, and NH_4^+ concentrations of 20 μN and Cl^-, Na^+, K^+, Ca^{2+}, and Mg^{2+} concentrations of 10 μN) was injected. Approximately every 10 days, the PILS was cleaned and the ion chromatographs recalibrated. A sample flow rate of 16.7 L/min for the PILS/IC was controlled by a critical orifice with a vacuum regulator. 20-min data were aggregated into hourly concentrations, where all three 20-min measurements within an hour were required to accept an hourly average.

2.3. Source Apportionment Methods

EPA's Positive Matrix Factorization (PMF) tool, EPA PMF [61], was used to apportion organic matter (OM) as measured by the HR-AMS, and is further described elsewhere [47,62]. Briefly, four factors were found with the PMF analysis: hydrocarbon-like organic aerosol (HOA), low-volatility oxygenated organic aerosol (LV-OOA), biomass burning organic aerosol (BBOA), and semi-volatile oxygenated organic aerosol (SV-OOA). These factors are typical of PMF deconvolution of HR-AMS data, and represent a spectrum of OA [51,63]. On average in this study, HOA made up 26% of the OM, while LV-OOA was highest in the afternoon and accounted for 26% of the OM. PMF-derived BBOA (PMF-BBOA) occurred in the evening hours, was transported predominantly from the residential area to the north, and on average constituted 12% of the OM; SV-OOA accounted for the remaining one-third of the OM.

3. Results

3.1. Ambient Concentrations and Temporal Variability of Biomass Burning Markers

Concentrations of organic matter, black carbon, and biomass burning indicators (levoglucosan, $C_2H_4O_2^+$, K^+, and UV-BC difference) varied widely during January 2008, typically reaching a peak in the early evening (*i.e.*, 1900 through 2100 LST). Figure 1 shows a time series for these species. OM at our roadside site was 3.3 μg/m^3 on average, while BC was 1.8 μg/m^3. Other aerosol and gaseous species were also measured and are summarized elsewhere [47]; in January 2008, OM and BC comprised 74% of the PM_1 mass measured via the AMS and Aethalometer (excluding metals and crustal material which were not measured). $C_2H_4O_2^+$ concentrations averaged 0.018 μg/m^3, and between 1800 and 0000 LST were nearly three times higher at 0.040 μg/m^3. PILS K^+ concentrations averaged 0.033 μg/m^3 across the month of measurements, while levoglucosan concentrations during the 12-h overnight samples averaged 0.14 μg/m^3. For comparison, the concentrations of elemental potassium at the Chemical Speciation Network (CSN) site in Las Vegas were 0.03 μg/m^3 across five measurement days that fell within our study period.

Figure 2 summarizes the typical diurnal pattern of the semi-continuous measurements. BC concentrations were similar in the morning and evening, during the rush hour commute times. OM showed a minor peak in the morning, and was on average three times higher in the evening than in the morning. See Supplementary Materials Figures S1 and S2 for diurnal box plots of OM and BC. $C_2H_4O_2^+$, K^+, and UV-BC difference all show a similar average diurnal pattern with a concentration peak extending from early evening through late night. K^+ concentrations decrease more slowly than $C_2H_4O_2^+$ after midnight, possibly suggesting that $C_2H_4O_2^+$ is being lost via other mechanisms (such as partitioning from particle to gas phase or atmospheric reactions) than those affecting the nonvolatile and non-reactive species K^+. K^+, $C_2H_4O_2^+$, and UV-BC difference are all lowest in the midday, when emissions of residential biomass burning are low, wind speeds and dispersion are higher, and OM is lower.

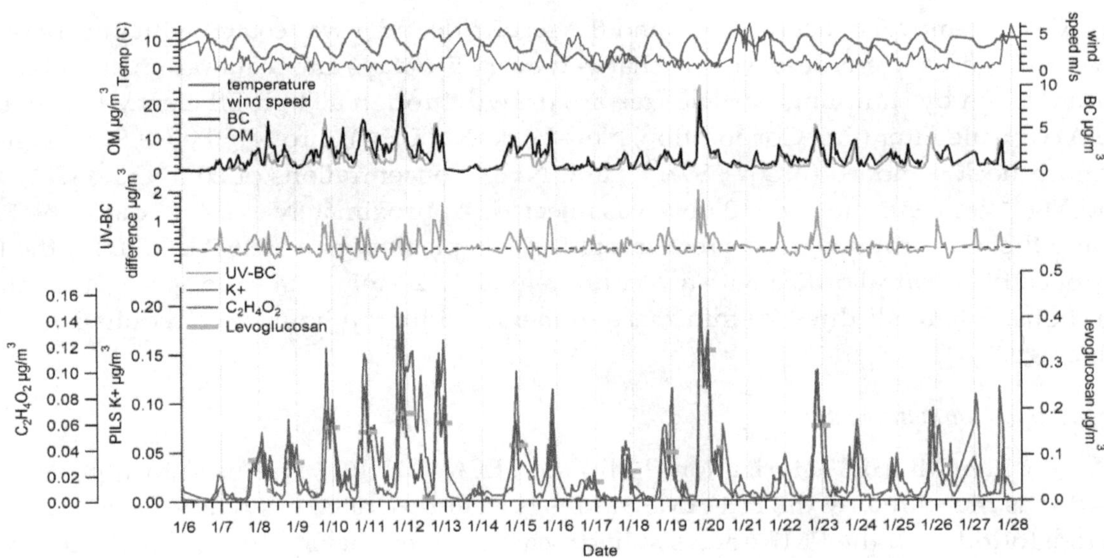

Figure 1. Time series of temperature, wind speed, Aethalometer black carbon (BC), Aerosol Mass Spectrometer (AMS) organic matter (OM), Aethalometer UV-BC difference, PILS K^+, AMS $C_2H_4O_2^+$, and levoglucosan from quartz fiber filters at Fyfe during January 2008 (all units in µg/m^3 except temperature in degrees C and wind speed in m/s).

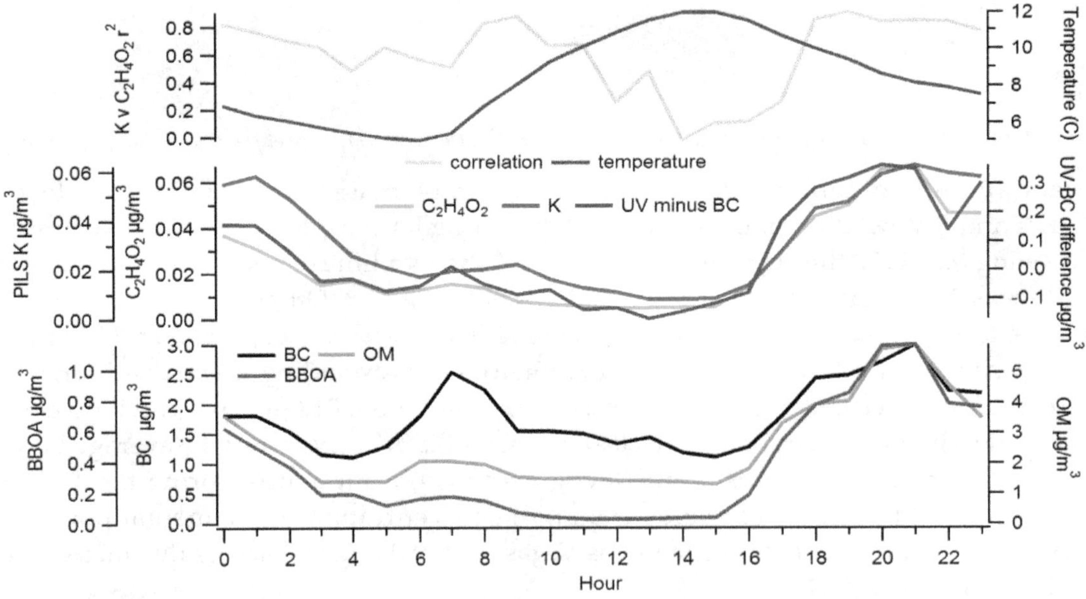

Figure 2. Average concentration by hour (LST) for Aethalometer BC, AMS OM, PMF-BBOA, UV-BC difference, PILS K^+, and AMS $C_2H_4O_2^+$ (all units µg/m^3), plus correlation (r^2) by hour of PILS K^+ *vs.* AMS $C_2H_4O_2^+$ and temperature (degrees C).

OM has a similar pattern as these BB indicators, while BC has a different pattern; concentrations of BC reach comparable average maxima in the morning and evening. The diurnal pattern of BC indicates that mobile source emissions related to rush hour traffic are likely the most important source of BC. The diurnal OM pattern—low concentrations in the midday and a steep rise in concentrations in the evening—is likely due to a mix of fresh emissions in the morning and evening with the rush hour and other activities, plus an additional evening source of biomass burning. This was further demonstrated with PMF analyses on the AMS data [47], which showed that fresh, hydrocarbon-like organic aerosol (HOA) was present in the morning and evening, and that additional semi-volatile oxidized organic aerosol (SV-OOA) and BBOA were evident during the evening peak.

3.2. Comparison among Biomass Burning Markers

3.2.1. Comparisons with Levoglucosan

There was a range in how well the potential biomass burning indicators correlated with each other. Filter-based levoglucosan was available only for a subset of times during the study, at varying intervals. Correlations of filter-based levoglucosan with other measurements are summarized in Figure 3, while correlation among semi-continuous measurements from other instruments is discussed in the next section.

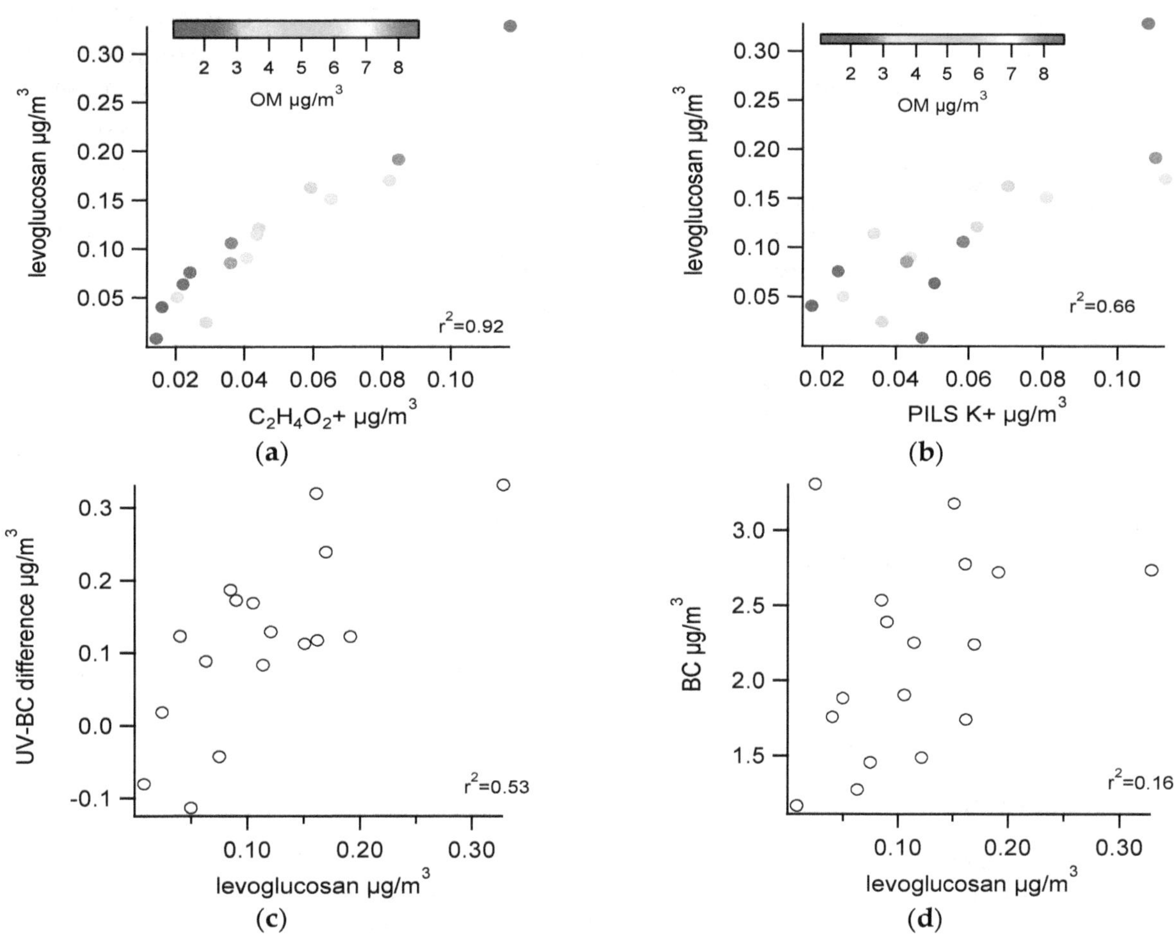

Figure 3. Scatter plots of levoglucosan concentrations (µg/m^3) correlated with (**a**) AMS $C_2H_4O_2^+$ (µg/m^3); (**b**) PILS K^+ (µg/m^3); (**c**) UV-BC difference (µg/m^3); and (**d**) BC (µg/m^3).

Levoglucosan concentrations measured from filters had high correlations with AMS $C_2H_4O_2^+$ (r^2 = 0.92). This is expected, since $C_2H_4O_2^+$ is a fragment from levoglucosan and other co-emitted anhydrous sugars; pure levoglucosan introduced into an AMS yields a suite of ions that has $C_2H_4O_2^+$ as 13%–14% of the total ion fragment pattern [25]. In contrast, there was only a moderate correlation of PILS K^+ (r^2 = 0.66) or UV-BC difference (r^2 = 0.53) with levoglucosan; no correlation was seen between levoglucosan and BC (r^2 = 0.16). The lower correlations are perhaps not surprising, as both BC and K^+ have other non-biomass burning sources; further, levoglucosan may be depleted during the 12-h sampling period via atmospheric reactions or phase partitioning to the gas phase, while BC and K^+ would not undergo similar processes. BC and K^+ are emitted primarily during flaming combustion, while levoglucosan is emitted more during smoldering combustion [25], which may also cause the lower correlation. The very low correlation with BC is likely because BC next to a roadway is predominantly from mobile sources, rather than from biomass burning. The modest correlation of levoglucosan with UV-BC difference, in the context of no correlation with BC, indicates that the UV-BC difference can be indicative of biomass burning aerosol, even when total BC is overwhelmingly from traffic-related sources.

3.2.2. Comparisons among Semi-Continuous Biomass Burning Markers

While there are a limited number of multi-hour samples of levoglucosan, the high correlation between levoglucosan and $C_2H_4O_2^+$ confirms that $C_2H_4O_2^+$ is an excellent tracer for levoglucosan and biomass burning emissions. We next examined correlations of hourly averaged $C_2H_4O_2^+$ concentrations with K^+, UV-BC difference values, and BC. Scatter plots of semi-continuous measurements are provided in Supplementary Materials Figure S2. As indicated by similar diurnal patterns, the measurements of biomass burning indicators were somewhat correlated, with some differences between species and time of day. The overall correlation coefficient (r^2) between K^+ and $C_2H_4O_2^+$ was 0.56, but if the correlations are examined by time of day, there is a large range in this correlation coefficient (Figure 2). Between 1800 and 0000 LST, when fresh biomass burning emissions are most likely, the correlation coefficient between K^+ and $C_2H_4O_2^+$ was 0.84; it slowly decreased through the morning until 1200 through 1600 LST, when the correlation coefficient was 0.19. Midday, when the correlation is lowest, is also when concentrations are lowest and approaching the detection limit; the lower correlations may simply be due to increased measurement noise closer to the detection limits.

UV-BC difference had only a modest correlation with $C_2H_4O_2^+$ ($r^2 = 0.43$), similar to the correlation between UV-BC difference and levoglucosan ($r^2 = 0.53$). BC and $C_2H_4O_2^+$ have a good correlation during the evening ($r^2 = 0.80$) during the period of strong residential wood combustion, but only a modest correlation in the morning ($r^2 = 0.35$). K^+ correlated poorly with both BC and UV-BC difference.

Overall, these results suggest that K^+ and UV-BC difference are only modestly good indicators of biomass burning in Las Vegas next to a roadway, probably at least in part because there are other sources of K^+ and BC at the monitoring site. It is clearly plausible that the majority of BC is from traffic-related emissions, which may complicate the relationship between UV-BC difference and levoglucosan or $C_2H_4O_2^+$. For K^+, the modest correlation with levoglucosan or $C_2H_4O_2^+$ may be due to differences in emissions of these species during flaming and smoldering processes, or to minor sources of K^+ confounding the relationship.

3.2.3. Urban Background Levels of $C_2H_4O_2^+$

There is a clear, strong relationship of levoglucosan with $C_2H_4O_2^+$ in the data here and in prior studies [25]. However, non-biomass burning sources, including organic acids, also can contribute to $C_2H_4O_2^+$ [28]. Lee *et al.* suggested that there is a background level of $C_2H_4O_2^+$, so that even when biomass burning is null, there is still some small concentration of $C_2H_4O_2^+$, approximately 0.3% of OA. This background $C_2H_4O_2^+$ can be seen in Figure 4, which shows the fraction of OM from $C_2H_4O_2^+$ ($fC_2H_4O_2^+$) *vs.* the fraction from m/z 44 (f44). During the morning and midday hours, a background of $C_2H_4O_2^+$ is evident of approximately 0.25% of the OM; during the evening, the fraction of mass from m/z 44 is much lower and the fraction from $C_2H_4O_2^+$ is much higher. Figure 4b shows how the relationship between m/z 44 and $C_2H_4O_2^+$ progresses, with a low m/z 44 fraction and $C_2H_4O_2^+$ fraction in the morning, followed by a midday increase in m/z 44 fraction, and an evening increase in $C_2H_4O_2^+$ fraction and decrease of m/z 44 fraction. This further shows the important contribution of biomass burning during the evening only, while other primary and secondary sources contribute to OM throughout the day.

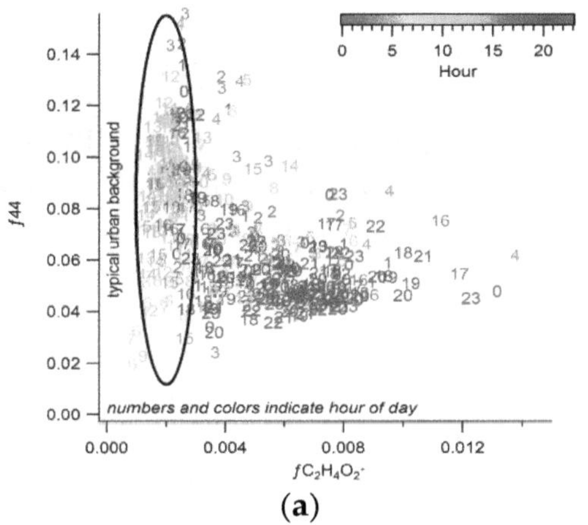

(a)

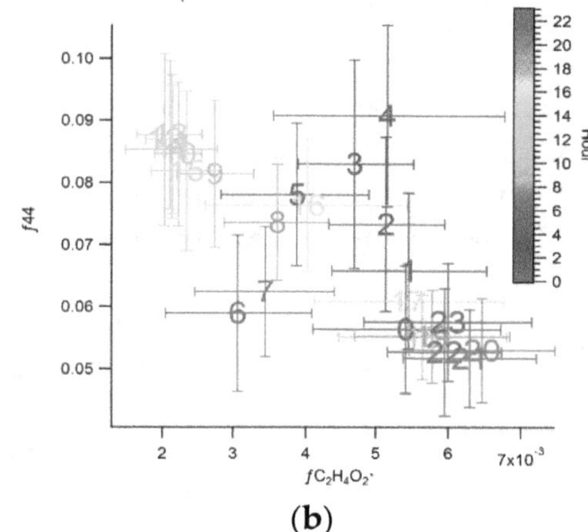

(b)

Figure 4. Scatter plot of the fraction of OM from m/z 44 and $C_2H_4O_2^+$, with color and numbers indicating hour of the day, for: (**a**) all data (hourly averages); and (**b**) data averaged by hour during the study.

3.3. Apportioning Biomass Burning via Multiple Methods

With the suite of biomass burning tracers observed here, multiple methods are available to apportion the contribution of biomass burning to OM: (1) use PMF-AMS, applying PMF to the AMS data to determine contributing factors, including BBOA [63,64], and comparing to the $(C_2H_4O_2^+ \times OM)^{-1}$ ratio reported in laboratory source experiments for biomass burning fuels [25]; (2) use a $(\text{levoglucosan/OC})^{-1}$ ratio as reported in filter-based source profiles, estimating the amount of OC from the levoglucosan concentrations and using an assumed OM/OC ratio to estimate BBOA contributions to OM; and (3) use the same process as for Method 2, but using potassium from PILS and source profiles. Since Methods 2 and 3 rely on source profiles, these methods should estimate primary emissions, if the source profiles represent only primary emissions. The PMF factor approach in Method 1 may include primary and some secondary aerosol formation associated with BBOA. However, the PMF method could underestimate secondary OA from biomass burning in the obtained BBOA factor if the secondary OA is chemically more similar to SV-OOA than to primary BBOA. Since SV-OOA concentrations observed here are typically concurrent with and higher than BBOA, secondary OA associated with biomass burning emissions may not be fully captured in the BBOA factor. With only three PMF factors, *i.e.*, with no SV-OOA factor, BBOA is higher than when four factors are used; it may be that with three factors, more of the secondary OA associated with biomass burning is contained in the BBOA factor. Table 1 summarizes the fraction of OM apportioned via each method. Brown *et al.* report the results from the PMF-AMS method where, using unit mass resolution AMS data and the EPA PMF program, on average 12% of the OM was attributable to biomass burning organic aerosol (BBOA) [47]. During overnight periods, BBOA was on average 26% of the OM.

Table 1. Fraction and standard deviation of OM (fOM) apportioned during 12 overnight (1700–0500 LST) periods, January evenings, and over all hours, via PMF-AMS, levoglucosan, and K^+. Apportionment by levoglucosan is available only for the 12 overnight filter sample periods.

Sample Range	% OM from BB via Levoglucosan	% OM from BB via PMF-AMS (BBOA)	% OM via K^+
12 12-h overnight periods	33% +/− 7%	26% +/− 9%	44% +/− 18%
All evenings (1800–2300 LST)	n/a	15% +/− 9%	26% +/− 24%
All hours	n/a	9% +/− 8%	25% +/− 25%

A number of studies have reported a range of $OM/C_2H_4O_2^+$ ratios from source experiments. Lee *et al.* report a value of 34.5 for the $OM/C_2H_4O_2^+$ ratio generated in biomass burning experiments [25]. Alfarra *et al.* used a combination of PMF and ^{14}C analyses to determine a similar ratio for OM to m/z 60, equal to 36, for wintertime wood combustion in Zurich, and suggested this ratio as a conservative estimate for apportioning BBOA [22]. The ratio of 34.5 is very close to the $OM/C_2H_4O_2^+$ ratio in the BBOA factor found here, which is 34.1, indicating that the BBOA factor is consistent with BBOA found in specific experiments where biomass burning is the main source of OA. In our PMF-AMS results, when using just three PMF factors, BBOA also comprises an average of 15% of the OM.

Fine *et al.* report an OC/levoglucosan ratio of 7.35 and an OC/K ratio of 20.83 for residential biomass burning emissions, used here to apportion BB OC based on our filter levoglucosan and PILS potassium measurements [38]. Recent studies have reported a wider range of levoglucosan/OC and K/OC emission ratios depending on biomass fuel type and burn conditions [16,29,65]. The 7.35 value for OC/levoglucosan is representative of fireplace combustion of hardwoods, which is likely appropriate for the Las Vegas area. Schmidl *et al.* developed a similar factor for Austrian fuels of 7.1 based on test burns in a tiled stove [66]. Puxbaum *et al.* [65] suggest an OM/OC conversion factor of 1.4 based on their calculations from the data reported in Fine *et al.* [11]. During the wintertime evening in Las Vegas, when biomass burning is most prevalent, the average OM/OC ratio is 1.46 [67], so a value of 1.4 for biomass burning appears reasonable. This yields a conversion of biomass burning OM equal to levoglucosan $\times$ 7.35 $\times$ 1.4. For potassium, biomass burning OM is calculated as $K^+ \times 20.83 \times 1.4$.

Figure 5 shows the fraction of OM by each method for periods when levoglucosan data are available. Figure 6 compares PMF-BBOA with levoglucosan measurements and levoglucosan-BBOA apportionment. Using conversions from levoglucosan, 33% of the OM is from biomass burning during the overnight periods. This range is slightly higher than the 26% apportioned via 4-factor PMF-AMS. Estimates of K-based BBOA are higher, with 44% of the OM apportioned to BBOA during the 12-h overnight periods with levoglucosan data, and 26% on all evenings. All but the highest levoglucosan concentration data points fall about the 1:1 line between PMF-BBOA and levoglucosan-BBOA in Figure 6, and on most evenings the PMF and levoglucosan apportionment methods yield a similar result. Apportionment via K^+ is consistently higher than all other methods. Each of the methods used has underlying uncertainties, in particular the selection of source profiles, since emissions of levoglucosan, K^+, and OM vary by wood type, flaming *vs.* smoldering *etc.* Results in Table 1 capture some of this uncertainty, showing that the K^+ method is the most uncertain compared to levoglucosan and AMS PMF.

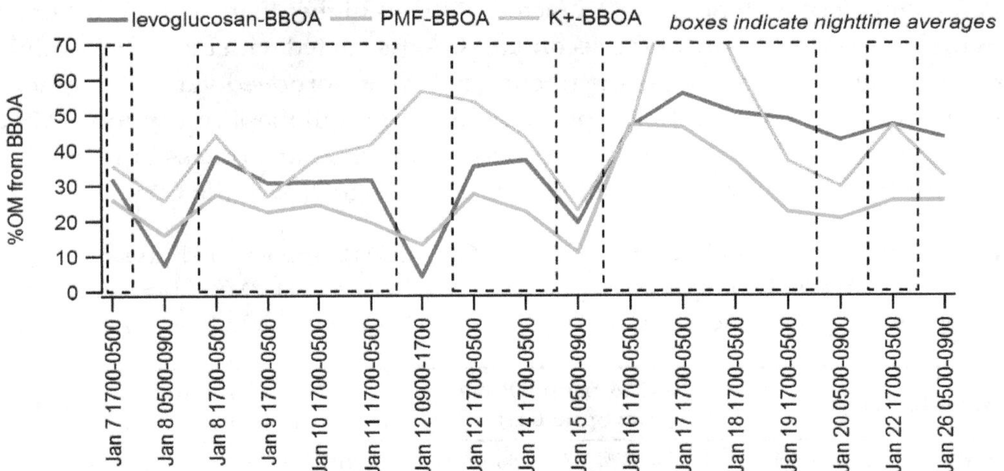

Figure 5. Percentage of OM apportioned by four methods for each time period where levoglucosan was quantified; boxes indicate nighttime averages (1700–0500 LST). Not shown is K-BBOA value of 100% apportioned OM on 17 January 1700-0500.

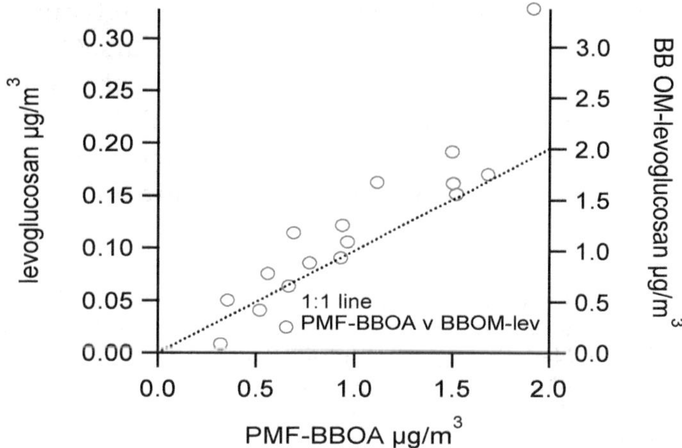

Figure 6. Comparison of PMF-BBOA concentrations *vs.* levoglucosan (primary *y*-axis) and BB OM by levoglucosan (secondary *y*-axis); all units are µg/m^3.

One difficulty with comparing 12-h average apportionments via levoglucosan and K^+ is that they may be lost at different rates by atmospheric processes [28], or emitted at varying rates as burning goes from flaming to smoldering [25]. In addition, as emissions age, gas/particle partitioning of semivolatile material may mean that the relationship of levoglucosan to OM emitted changes over time, as organic material is either condensed into the particle phase or partitioned in the vapor phase [17,28,68]. We examined the hourly average ratio and correlation between K^+ and $C_2H_4O_2^+$ to understand how the relationship varies during the night (Figure 2). Between 1900 and 2300 LST, the ratio (0.625) and correlation (r^2 = 0.80) between K^+ and $C_2H_4O_2^+$ is consistent, but it degrades after 0000 LST, which is likely around the time that emissions have nearly stopped and levoglucosan may be lost via atmospheric reactions.

The sample with the highest disagreement between methods was the night of 19 January, when levoglucosan-BBOA and K-BBOA are similar (37%–39% of the OM) but are 1.6 times higher than PMF-BBOA. As seen in Figure 1, this was not only the evening of the highest levoglucosan and $C_2H_4O_2^+$ concentrations but also highest OM. It may be that PMF is under-predicting the amount of BBOA, since the OM concentration and possibly composition is quickly changing. The $Q/Q_{expected}$ ratio and scaled residuals from PMF during this evening are low, indicating a good fit, but SV-OOA is also very high during this evening, so it is likely that some mass assigned by PMF as SV-OOA is actually BBOA. Since a constant profile is needed in PMF, differences in the BBOA composition between evenings mean that a "typical" or average profile is found; deviations from this profile suggest that mass appears to be apportioned to SV-OOA. However, as the results are consistent for all the other data points, our conclusion is that the three BBOA methods using levoglucosan and AMS data compare rather well most of the time. As there are other non-biomass burning sources of K^+ in the area, and the amount of K^+ emitted depends on the amount of flaming *vs.* smoldering emissions, assuming all the K^+ is from biomass burning yields an upper limit of BBOA that is likely less accurate than the other methods used here.

4. Conclusions

Urban-scale biomass burning was found to be a surprising source of aerosol at Fyfe Elementary School in Las Vegas, even though the monitoring site was located next to a major freeway in a city with no tradition of home heating from wood stoves or fireplaces. Multiple methods of estimating the contribution of this source to fine PM were compared; HR-AMS measurements correlated with levoglucosan measurements, and both yielded similar estimates of total biomass burning organic aerosol (BBOA). Water-soluble potassium correlated with AMS $C_2H_4O_2^+$ only during evening hours, when biomass burning was relatively high; during other hours, there was little correlation, indicating that although K^+ can be a useful biomass burning indicator when biomass burning is high, other sources tend to overwhelm the K^+ concentrations during other hours. On average, BBOA comprised 9%–14% of the organic matter (OM), but was only significant during the evening hours, when OM was highest. During the overnight period between 1700 and 0500 LST, BBOA contributed between 26% and 33% of the OM (range derived from different analysis/measurement techniques). Thus, residential biomass burning is an unexpected, but relatively important, source of $PM_{2.5}$ in Las Vegas.

Acknowledgments: Gary Norris and David Olson at EPA's Office of Research and Development provided the chemical analysis of levoglucosan. Funding for the core measurements, including BC and meteorology, as well as for site operations, was provided by Nevada Department of Transportation. Sonoma Technology, Inc. (Petaluma, CA, USA), provided supplemental internal funding for the HR-AMS measurements. Lastly, we appreciate the review and comments from Steven G. Brown's PhD committee on this work, including Sonia Kreidenweis, Colette Heald, and Anthony Marchese.

Author Contributions: Steven Brown, Paul Roberts, and Jeffrey Collett conceived and designed the experiments; Taehyoung Lee and Steven Brown performed the experiments; Steven Brown analyzed the data; Taehyoung Lee contributed analysis tools; Steven Brown wrote the paper.

References

1. Bergauff, M.A.; Ward, T.J.; Noonan, C.W.; Palmer, C.P. The effect of a woodstove changeout on ambient levels of $PM_{2.5}$ and chemical tracers for woodsmoke in Libby, Montana. *Atmos Environ.* **2009**, *43*, 2938–2943. [CrossRef]
2. Ward, T.; Noonan, C. Results of a residential indoor $PM_{2.5}$ sampling program before and after a woodstove changeout. *Indoor Air* **2008**, *18*, 408–415. [CrossRef] [PubMed]
3. Allen, G.A.; Miller, P.J.; Rector, L.J.; Brauer, M.; Su, J.G. Characterization of valley winter woodsmoke concentrations in Northern NY using highly time-resolved measurements. *Aerosol Air Qual. Res.* **2011**, *11*, 519–530. [CrossRef]
4. Wang, Y.; Hopke, P.K.; Utell, M.J. Urban-scale spatial-temporal variability of black carbon and winter residential wood combustion particles. *Aerosol Air Qual. Res.* **2011**, *11*, 473–481. [CrossRef]
5. Lighty, J.S.; Veranth, J.M.; Sarofim, A.F. Combustion aerosols: Factors governing their size and composition and implications to human health. *J. Air Waste Manag. Assoc.* **2000**, *50*, 1565–1618. [CrossRef] [PubMed]
6. Barregard, L.; Sallsten, G.; Andersson, L.; Almstrand, A.-C.; Gustafson, P.; Andersson, M.; Olin, A.-C. Experimental exposure to wood smoke: Effects on airway inflammation and oxidative stress. *Occup. Environ. Med.* **2007**, *65*, 319–324. [CrossRef] [PubMed]
7. Freeman, L.; Stiefer, P.S.; Weir, B.R. *Carcinogenic Risk and Residential Wood Smoke*; Systems Applications International: San Rafael, CA, USA, 1992.
8. Seagrave, J.; McDonald, J.D.; Bedrick, E.; Edgerton, E.S.; Gigliotti, A.P.; Jansen, J.J.; Ke, L.; Naeher, L.P.; Seilkop, S.K.; Zheng, M.; *et al.* Lung toxicity of ambient particulate matter from southeastern U.S. sites with different contributing sources: Relationships between composition and effects. *Environ. Health Perspect.* **2006**, *114*, 1387–1393. [CrossRef] [PubMed]
9. Travis, C.C.; Etnier, E.L.; Meyer, H.R. Health risks of residential wood heat. *Environ. Manag.* **1985**, *9*, 209–215. [CrossRef]
10. Naeher, L.P.; Brauer, M.; Lipsett, M.; Zelikoff, J.T.; Simpson, C.D.; Koenig, J.Q.; Smith, K.R. Woodsmoke health effects: A review. *Inhal. Toxicol.* **2007**, *19*, 67–106. [CrossRef] [PubMed]
11. Fine, P.M.; Cass, G.R.; Simoneit, B.R.T. Chemical characterization of fine particle emissions from the fireplace combustion of wood types grown in the Midwestern and Western United States. *Environ. Eng. Sci.* **2004**, *21*, 387–409. [CrossRef]

12. Simoneit, B.R.T.; Schauer, J.J.; Nolte, C.G.; Oros, D.R.; Elias, V.O.; Fraser, M.P.; Rogge, W.F.; Cass, G.R. Levoglucosan, a tracer for cellulose in biomass burning and atmospheric particles. *Atmos. Environ.* **1999**, *33*, 173–182. [CrossRef]
13. Simoneit, B.R.T. Biomass burning—A review of organic tracers for smoke from incomplete combustion. *Appl. Geochem.* **2002**, *17*, 129–162. [CrossRef]
14. Engling, G.; Herckes, P.; Kreidenweis, S.M.; Malm, W.C.; Collett, J.L. Composition of the fine organic aerosol in Yosemite National Park during the 2002 Yosemite Aerosol Characterization Study. *Atmos. Environ.* **2006**, *40*, 2959–2972. [CrossRef]
15. Schauer, J.J.; Kleeman, M.J.; Cass, G.R.; Simoneit, B.R.T. Measurement of emissions from air pollution sources. 3. C_1 through C_{29} organic compounds from fireplace combustion of wood. *Environ. Sci. Technol.* **2001**, *35*, 1716–1728. [CrossRef] [PubMed]
16. Sullivan, A.P.; Holden, A.S.; Patterson, L.A.; McMeeking, G.R.; Kreidenweis, S.M.; Malm, W.C.; Hao, W.M.; Wold, C.E.; Collett, J.L., Jr. A method for smoke marker measurements and its potential application for determining the contribution of biomass burning from wildfires and prescribed fires to ambient $PM_{2.5}$ organic carbon. *J. Geophys. Res.* **2008**. [CrossRef]
17. Hennigan, C.J.; Miracolo, M.A.; Engelhart, G.J.; May, A.A.; Presto, A.A.; Lee, T.; Sullivan, A.P.; McMeeking, G.R.; Coe, H.; Wold, C.E.; *et al.* Chemical and physical transformations of organic aerosol from the photo-oxidation of open biomass burning emissions in an environmental chamber. *Atmos. Chem. Phys.* **2011**, *11*, 7669–7686. [CrossRef]
18. Hoffmann, D.; Tilgner, A.; Iinuma, Y.; Herrmann, H. Atmospheric stability of levoglucosan: A detailed laboratory and modeling study. *Environ. Sci. Technol.* **2010**, *44*, 694–699. [CrossRef] [PubMed]
19. Slade, J.H.; Knopf, D.A. Multiphase OH oxidation kinetics of organic aerosol: The role of particle phase state and relative humidity. *Geophys. Res. Lett.* **2014**, *41*, 5297–5306. [CrossRef]
20. Kessler, S.H.; Smith, J.D.; Che, D.L.; Worsnop, D.R.; Wilson, K.R.; Kroll, J.H. Chemical sinks of organic aerosol: Kinetics and products of the heterogeneous oxidation of erythritol and levoglucosan. *Environ. Sci. Technol.* **2010**, *44*, 7005–7010. [CrossRef] [PubMed]
21. Slade, J.H.; Knopf, D.A. Heterogeneous OH oxidation of biomass burning organic aerosol surrogate compounds: Assessment of volatilisation products and the role of OH concentration on the reactive uptake kinetics. *Phys. Chem. Chem. Phys.* **2013**, *15*, 5898–5915. [CrossRef] [PubMed]
22. Alfarra, M.R.; Prévôt, A.S.H.; Szidat, S.; Sandradewi, J.; Weimer, S.; Lanz, V.A.; Schreiber, D.; Mohr, M.; Baltensperger, U. Identification of the mass spectral signature of organic aerosols from wood burning emissions. *Environ. Sci. Technol.* **2007**, *41*, 5770–5777. [CrossRef] [PubMed]
23. Weimer, S.; Alfarra, M.R.; Schreiber, D.; Mohr, M.; Prévôt, A.S.H.; Baltensperger, U. Organic aerosol mass spectral signatures from wood-burning emissions: Influence of burning conditions and wood type. *J. Geophys. Res. Atmos.* **2008**. [CrossRef]
24. Schneider, J.; Weimer, S.; Drewnick, F.; Borrmann, S.; Helas, G.; Gwaze, P.; Schmid, O.; Andreae, M.O.; Kirchner, U. Mass spectrometric analysis and aerodynamic properties of various types of combustion-related aerosol particles. *Int. J. Mass Spec.* **2006**, *258*, 37–49. [CrossRef]
25. Lee, T.; Sullivan, A.P.; Mack, L.; Jimenez, J.L.; Kreidenweis, S.M.; Onasch, T.B.; Worsnop, D.R.; Malm, W.; Wold, C.E.; Hao, W.M.; *et al.* Chemical smoke marker emissions during flaming and smoldering phases of laboratory open burning of wildland fuels. *Aerosol Sci. Technol.* **2010**. [CrossRef]
26. Mohr, C.; Huffman, J.A.; Cubison, M.; Aiken, A.C.; Docherty, K.S.; Kimmel, J.R.; Ulbrich, I.M.; Hannigan, M.; Jimenez, J.L. Characterization of primary organic aerosol emissions from meat cooking, trash burning, and motor vehicles with high-resolution aerosol mass spectrometry and comparison with ambient and chamber observations. *Environ. Sci. Technol.* **2009**, *43*, 2443–2449. [CrossRef] [PubMed]
27. Takegawa, N.; Miyakawa, T.; Kawamura, K.; Kondo, Y. Contribution of selected dicarboxylic and omega-oxocarboxylic acids in ambient aerosol to the m/z 44 signal of an aerodyne aerosol mass spectrometer. *Aerosol Sci. Technol.* **2007**, *41*, 418–437. [CrossRef]
28. Cubison, M.J.; Ortega, A.M.; Hayes, P.L.; Farmer, D.K.; Day, D.; Lechner, M.J.; Brune, W.H.; Apel, E.; Diskin, G.S.; Fisher, J.A.; *et al.* Effects of aging on organic aerosol from open biomass burning smoke in aircraft and laboratory studies. *Atmos. Chem. Phys.* **2011**, *11*, 12049–12064. [CrossRef]

29. Minguillon, M.C.; Perron, N.; Querol, X.; Szidat, S.; Fahrni, S.M.; Alastuey, A.; Jimenez, J.L.; Mohr, C.; Ortega, A.M.; Day, D.A.; *et al.* Fossil *versus* contemporary sources of fine elemental and organic carbonaceous particulate matter during the DAURE campaign in Northeast Spain. *Atmos. Chem. Phys.* **2011**, *11*, 12067–12084. [CrossRef]
30. Kim, E.; Larson, T.V.; Hopke, P.K.; Slaughter, C.; Sheppard, L.E.; Claiborn, C. Source identification of $PM_{2.5}$ in an arid northwest U.S. city by positive matrix factorization. *Atmos. Res.* **2003**, *66*, 291–305. [CrossRef]
31. Poirot, R. Tracers of Opportunity: Potassium. Available online: http://capita.wustl.edu/PMFine/Workgroup/SourceAttribution/Reports/In-progress/potass/Kcover.htm (accessed on 12 April 2016).
32. Liu, W.; Wang, Y.; Russell, A.; Edgerton, E.S. Atmospheric aerosol over two urban-rural pairs in the southeastern United States: Chemical composition and possible sources. *Atmos. Environ.* **2005**, *39*, 4453–4470. [CrossRef]
33. Brown, S.G.; Frankel, A.; Raffuse, S.M.; Roberts, P.T.; Hafner, H.R.; Anderson, D.J. Source apportionment of fine particulate matter in Phoenix, Arizona, using positive matrix factorization. *J. Air Waste Manag. Assoc.* **2007**, *57*, 741–752. [PubMed]
34. Aiken, A.C.; de Foy, B.; Wiedinmyer, C.; DeCarlo, P.F.; Ulbrich, I.M.; Wehrli, M.N.; Szidat, S.; Prévôt, A.S.H.; Noda, J.; Wacker, L.; *et al.* Mexico City aerosol analysis during MILAGRO using high resolution aerosol mass spectrometry at the urban supersite (T0). Part 2: Analysis of the biomass burning contribution and the non-fossil carbon fraction. *Atmos. Chem. Phys.* **2010**, *10*, 5315–5341. [CrossRef]
35. Zhang, X.; Hecobian, A.; Zheng, M.; Frank, N.H.; Weber, R.J. Biomass burning impact on $PM_{2.5}$ over the southeastern US during 2007: Integrating chemically speciated FRM filter measurements, MODIS fire counts and PMF analysis. *Atmos. Chem. Phys.* **2010**, *10*, 6839–6853. [CrossRef]
36. Schauer, J.J.; Kleeman, M.J.; Cass, G.R.; Simoneit, B.R.T. Measurement of emissions from air pollution sources. 1. C_1 through C_{29} organic compounds from meat charbroiling. *Environ. Sci. Technol.* **1999**, *33*, 1566–1577. [CrossRef]
37. Fine, P.M.; Cass, G.R.; Simoneit, B.R.T. Chemical characterization of fine particle emissions from the fireplace combustion of woods grown in the southern United States. *Environ. Sci. Technol.* **2002**, *36*, 1442–1451. [CrossRef] [PubMed]
38. Fine, P.M.; Cass, G.R.; Simoneit, B.R.T. Organic compounds in biomass smoke from residential wood combustion: Emissions characterization at a continental scale. *J. Geophys. Res. Atmos.* **2002**. [CrossRef]
39. Fine, P.M.; Cass, G.R.; Simoneit, B.R.T. Chemical characterization of fine particle emissions from fireplace combustion of woods grown in the northeastern United States. *Environ. Sci. Technol.* **2001**, *35*, 2665–2675. [CrossRef] [PubMed]
40. Allen, G.A.; Lawrence, J.; Koutrakis, P. Field validation of a semi-continuous method for aerosol black carbon (Aethalometer) and termporal patterns of summertime hourly black carbon measurements in Southwestern Pennsylvania. *Atmos. Environ.* **1999**, *33*, 817–823. [CrossRef]
41. Sandradewi, J.; Prévôt, A.S.H.; Weingartner, E.; Schmidhauser, R.; Gysel, M.; Baltensperger, U. A study of wood burning and traffic aerosols in an Alpine valley using a multi-wavelength Aethalometer. *Atmos. Environ.* **2008**, *42*, 101–112. [CrossRef]
42. Harrison, R.M.; Beddowsa, D.C.S.; Jones, A.M.; Calvo, A.; Alves, C.; Pio, C. An evaluation of some issues regarding the use of aethalometers to measure woodsmoke concentrations. *Atmos. Environ.* **2013**, *80*, 540–548. [CrossRef]
43. Sandradewi, J.; Prévôt, A.S.H.; Alfarra, M.R.; Szidat, S.; Wehrli, M.N.; Ruff, M.; Weimer, S.; Lanz, V.A.; Weingartner, E.; Perron, N.; *et al.* Comparison of several wood smoke markers and source apportionment methods for wood burning particulate mass. *Atmos. Chem. Phys. Discuss.* **2008**, *8*, 8091–8118. [CrossRef]
44. Favez, O.; El Haddad, I.; Piot, C.; Boréave, A.; Abidi, E.; Marchand, N.; Jaffrezo, J.-L.; Besombes, J.-L.; Personnaz, M.-B.; Sciare, J.; *et al.* Inter-comparison of source apportionment models for the estimation of wood burning aerosols during wintertime in an Alpine city (Grenoble, France). *Atmos. Chem. Phys.* **2010**, *10*, 5295–5314. [CrossRef]
45. Green, M.C.; Chow, J.C.; Hecobian, A.; Etyemezian, V.; Kuhns, H.; Watson, J.G. *Las Vegas Valley Visibility and $PM_{2.5}$ Study*; Final Report Prepared for the Clark County Department of Air Quality Management, Las Vegas, NV; Desert Research Institute: Las Vegas, NV, USA, 2002.

46. Watson, J.G.; Barber, P.W.; Chang, M.C.O.; Chow, J.C.; Etyemezian, V.R.; Green, M.C.; Keislar, R.E.; Kuhns, H.D.; Mazzoleni, C.; Moosmüller, H.; *et al. Southern Nevada Air Quality Study*; Final Report Prepared for the U.S. Department of Transportation, Washington, DC; Desert Research Institute: Reno, NV, USA, 2007.
47. Brown, S.G.; Lee, T.; Norris, G.A.; Roberts, P.T.; Collett, J.L., Jr.; Paatero, P.; Worsnop, D.R. Receptor modeling of near-roadway aerosol mass spectrometer data in Las Vegas, Nevada, with EPA PMF. *Atmos. Chem. Phys.* **2012**, *12*, 309–325. [CrossRef]
48. Brown, S.G.; McCarthy, M.C.; DeWinter, J.L.; Vaughn, D.L.; Roberts, P.T. Changes in air quality at near-roadway schools after a major freeway expansion in Las Vegas, Nevada. *J. Air Waste Manag. Assoc.* **2014**, *64*, 1002–1012. [CrossRef]
49. DeCarlo, P.; Kimmel, J.R.; Trimborn, A.; Northway, M.; Jayne, J.T.; Aiken, A.C.; Gonin, M.; Fuhrer, K.; Horvath, T.; Docherty, K.S.; *et al.* Field-deployable, high-resolution, time-of-flight aerosol mass spectrometer. *Anal. Chem.* **2006**, *78*, 8281–8289. [CrossRef] [PubMed]
50. Jimenez, J.L.; Jayne, J.T.; Shi, Q.; Kolb, C.E.; Worsnop, D.R.; Yourshaw, I.; Seinfeld, J.H.; Flagan, R.C.; Zhang, X.F.; Smith, K.A.; *et al.* Ambient aerosol sampling using the Aerodyne Aerosol Mass Spectrometer. *J. Geophys. Res. Atmos.* **2003**. [CrossRef]
51. Zhang, Q.; Jimenez, J.L.; Canagaratna, M.R.; Ulbrich, I.M.; Ng, N.L.; Worsnop, D.R.; Sun, Y. Understanding atmospheric organic aerosols via factor analysis of aerosol mass spectrometry: A review. *Anal. Bioanal. Chem.* **2011**, *401*, 3045–3067. [CrossRef] [PubMed]
52. Sun, Y.; Zhang, Q.; MacDonald, A.M.; Hayden, K.; Li, S.M.; Liggio, J.; Liu, P.S.K.; Anlauf, K.G.; Leaitch, W.R.; Steffen, A.; *et al.* Size-resolved aerosol chemistry on Whistler Mountain, Canada with a high-resolution aerosol mass spectrometer during INTEX-B. *Atmos. Chem. Phys.* **2009**, *9*, 3095–3111. [CrossRef]
53. Canagaratna, M.R.; Jayne, J.T.; Jimenez, J.L.; Allan, J.D.; Alfarra, M.R.; Zhang, Q.; Onasch, T.B.; Drewnick, F.; Coe, H.; Middlebrook, A.; *et al.* Chemical and microphysical characterization of ambient aerosols with the aerodyne aerosol mass spectrometer. *Mass Spectrom. Rev.* **2007**, *26*, 185–222. [CrossRef] [PubMed]
54. Drewnick, F.; Hings, S.S.; Alfarra, M.R.; Prevot, A.S.H.; Borrmann, S. Aerosol quantification with the Aerodyne Aerosol Mass Spectrometer: Detection limits and ionizer background effects. *Atmos. Measure. Tech.* **2009**, *2*, 33–46. [CrossRef]
55. Olson, D.A.; Vedantham, R.; Norris, G.A.; Brown, S.G.; Roberts, P. Determining source impacts near roadways using wind regression and organic source markers. *Atmos. Environ.* **2012**, *47*, 261–268. [CrossRef]
56. Orsini, D.A.; Ma, Y.L.; Sullivan, A.; Sierau, B.; Baumann, K.; Weber, R.J. Refinements to the particle-into-liquid sampler (PILS) for ground and airborne measurements of water soluble aerosol composition. *Atmos. Environ.* **2003**, *37*, 1243–1259. [CrossRef]
57. Weber, R.J.; Orsini, D.; Daun, Y.; Lee, Y.N.; Klotz, P.J.; Brechtel, F. A particle-into-liquid collector for rapid measurement of aerosol bulk chemical composition. *Aerosol Sci. Technol.* **2001**, *35*, 718–727. [CrossRef]
58. Weber, R.; Orsini, D.; Duan, Y.; Baumann, K.; Kiang, C.S.; Chameides, W.; Lee, Y.N.; Brechtel, F.; Klotz, P.; Jongejan, P.; *et al.* Intercomparison of near real time monitors of $PM_{2.5}$ nitrate and sulfate at the U.S. Environmental Protection Agency Atlanta Supersite. *J. Geophys. Res. Atmos.* **2003**. [CrossRef]
59. Sorooshian, A.; Brechtel, F.J.; Ma, Y.L.; Weber, R.J.; Corless, A.; Flagan, R.C.; Seinfeld, J.H. Modeling and characterization of a particle-into-liquid sampler (PILS). *Aerosol Sci. Technol.* **2006**, *40*, 396–409. [CrossRef]
60. Lee, T.; Yu, X.-Y.; Kreidenweis, S.M.; Malm, W.C.; Collett, J.L. Semi-continuous measurement of $PM_{2.5}$ ionic composition at several rural locations in the United States. *Atmos. Environ.* **2008**, *42*, 6655–6669. [CrossRef]
61. Norris, G.; Duvall, R.; Brown, S.; Bai, S. *EPA Positive Matrix Factorization (PMF) 5.0 Fundamentals and User Guide*; Prepared for the U.S. Environmental Protection Agency Office of Research and Development: Washington, DC, USA, 2014.
62. Brown, S.G.; Eberly, S.; Paatero, P.; Norris, G.A. Methods for estimating uncertainty in PMF solutions: Examples with ambient air and water quality data and guidance on reporting PMF results. *Sci. Total Environ.* **2015**, *518*, 626–635. [CrossRef] [PubMed]
63. Ulbrich, I.M.; Canagaratna, M.R.; Zhang, Q.; Worsnop, D.R.; Jimenez, J.L. Interpretation of organic components from Positive Matrix Factorization of aerosol mass spectrometric data. *Atmos. Chem. Phys.* **2009**, *9*, 2891–2918. [CrossRef]
64. Lanz, V.A.; Alfarra, M.R.; Baltensperger, U.; Buchmann, B.; Hueglin, C.; Prévôt, A.S.H. Source apportionment of submicron organic aerosols at an urban site by factor analytical modelling of aerosol mass spectra. *Atmos. Chem. Phys.* **2007**, *7*, 1503–1522. [CrossRef]

65. Puxbaum, H.; Caseiro, A.; Sanchez-Ochoa, A.; Kasper-Giebl, A.; Claeys, M.; Gelencser, A.; Legrand, M.; Preunkert, S.; Pio, C. Levoglucosan levels at background sites in Europe for assessing the impact of biomass combustion on the European aerosol background. *J. Geophys. Res.* **2007**. [CrossRef]
66. Schmidl, C.; Marr, I.L.; Caseiro, A.; Kotianova, P.; Berner, A.; Bauer, H.; Kasper-Giebl, A.; Puxbaum, H. Chemical characterisation of fine particle emissions from wood stove combustion of common woods growing in mid-European Alpine regions. *Atmos. Environ.* **2008**, *42*, 126–141. [CrossRef]
67. Brown, S.G.; Lee, T.; Roberts, P.T.; Collett, J.L., Jr. Variations in the OM/OC ratio of urban organic aerosol next to a major roadway. *J. Air Waste Manag. Assoc.* **2013**, *63*, 1422–1433. [CrossRef] [PubMed]
68. Oja, V.; Suuberg, E.M. Vapor pressures and enthalpies of sublimation of D-glucose, D-xylose, cellobiose, and levoglucosan. *J. Chem. Eng. Data* **1999**, *44*, 26–29. [CrossRef]

7

Source Apportionment of Sulfate and Nitrate over the Pearl River Delta Region in China

Xingcheng Lu [1] and Jimmy C. H. Fung [1,2,*]

[1] Division of Environment, Hong Kong University of Science & Technology, Clear Water Bay, Hong Kong, China; xluad@ust.hk

[2] Department of Mathematics, Hong Kong University of Science & Technology, Clear Water Bay, Hong Kong, China

* Correspondence: majfung@ust.hk

Abstract: In this work, the Weather Research Forecast (WRF)–Sparse Matrix Operator Kernel Emission (SMOKE)–Comprehensive Air Quality Model with Extensions (CAMx) modeling system with particulate source apportionment technology (PSAT) module was used to study and analyze the source apportionment of sulfate and nitrate particulate matter in the Pearl River Delta region (PRD). The results show that superregional transport was an important contributor for both sulfates and nitrates in all 10 cities in this region in both February (winter) and August (summer). Especially in February, the average super-regional contribution of sulfate and nitrate reached up to 80% and 56% respectively. For the local and regional source category, power plant emissions (coal-fired and oil-fired) and industry emissions were important for sulfate formation in this region. Industry emissions and mobile emissions are important for nitrate formation in this region. In August, the sum of these two sources contributed around over 60% of local and regional nitrate. The contributions from power plant emissions and marine emissions became important in August due to the southerly prevailing wind direction. Area sources and biogenic emissions were negligible for sulfate and nitrate formation in this region. Our results reveal that cross-province cooperation is necessary for control of sulfates and nitrates in this region.

Keywords: source apportionment; sulfate; nitrate; CAMx; PSAT

---

1. Introduction

Rapid and continuous economic growth has brought great wealth to China, and the material living conditions of its citizens have improved greatly. However, the large-scale urbanization process is modifying the landscape and turning more and more forests and wetlands into concrete surfaces. The urbanization process has also played a role in clustering vehicles and industrial factories, which leads to worsening environmental conditions. Since the implementation of the openness policy, the Pearl River Delta (PRD) region has become China's major engine for economic growth and one of the world's main manufacturing hubs. Although the incomes and convenience of living of the local residents have greatly improved, more and more people have complained about the smell of waste water, poor visibility, and inhalation of high levels of air pollutants. $PM_{2.5}$, NO_2, and O_3 are the three major ambient pollutants in this region. According to the observation data, the peak ozone concentration exceeded 100 ppb in 15 days during August 2011 at the Guangzhou Luhu station. In Hong Kong, the average NO_2 concentration surpassed 80 ppb in February 2011 at Causeway Bay, a roadside station. At the Nansha station, the $PM_{2.5}$ concentrations reached 150 $\mu g/m^3$ during four episodes with high levels of particulate matter (PM) in January 2011. Once emitted or formed in the atmosphere, these pollutants can also be deposited onto the ground via wet deposition and dry deposition. Acid rain is another important problem caused by substantial emissions of SO_2 and NO_x in this region [1].

The air quality issues mentioned above have inspired many studies of these problems over this region. In the meteorological model MM5 simulation study, Lo et al. [2] found that urbanization played a significant role in trapping pollutants by influencing the land-sea breezes around the PRD region. Li et al. [3] applied the CAMx model coupled with ozone source apportionment technology (OSAT) to study the sources of ozone during episode days and found that superregional transport is an important source of ozone in this region. Wu et al. [4] also used the chemical transport model CAMx coupled with particulate source apportionment technology (PSAT) to study the source apportionment of fine PM and found that local mobile emissions and superregional transport were the dominant contributors of PM over this region. Yao et al. [5] found that the mountains to the north of the PRD region trapped pollutants and further worsened the air quality. Lu et al. [6] recently applied CAMx-OSAT to study the source apportionment of ambient NO_x in this region and found that heavy duty diesel vehicles are the major contributor to this pollutant. Many observation-based studies have also been launched in this region. For example, Yuan et al. [7] applied a positive matrix factorization method to identify the major sources of PM_{10} in Hong Kong and found that vehicle emissions were the greatest contributor. Xue et al. [8] claimed that the liquid water content could determine the sulfate and nitrate abundance in $PM_{2.5}$ at polluted sites in Hong Kong. With all of these studies, the generation mechanism and cause of episodes for related pollutants have been relatively well described in the PRD region.

$PM_{2.5}$ is liquid or solid matter suspended in the air with a particle aerodynamic diameter of less than 2.5 μm. Long-term exposure to this pollutant may increase the risk of cardiovascular disease, respiratory disease, lung cancer, and other disorders. $PM_{2.5}$ has several components, including sulfates, nitrates, biogenic components, and crustal dust. Wu et al. [4] performed a detailed source apportionment analysis for this pollutant over this region during April and December. However, this pollutant has several important components that may come from different regions or processes. Therefore, to further understand these sources, it is necessary to study the sources of its major components. Unlike some primary gaseous pollutants (NO_x and SO_2), $PM_{2.5}$ cannot be merely reduced by simply controlling specific single source. Sulfates and nitrates are the two important anthropogenic components of $PM_{2.5}$ in this region and they can be controlled effectively once their sources are identified. Therefore, the $PM_{2.5}$ concentration can be reduced gradually if the local government can first focus on controlling sulfate and nitrate. According to Huang et al. [9], $PM_{2.5}$ in western Hong Kong consisted of 31% secondary sulfate and 13% secondary nitrate. Zhang et al. [10] studied sulfate and nitrate sources throughout China with the CMAQ model at a 36-km grid resolution and found that power plant and mobile emissions were the dominant sources of these two components. In this study, we applied a Weather Research Forecast (WRF)–Sparse Matrix Operator Kernel Emission (SMOKE)–Comprehensive Air Quality Model with Extensions (CAMx) modeling system and PSAT to study the source apportionment of sulfates and nitrates in the PRD region with a 3-km model resolution, which is sufficiently fine to analyze local contribution, regional transport (from other cities within the PRD region), and superregional transport (from outside the PRD region). We chose February and August 2011 to represent both winter (northerly prevailing winds) and summer (southerly prevailing winds) conditions in this region.

The remainder of this article is organized as follows. Section 2 describes the model domain setting, the choice of the parameterization scheme, the emission inventory, and the PSAT module. Section 3 contains the model evaluation and a discussion of the source apportionment results; and the overall study is summarized in Section 4.

2. Model and Methods

2.1. Model Description

Weather Research Forecast v3.2 (WRF) was used to simulate the meteorology field in this study. The WRF scheme selection is listed as follows. We chose Grell-Devenyi ensemble cumulus parameterization for cumulus scheme, WRF single-moment six-class scheme for microphysics, the Yonsei University PBL scheme, Dudhia's shortwave radiation scheme, the rapid radiative transfer model for long-wave radiation and the Noah land-surface model. The observation data (wind and temperature) from Hong Kong Observatory were nudged to domain 3. The ambient pollutant concentration was simulated by CAMx v6.00, and Euler backward iterative (EBI) was used for chemical solver, the RADM scheme for aqueous phase chemistry, the K-theory for vertical diffusion, CB05 for gas phase chemistry, ISORROPIA v1.7 for the inorganic aerosol scheme, and SOAP for the secondary organic scheme.

The domain setting is shown in Figure 1. In general, our simulation had three nested domains with resolutions of 27 km, 9 km, and 3 km. The boundary conditions for domain 1 were generated from GEOS-Chem to better match the Asian pollutant background [11]. The outer domain covered a large part of China and some other countries, such as Korea, Japan, and Thailand. Domain 2 covered the entire Guangdong province, and domain 3 included all of the important cities in the PRD region. One should note that the domain for the meteorology model (in black) was intentionally larger than that for the air quality model (in red) because it can help to minimize the boundary effect for air quality simulation.

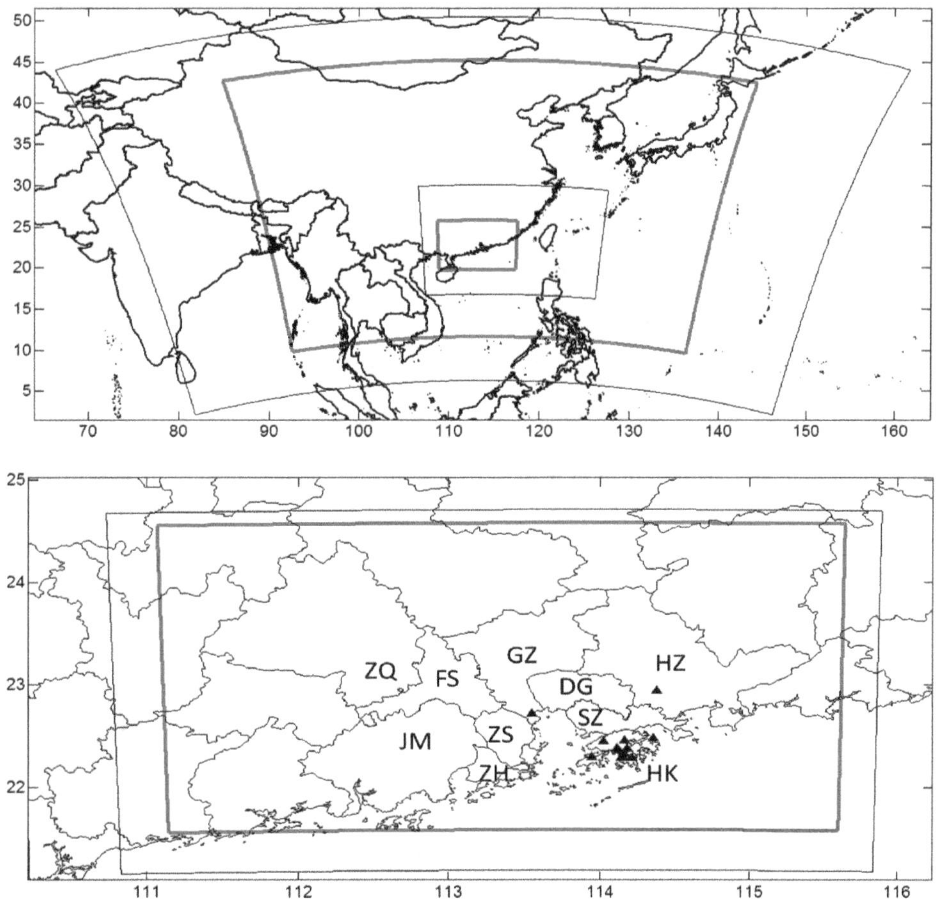

Figure 1. Model domain setting and air quality station location (triangles). HK, Hong Kong; SZ, Shenzhen; DG, Dongguan; HZ, Huizhou; GZ, Guangzhou; FS, Foshan; ZQ, Zhaoqing; JM, Jiangmen; ZH, Zhuhai; ZS, Zhuhai.

The emission inventory for domains 1 and 2 was based on the INTEX-B 2006 regional emission inventory with some updates according to the study from Zhang et al. [12]. For domain 3, the PRD region, a highly resolved emission inventory in this region for 2006 was implemented [13]. The PRD emission mapping for SO_2 and NO_x is shown in Figure 2. The emissions in this region are clustered in Shenzhen, Hong Kong, and Guangzhou. The emissions in some cities, such as Huizhou, Jiangmen, and Zhaoqing (see Figure 2), were much lower than those in the center of this region. The biogenic emissions were generated with the Model of Emission of Gases and Aerosols from Nature (MEGAN v2.04). The MEGAN inputs, such as the leaf area index and plant functional type, were all generated from MODIS satellite data. All emissions were processed and combined with the SMOKE (v2.4) system.

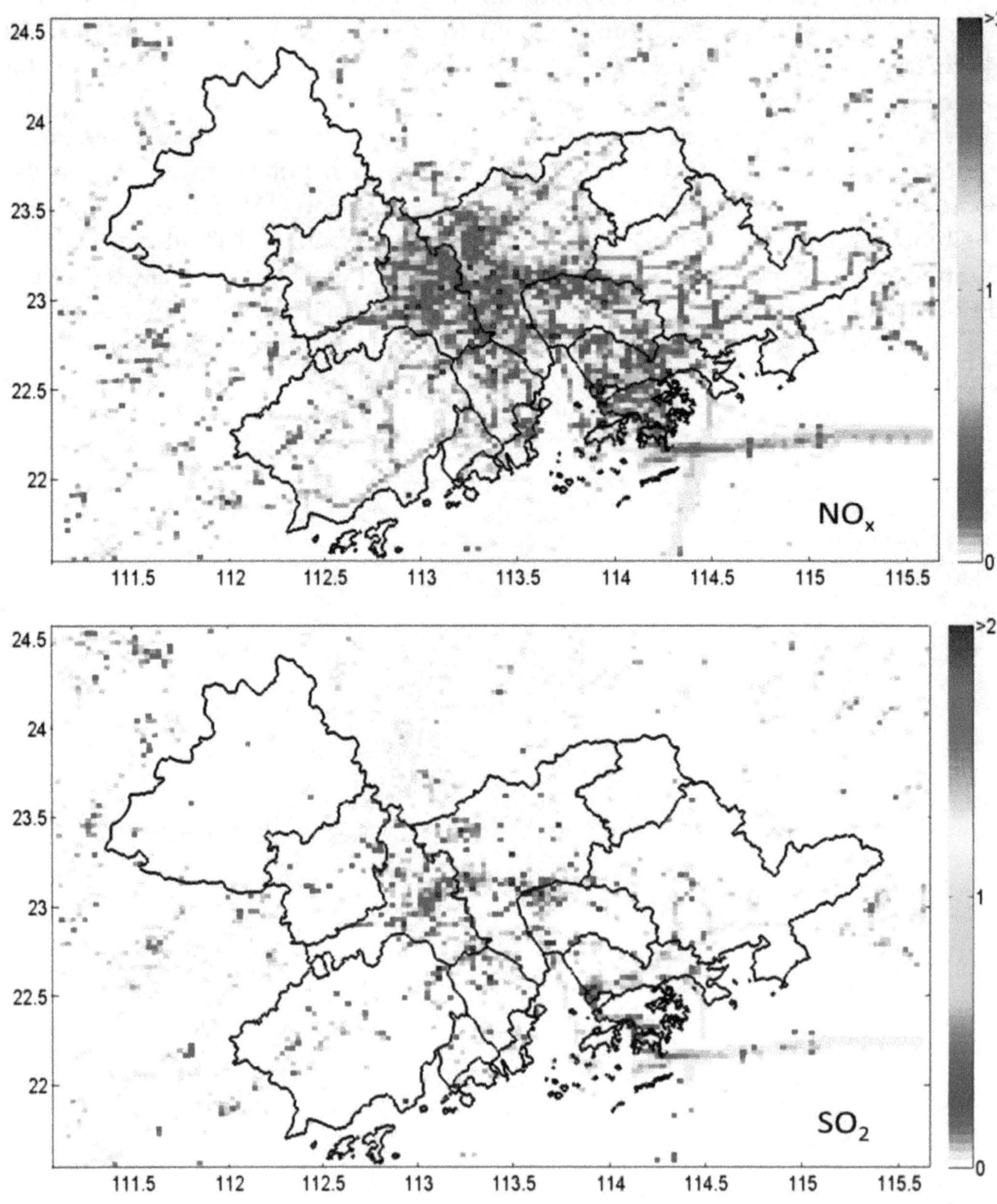

Figure 2. August and February averaged SO_2 and NO_x emissions in the Pearl River Delta (PRD) region. Unit: moles/s.

2.2. Particulate Source Apportionment Technology (PSAT)

As a CAMx module, PSAT has been developed to track the sources (geographic regions and source categories) of PM components. PSAT can be used to track particulate sulfates, nitrates, ammonium, secondary organic aerosols, and six categories of primary PM. The locations at which secondary PM forms may not be the same as the locations at which the precursors are emitted into the atmosphere, and PSAT is able to track such process [14].

In this study, we separated the source into six categories and 10 regions. The categories include mobile sources, industrial point sources, power plant point sources, area sources, marine sources, and biogenic sources in the PRD region. As shown in Figure 3, we separated the region into 10 cities: Guangzhou (GZ), Shenzhen (SZ), Huizhou (HZ), Dongguan (DG), Jiangmen (JM), Zhuhai (ZH), Zhongshan (ZS), Zhaoqing (ZQ), Foshan (FS), and Hong Kong (HK). In addition to the source categories and source regions, the PSAT can automatically track pollutants at the southern, northern, western, and eastern boundaries. The source apportionment of sulfates and nitrates in the region can be expressed by Equation (1).

$$Sulfate/Nitrate(n) = \sum_{i=1}^{10}\sum_{j=1}^{6} S(i,\ j) + BC_{E+W+S+N} + IC \tag{1}$$

where sulfate/nitrate (n) is the sulfate and nitrate concentration in city n, $S(i,j)$ represents the sulfates or nitrates from city i and source j, BC is the boundary source from each of the four directions, and IC is the initial condition. We used local sources, regional sources, and superregional sources to analyze the results. A local source indicates that the source is a local city, a regional source indicates that the source is another city within the PRD region, and a superregional source indicates that the source is outside the PRD region.

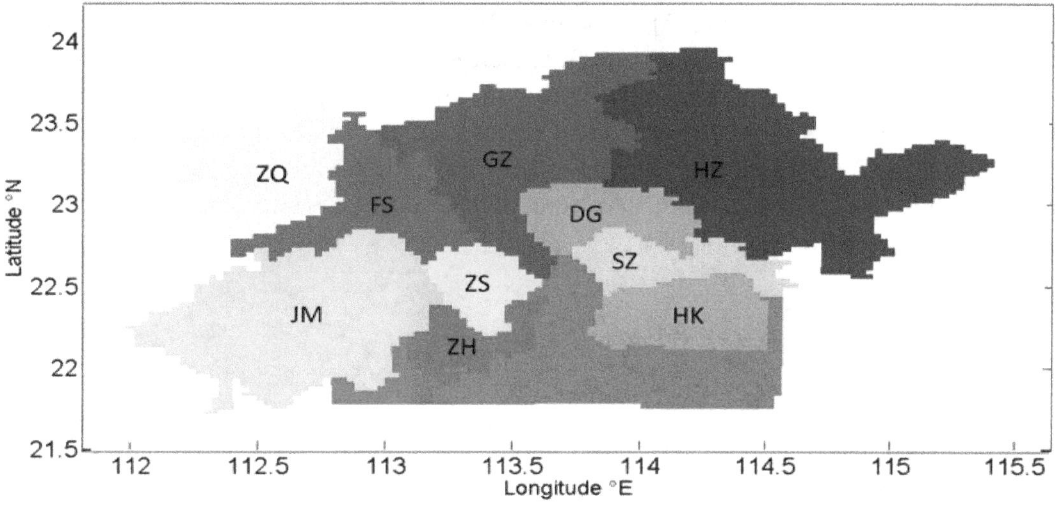

Figure 3. The source regions defined in the PRD region.

3. Results and Discussion

3.1. Model Evaluation

Observation data from 37 meteorology stations in the PRD region were used in our evaluation. Table 1 presents the evaluation (37 stations on average) of the 2-m temperature, wind speed, and wind direction as simulated by WRF. The index of agreement (IOA) formula used for wind direction in this study differs from the formulation for scalar variables; it follows that introduced by Kwok et al. [9]. For the wind speed simulation results, the root-mean-square error (RMSE) ranged from 1.4 to 1.8, the normalized mean bias (NMB) ranged from 0.29 to 0.37, and the IOA ranged from 0.68 to 0.7. The IOA for wind speed was better than 0.88 for both February and August. For the 2-m temperature, the NMB was 0.12 for February and −0.009 for August, which indicates that the temperature was slightly overestimated in winter and underestimated in summer. The simulated meteorology field can reveal the difference between rural area and urban area. The spatial wind mapping can be found in [6]. The results were good and acceptable for further use to drive the air quality model.

Table 1. Evaluation of hourly Weather Research Forecast (WRF) meteorology simulation.

		RMSE	NMB	IOA
February	Wind speed	1.8	0.29	0.70
	Wind direction	-	−0.16	0.88
	Temperature (2m)	3	0.12	0.81
August	Wind speed	1.4	0.37	0.68
	Wind direction	-	0.05	0.88
	Temperature (2m)	2.2	−0.009	0.73

Table 2 shows the model evaluation statistics matrix for the hourly CAMx results for $PM_{2.5}$, sulfates, and nitrates. We used 14 stations in the PRD region to evaluate the $PM_{2.5}$ simulation; the station locations are shown in Figure 1. For sulfates and nitrates, we have only the hourly data from the Hong Kong University of Science and Technology supersite (22.34°N, 114.27°E). Figure 4 shows the $PM_{2.5}$ time series comparison between the simulation and the observations. In general, the model yielded a reasonable $PM_{2.5}$ simulation; the RMSE ranged from 17.1 to 20.1, the IOA ranged from 0.68 to 0.76, and the NMB ranged from −0.24 to −0.012. The sulfate simulation was also acceptable; the RMSE, IOA, and NMB ranged from 4.9 to 6.1, 0.60 to 0.81, and −0.35 to −0.19, respectively. However, the nitrate simulation was not as good as those for sulfates and $PM_{2.5}$. This nitrate simulation discrepancy has also been noted in other studies [15], probably this is due to (1): the discrepancy in HNO_3 and NH_3 dry deposition velocity simulation [16]; (2): the emission of ammonium and NO_x had significant room for improvement from the spatial and temporal aspects; (3): the imperfect simulation for the chemical reaction involving Ca^{2+}, Na^+ and HNO_3 since these reactions are important for the coarse mode nitrate formation [17]. Nonetheless, the nitrate simulation by the CAMx model can catch the magnitude of the observation data in both August and February, as shown in Table 2.

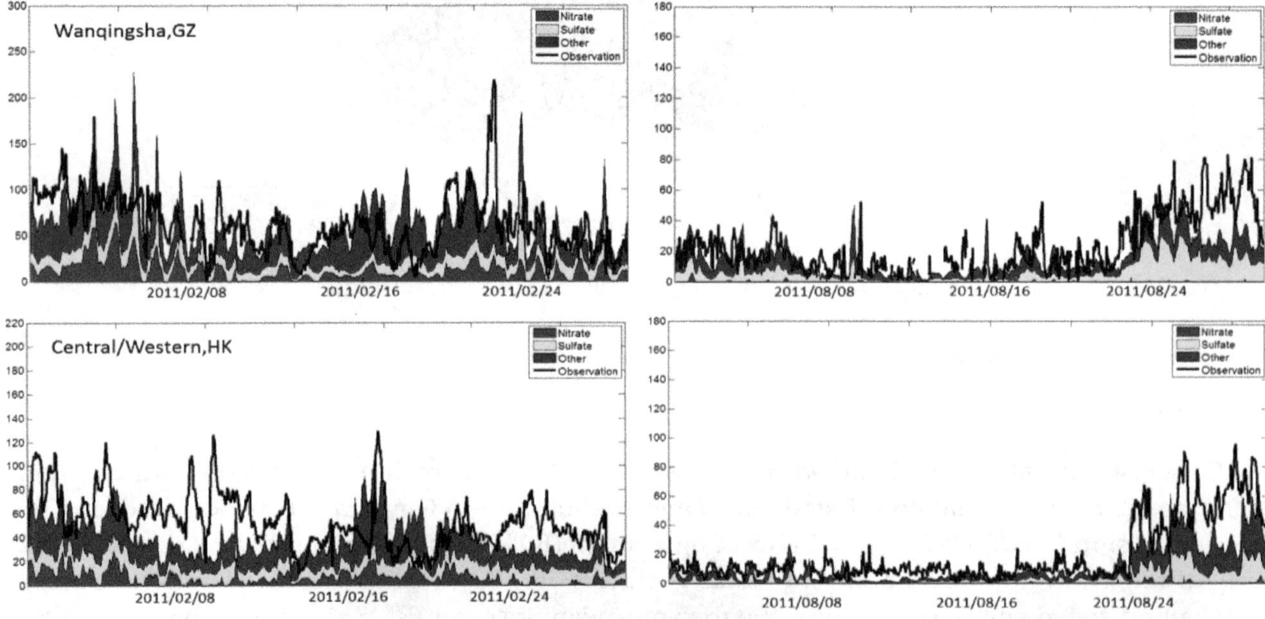

Figure 4. Model and observation comparison for two stations in the PRD region.

Table 2. Evaluation of hourly particulate matter ($PM_{2.5}$), sulfate and nitrate simulated by Comprehensive Air Quality Model with Extensions (CAMx).

$PM_{2.5}$	RMSE	IOA	NMB	Mean-Sim	Mean-OBS
February	20.1	0.68	−0.012	44.2	43.5
August	17.1	0.76	−0.24	19.9	26.0
Sulfate	**RMSE**	**IOA**	**NMB**	**Mean-Sim**	**Mean-OBS**
February	6.1	0.60	−0.19	10.6	13.2
August	4.9	0.81	−0.35	4.2	6.6
Nitrate	**RMSE**	**IOA**	**NMB**	**Mean-Sim**	**Mean-OBS**
February	6.0	0.43	0.62	5.2	3.2
August	1.5	0.29	−0.51	0.4	0.8

3.2. Local, Regional and Super-Regional Contribution

Table 3 shows the local, regional, and superregional contributions of sulfates and nitrates in the 10 PRD cities during February and August. Table 4 shows the regional contribution to a specific city by other cities (the top 3 contributors are shown) in the PRD region. The superregional contribution is the dominant sulfate source in February over the 10 cities, ranging from 66.8% (Foshan) to 94.0% (Huizhou), mainly because the northerly prevailing winds blew this pollutant from northern China into the PRD region during winter. Huizhou is in the northwestern part of the PRD region and has low local emissions. It is the first station reached by pollutants entering the PRD region; as a result, the superregional contribution of sulfates reached almost 100% in this city. The local contribution and regional contribution in Foshan were relatively higher than those in the other nine cities, mainly due to the substantial emissions from this city, as shown in Figure 2, and because Foshan is immediately adjacent to Guangzhou, which is the largest city in the region and also has substantial emissions. As shown in Table 4, Guangzhou contributed 49% of the regional sulfate to Foshan in February. The superregional contributions for Shenzhen and Hong Kong were 82.4% and 91.6%, respectively. During August, the superregional contribution of sulfate decreased to half the level in February, mainly due to the southerly prevailing winds during summer in this region. In August, the superregional contributions ranged from 59.9% (Shenzhen) to 77.4% (Zhuhai). The local contributions from Shenzhen and Guangzhou were the highest—21.6% and 18.6%, respectively—partly due to the large number of vehicles in these two major cities. In 2014, the number of vehicles in Guangzhou and Shenzhen reached 2.7 million and 2.9 million, respectively. From the analysis mentioned above and from Table 3, it can be seen that sulfates came mainly from sources outside the local city; hence, the pollutant issue cannot be solved by local government alone.

Table 3. Local, regional and super-regional contribution of sulfate and nitrate in 10 cities over the PRD region (in μg/m^3).

	Sulfate					
	February			August		
	Local	Regional	S-Regional	Local	Regional	S-Regional
HZ	0.4 (4%)	0.2 (2%)	10.0 (94%)	0.3 (6%)	1.1 (23%)	3.4 (71%)
GZ	1.5 (12%)	1.3 (10%)	9.5 (78%)	1.3 (19%)	1.6 (23%)	4.0 (58%)
FS	1.6 (11%)	3.0 (22%)	9.3 (67%)	1.0 (14%)	1.8 (26%)	4.1 (60%)
DG	1.2 (10%)	1.3 (11%)	9.7 (79%)	0.9 (16%)	1.4 (24%)	3.6 (60%)
JM	1.0 (8%)	1.9 (16%)	9.0 (76%)	0.6 (12%)	0.8 (15%)	3.9 (74%)
SZ	1.3 (11%)	0.7 (7%)	9.8 (82%)	1.2 (22%)	1.0 (19%)	3.3 (60%)
ZS	0.7 (6%)	2.5 (20%)	9.3 (74%)	0.4 (8%)	1.2 (23%)	3.5 (69%)
ZQ	0.8 (6%)	2.7 (21%)	9.2 (73%)	0.4 (6%)	1.9 (27%)	4.7 (67%)
HK	0.4 (4%)	0.5 (5%)	10.0 (92%)	0.7 (16%)	0.5 (12%)	3.1 (73%)
ZH	0.6 (5%)	1.6 (14%)	9.3 (81%)	0.3 (7%)	0.7 (15%)	3.3 (77%)

Table 3. *Cont.*

	Nitrate					
	February			August		
	Local	Regional	S-Regional	Local	Regional	S-Regional
HZ	0.7 (12%)	0.2 (4%)	5.0 (85%)	0.1 (13%)	0.4 (49%)	0.3 (38%)
GZ	1.7 (17%)	3.2 (33%)	4.8 (50%)	0.2 (22%)	0.5 (54%)	0.2 (24%)
FS	1.7 (14%)	5.4 (44%)	5.2 (42%)	0.2 (23%)	0.4 (49%)	0.2 (28%)
DG	0.8 (10%)	2.7 (34%)	4.3 (55%)	0.04 (11%)	0.2 (58%)	0.1 (31%)
JM	0.9 (9%)	4.3 (47%)	4.1 (44%)	0.2 (25%)	0.2 (31%)	0.3 (44%)
SZ	1.0 (13%)	1.4 (19%)	5.0 (68%)	0.1 (14%)	0.2 (51%)	0.1 (35%)
ZS	0.6 (5%)	5.4 (50%)	4.9 (45%)	0.1 (11%)	0.4 (64%)	0.2 (25%)
ZQ	0.8 (7%)	5.5 (48%)	5.2 (45%)	0.1 (8%)	0.6 (60%)	0.3 (32%)
HK	0.6 (10%)	0.8 (14%)	4.6 (76%)	0.1 (30%)	0.2 (40%)	0.1 (30%)
ZH	0.5 (6%)	3.9 (49%)	3.6 (45%)	0.1 (15%)	0.3 (59%)	0.1 (26%)

Table 4. Regional contribution of sulfates and nitrates by other cities (top 3 for each).

	February					
	Sulfate			Nitrate		
HZ	HK (24%)	SZ (21%)	DG (20%)	SZ (34%)	DG (22%)	GZ (21%)
GZ	DG (28%)	HK (21%)	SZ (18%)	DG (24%)	SZ (24%)	HK (19%)
FS	GZ (49%)	HK (12%)	JM (9%)	GZ (38%)	HK (12%)	DG (11%)
DG	SZ (31%)	HZ (31%)	HK (18%)	HZ (34%)	SZ (34%)	HK (20%)
JM	GZ (27%)	FS (21%)	HK (15%)	GZ (29%)	FS (22%)	DG (12%)
SZ	HK (36%)	HZ (30%)	DG (23%)	HK (40%)	HZ (39%)	DG (17%)
ZS	HK (23%)	SZ (17%)	ZH (14%)	HK (24%)	GZ (20%)	SZ (20%)
ZQ	FS (34%)	GZ (19%)	JM (14%)	FS (28%)	GZ (26%)	JM (15%)
HK	SZ (55%)	HZ (20%)	DG (11%)	SZ (47%)	HZ (34%)	DG (11%)
ZH	HK (26%)	DG (15%)	SZ (14%)	HK (21%)	GZ (20%)	SZ (20%)
	August					
	Sulfate			Nitrate		
HZ	HK (36%)	DG (21%)	SZ (18%)	HK (39%)	SZ (27%)	DG (17%)
GZ	DG (30%)	HK (16%)	FS (14%)	DG (21%)	HK (16%)	FS (14%)
FS	GZ (27%)	JM (16%)	DG (15%)	JM (37%)	GZ (17%)	ZH (15%)
DG	SZ (32%)	HK (27%)	GZ (15%)	SZ (27%)	HK (25%)	GZ (13%)
JM	HK (21%)	FS (17%)	ZH (16%)	FS (27%)	ZH (27%)	GZ (20%)
SZ	HK (62%)	DG (13%)	GZ (7%)	HK (51%)	GZ (13%)	DG (10%)
ZS	ZH (23%)	HK (19%)	GZ (18%)	HK (20%)	ZH (20%)	GZ (19%)
ZQ	FS (23%)	GZ (16%)	JM (16%)	JM (38%)	FS (16%)	ZH (12%)
HK	SZ (41%)	DG (20%)	GZ (15%)	SZ (39%)	GZ (16%)	DG (14%)
ZH	HK (25%)	GZ (18%)	DG (16%)	GZ (25%)	SZ (21%)	DG (14%)

The average nitrate concentration in August in the PRD region was only 0.6 $\mu g/m^3$. This low concentration was mainly a result of the partitioning of particulate nitrates to a gas phase at high temperatures. Due to the southerly prevailing winds, the regional contribution of nitrates from other cities in the PRD region was greater than the superregional contribution, except for in Jiangmen. In February, the superregional contribution of nitrate ranged from 42.2% in Foshan to 84.5% in Huizhou. As with sulfates, the high superregional nitrate contribution in Huizhou was mainly a result of its geographic location and the wind direction. There were relatively fewer nitrates from superregional sources. One reason for this finding is the rapid transformation of NO_x into HNO_3, which increased the levels of locally and regionally generated nitrates [18]. The regional contribution ranged from 3.6% in Huizhou to 49.5% in Zhongshan. The northerly prevailing winds prevent much of the nitrates generated in the PRD (Shenzhen and Hong Kong) from entering Huizhou. In February, Hong Kong is downwind from Shenzhen and Huizhou; hence, the regional nitrates were mainly contributed by these two cities, whose contributions were 47% and 34%, respectively, as shown in Table 4. As with sulfates, a substantial amount of nitrates came mainly from regional and superregional sources; therefore, the control policy for this pollutant requires intergovernmental cooperation. However, one should note that there exists uncertainty for the model simulation results. The uncertainty derived from emission inventory, meteorology field, and the chemical reaction mechanism. The overestimation of wind speed may lead to the possible overestimation of regional and super-regional transport contribution. Since not all the chemical reactions are included in the model, the uncertainty is not small for the source apportionment results. In the future, more observation data is needed to quantify the gap between the model simulation and true condition.

3.3. Source Category Contribution

Figure 5 shows the source categories of sulfates and nitrates for local and regional sources over the PRD region in August. As shown in Table 3, in February, most of the sulfate and nitrate are from the super-regional source. Therefore, we only show the source category for August in this part. In this month, power plants were the dominant regional and local sulfate source, followed by industrial sources. Because of the seasonal wind effect, the pollutants emitted from the power plants are blown to the north by the southerly winds, and the sulfate contribution from this source therefore became important. Contini et al. [19] also found that power plant emission can contribute to the sulfate emission in Italy. At the same time, the wind pattern can also influence the contribution of the marine sources. For example, as seen in Figure 2, the ocean channel is located at the southeastern part of this region, and the marine contribution of sulfates in Hong Kong exceeded 30% under southerly wind conditions. The importance of marine emissions for the sulfate formation can also be found in European cities [20,21]. The sulfate contribution from mobile emissions is shown to be relatively large in this work when compared to another study which used the INTEX-B as the only emission inventory [10]. The reason for this is because the SO_2 emitted by vehicles in the emission inventory we used [13] was relatively larger than that in INTEX-B [12]. In the PRD 2006 emission inventory, the ratio of mobile emitted SO_2 over the power plant emitted SO_2 was 10.5%; while in INTEX-B, this ratio was only around 0.7%. The area sources and biogenic sources contributed only a negligible amount of sulfates over this region in this month.

The main contributors of nitrates in August were mobile (average 33.9%) and power plant (average 35.7%) emissions. The sum of mobile and power plant contributions in all the cities is above 60%. In this month, industrial and marine emissions also contributed a substantial amount of nitrate. As with sulfates, biogenic emissions of nitrates were also negligible over this region. Due to the southerly winds, marine sources contributed over 15% of nitrate in Shenzhen and Hong Kong (including aged sea-spray sodium nitrate). In order to control the nitrate in this region, the local government should focus on two points: (1) take action to control mobile emissions (e.g., odd-even number restriction); (2) reduce the industrial emission by establishing stricter supervision regulation. However, as is the case with sulfate, super-regional contribution is of great importance and therefore cross regional cooperation is highly important.

3.4. Source Apportionment in City Center

The city center of an urban area is the commercial and geographical heart of the city. It has a high population density, and its air quality has important effects on the health of local citizens. Hence, to protect public health, it is also important for city government to formulate a specific control policy for these areas. In this study, we further analyzed the city centers in Guangzhou (23.13°N, 113.26°E), Shenzhen (22.55°N, 114.10°E) and Hong Kong (22.28°N, 114.16°E), which are the three most densely populated cities in the PRD region. Table 5 shows the source apportionment results for sulfates and nitrates in the city centers of the three major cities mentioned above. In February, the superregional contribution of sulfates in the city center of Hong Kong was similar to the average value for the whole city. However, the super-regional contribution in the city center of the other two cities was lower than the average value for the whole cities. In the INTEX-B emission inventory, the SO_2 from power plants and industrial emissions represented 59% and 31% of total SO_2 emission respectively. Hence, most of the sulfate in this region is coming from power plants and industrial sources. In Guangzhou, the contributions from regional and local sources made up 32% and 9% of total sulfates, respectively. In the city center of Shenzhen, the regional contribution of sulfate formation is 6% and the local sulfate contribution is 16%. The regional contribution is low in the city center during February. One of the main reasons for this was that in February the predominant wind is in north-east direction and an important power plant is located in the southern part of this region. In August, the superregional contribution of sulfates decreased due to the southerly prevailing wind direction. In this month, the super-regional contribution of sulfate in the city centers of Shenzhen, Guangzhou and Hong Kong were 55%, 47% and 67%. The regional and local source contribution increased due to the southern prevailing wind that brought the sulfate from the power plant to the city centers. In February, the superregional contribution of nitrates comprised more than 70% of the levels in the city centers of Hong Kong and Shenzhen, but only 39% of the level in Guangzhou. Compared to the sulfate in February, the sum of regional and local sources of nitrate is larger. One main reason for this was that a substantial amount of nitrate is from mobile emissions and large number of vehicles were clustering in the city center. Hence, it is anticipated that traffic control policy should be effective in controlling the urban nitrate level. In August, super regional contributions of nitrate in the three city centers were all below 50%, which is due to the southern prevailing wind, as noted above. As with the average value for the whole city, the superregional contribution was the most important source of sulfates and nitrates in the city center. This result further indicates that cooperation with other provinces is necessary for sulfate and nitrate control over the PRD region.

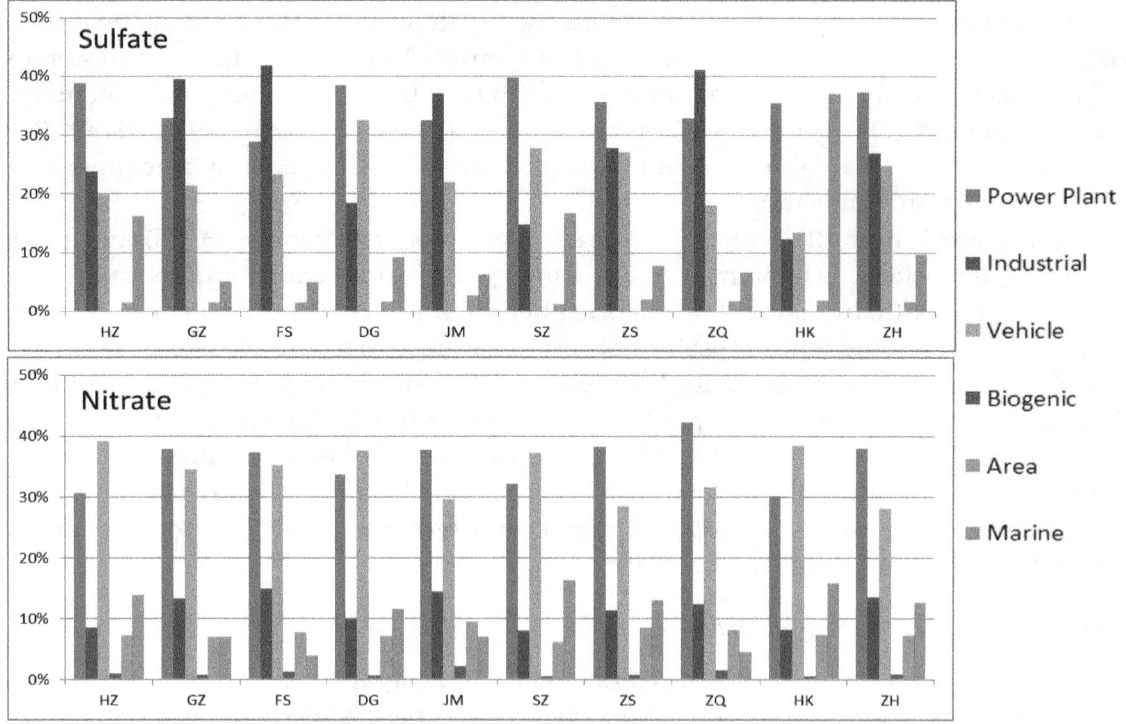

Figure 5. Source category of sulfates and nitrates over ten cities over the PRD region in August.

Table 5. Source apportionment of sulfates and nitrates in the city centers of Guangzhou, Shenzhen and Hong Kong.

February-Sulfates	GZ	SZ	HK
Local	32%	16%	6%
Regional	9%	6%	3%
Super-regional	59%	78%	91%
August-Sulfates	**GZ**	**SZ**	**HK**
Local	32%	25%	23%
Regional	18%	20%	10%
Super-regional	47%	55%	67%
February-Nitrates	**GZ**	**SZ**	**HK**
Local	27%	15%	12%
Regional	34%	14%	9%
Super-regional	39%	71%	79%
August-Nitrates	**GZ**	**SZ**	**HK**
Local	33%	24%	11%
Regional	47%	43%	49%
Super-regional	20%	33%	40%

4. Conclusions

In this study, we applied the WRF-SMOKE-CAMx modeling system with PSAT to study the source apportionment of sulfates and nitrates in the PRD region. In both August and February, the superregional contribution was the dominant source of sulfates and nitrates in this region, although it was lower in August due to the southerly prevailing wind direction. The level of nitrates from superregional sources was lower than that of sulfates. The process and influencing conditions for this include: (1) the excess ammonia amount that can be reacted with nitric acid; (2) the temperature that is low enough to prevent the HNO_3 vaporization [22]; (3) the wetness of the surface (HNO_3 can deposit quickly on the wet surface). Industrial sources and power plant emissions were the two major contributors among the regional and local sources. For nitrates, the contribution from mobile emissions was much greater than that from other sources and exceeded 40% in all 10 cities in this region during August. Given the high population density in the city centers, we also analyzed the source apportionment results in the city centers of Hong Kong, Guangzhou, and Shenzhen. As with the overall situation, the superregional contribution was very important in the city center, especially the city center of Hong Kong. With the exception of the superregional contribution, mobile sources contributed the most sulfates and nitrates in all three city centers during both months. Our results indicate that the sulfate and nitrate pollution issue cannot be solved by any single city or even by all 10 cities in the PRD region. A higher-level effort, such as interprovincial cooperation, is needed for a better control policy for this region. The $PM_{2.5}$ pollution problem can be alleviated only when sulfates and nitrates are controlled effectively. Because the superregional contribution is important for these two pollutants, future study of the source apportionment over a larger area (e.g., including Hunan, Jiangxi, and Fujian provinces) with fine model resolution is necessary to better understand the sources from other provinces and their influence on the sulfate and nitrate levels in the PRD region. One of the limitations of this work was that we did not classify the source category in the super-regional contribution. Therefore, further work is needed to provide a more detailed source contribution for sulfate and nitrate formation in this region. This work focuses on sulfate and nitrate, the sources of other $PM_{2.5}$ components, such as OC and EC, should be further studied once the sulfate and nitrate are controlled effectively over this region in the future.

Acknowledgments: This work was supported by the National Science Foundation of China (No. 41375103), NSFC/RGC Grant N_HKUST631/05, NSFC-GD Grant U1033001, and RGC Grant 16300715.

Author Contributions: Xingcheng Lu did the model simulation, analyzed the model results and wrote the paper. Xingcheng Lu and Jimmy Fung designed the framework of this paper and revised the paper.

Appendix

Model performance statistics formula:

$$RMSE = \sqrt{\frac{1}{n}\sum_{i=1}^{n}(S_i - O_i)^2} \tag{A1}$$

$$IOA = 1 - \frac{\sum_{i=1}^{n}(O_i - S_i)^2}{\sum_{i=1}^{n}\left(|O_i - \overline{O}| + |S_i - \overline{S}|\right)^2} \tag{A2}$$

$$NMB = \frac{\sum_{i=1}^{N}(S_i - O_i)}{\sum_{i=1}^{N}O_i} \times 100\% \tag{A3}$$

$$Mean - Sim = \frac{1}{n}\sum_{i=1}^{n}S_i \tag{A4}$$

$$Mean - Obs = \frac{1}{n}\sum_{i=1}^{n}O_i \tag{A5}$$

$$\begin{gathered} IOA_{wind} = 1 - \frac{\sum_{i=1}^{N} f(|\theta_i - \phi_i|)}{180^2},\ 0^\circ \leqslant \theta_i \leqslant 360^\circ \ and\ 0^\circ \leqslant \phi_i \leqslant 360^\circ \\ f(|\theta_i - \phi_i|) = |\theta_i - \phi_i|^2,\ |\theta_i - \phi_i| \leqslant 180^\circ \\ f(|\theta_i - \phi_i|) = (360^\circ - |\theta_i - \phi_i|)^2,\ |\theta_i - \phi_i| > 180^\circ \end{gathered} \tag{A6}$$

where "RMSE" represents root mean square error, "IOA" represents index of agreement, "IOAwind" represents the index of agreement for wind direction evaluation, "NMB" represents normalized mean bias, Mean-Sim represents the mean of the model simulation and Mean-Obs represents the mean of the observation results. S_i is the hourly simulation, O_i is the hourly observation value, θ_i is the observation wind direction and ϕ_i is the simulated wind direction at time i.

References

1. Lu, X.; Fung, J.C.H.; Wu, D. Modeling wet deposition of acid substances over the PRD region in China. *Atmosp. Environ.* **2015**, *122*, 819–828. [CrossRef]
2. Lo, J.C.; Lau, A.K.; Fung, J.C.; Chen, F. Investigation of enhanced cross-city transport and trapping of air pollutants by coastal and urban land-sea breeze circulations. *J. Geophys. Res. Atmos.* **2006**, *111*. [CrossRef]
3. Li, Y.; Lau, A.H.; Fung, J.H.; Zheng, J.Y.; Zhong, L.J.; Louie, P.K.K. Ozone source apportionment (OSAT) to differentiate local regional and super-regional source contributions in the Pearl River Delta region, China. *J. Geophys. Res. Atmos.* **2012**, *117*, D15–D16. [CrossRef]
4. Wu, D.; Fung, J.C.H.; Yao, T.; Lau, A.K.H. A study of control policy in the Pearl River Delta region by using the particulate matter source apportionment method. *Atmos. Environ.* **2013**, *76*, 147–161. [CrossRef]
5. Yao, T.; Fung, J.C.H.; Ma, H.; Lau, A.K.H.; Chan, P.W.; Yu, J.Z.; Xue, J. Enhancement in secondary particulate matter production due to mountain trapping. *Atmos. Res.* **2014**, *147*, 227–236. [CrossRef]
6. Lu, X.; Yao, T.; Li, Y.; Fung, J.C.H.; Lau, A.K.H. Source apportionment and health effect of NOx over the Pearl River Delta region in southern China. *Environ. Pollut.* **2016**, *212*, 135–146. [CrossRef] [PubMed]

7. Yuan, Z.; Lau, A.K.H.; Zhang, H.; Yu, J.Z.; Louie, P.K.; Fung, J.C. Identification and spatiotemporal variations of dominant PM_{10} sources over Hong Kong. *Atmos. Environ.* **2006**, *40*, 1803–1815. [CrossRef]
8. Xue, J.; Griffith, S.M.; Yu, X.; Lau, A.K.; Yu, J.Z. Effect of nitrate and sulfate relative abundance in $PM_{2.5}$ on liquid water content explored through half-hourly observations of inorganic soluble aerosols at a polluted receptor site. *Atmos. Environ.* **2014**, *99*, 24–31. [CrossRef]
9. Huang, X.H.; Bian, Q.; Ng, W.M.; Louie, P.K.; Yu, J.Z. Characterization of $PM_{2.5}$ major components and source investigation in suburban Hong Kong: A one year monitoring study. *Aerosol Air Qual. Res.* **2014**, *14*, 237–250. [CrossRef]
10. Zhang, H.; Li, J.; Ying, Q.; Yu, J.Z.; Wu, D.; Cheng, Y.; Jiang, J. Source apportionment of $PM_{2.5}$ nitrate and sulfate in China using a source-oriented chemical transport model. *Atmos. Environ.* **2012**, *62*, 228–242. [CrossRef]
11. Fu, J.; Lam, Y.; Gao, Y.; Jacob, D.; Carouge, C.; Dolwick, P.; Jang, C. Recent study of U.S. ozone background concentrations using GEOS-Chem. In Proceedings of the 5th GEOS-Chem Meeting, Cambridge, MA, USA, 2–5 May 2011.
12. Zhang, Q.; Streets, D.G.; Carmichael, G.R.; He, K.B.; Huo, H.; Kannari, A.; Klimont, Z.; Park, I.S.; Reddy, S.; Fu, J.S.; et al. Asian emissions in 2006 for the NASA INTEX-B mission. *Atmos. Chem. Phys.* **2009**, *9*, 5131–5153. [CrossRef]
13. Zheng, J.; Zhang, L.; Che, W.; Zheng, Z.; Yin, S. A highly resolved temporal and spatial air pollutant emission inventory for the Pearl River Delta region, China and its uncertainty assessment. *Atmos. Environ.* **2009**, *43*, 5112–5122. [CrossRef]
14. Yarwood, G.; Morris, R.E.; Wilson, G.M. Particulate matter source apportionment technology (PSAT) in the CAMx photochemical grid model. In *Air Pollution Modeling and Its Application XVII*; Springer US: New York, NY, USA, 2007; pp. 478–492.
15. Kwok, R.H.; Fung, J.C.; Lau, A.K.; Fu, J.S. Numerical study on seasonal variations of gaseous pollutants and particulate matters in Hong Kong and Pearl River Delta Region. *J. Geophys. Res. Atmos. (1984–2012)* **2010**, *115*. [CrossRef]
16. Shimadera, H.; Hayami, H.; Chatani, S.; Morino, Y.; Mori, Y.; Morikawa, T.; Ohara, T. Sensitivity analysis of influencing factors on $PM_{2.5}$ nitrate simulation. In Proceedings of the 11th Annual CMAS Conference, Chapel Hill, NC, USA, 15–17 October 2012.
17. Contini, D.; Cesari, D.; Genga, A.; Siciliano, M.; Ielpo, P.; Guascito, M.R.; Conte, M. Source apportionment of size-segregated atmospheric particles based on the major water-soluble components in Lecce (Italy). *Sci. Total Environ.* **2014**, *472*, 248–261. [CrossRef] [PubMed]
18. Ying, Q.; Kleeman, M. Regional contributions to airborne particulate matter in central California during a severe pollution episode. *Atmos. Environ.* **2009**, *43*, 1218–1228. [CrossRef]
19. Contini, D.; Cesari, D.; Conte, M.; Donateo, A. Application of PMF and CMB receptor models for the evaluation of the contribution of a large coal-fired power plant to PM_{10} concentrations. *Sci. Total Environ.* **2016**, *560*, 131–140. [CrossRef] [PubMed]
20. Becagli, S.; Sferlazzo, D.M.; Pace, G.; Sarra, A.D.; Bommarito, C.; Calzolai, G.; Ghedini, C.; Lucarelli, F.; Meloni, D.; Monteleone, F.; et al. Evidence for heavy fuel oil combustion aerosols from chemical analyses at the island of Lampedusa: A possible large role of ships emissions in the Mediterranean. *Atmos. Chem. Phys.* **2012**, *12*, 3479–3492. [CrossRef]
21. Cesari, D.; Genga, A.; Ielpo, P.; Siciliano, M.; Mascolo, G.; Grasso, F.M.; Contini, D. Source apportionment of $PM_{2.5}$ in the harbour–industrial area of Brindisi (Italy): Identification and estimation of the contribution of in-port ship emissions. *Sci. Total Environ.* **2014**, *497*, 392–400. [CrossRef] [PubMed]
22. Griffith, S.M.; Huang, X.H.; Louie, P.K.K.; Yu, J.Z. Characterizing the thermodynamic and chemical composition factors controlling $PM_{2.5}$ nitrate: Insights gained from two years of online measurements in Hong Kong. *Atmos. Environ.* **2015**, *122*, 864–875. [CrossRef]

8

The Spatial Variation of Dust Particulate Matter Concentrations during Two Icelandic Dust Storms in 2015

Pavla Dagsson-Waldhauserova [1,2,3,*], Agnes Ösp Magnusdottir [1], Haraldur Olafsson [2,4] and Olafur Arnalds [1]

1 Faculty of Environmental Sciences, Agricultural University of Iceland, Keldnaholt, Reykjavik 112, Iceland; agnes@lbhi.is (A.Ö.M.); oa@lbhi.is (O.A.)
2 Faculties of Physical and Earth Sciences, University of Iceland, Reykjavik 101, Iceland; haraldur@vedur.is
3 Faculty of Environmental Sciences, Czech University of Life Sciences, Prague 165 21, Czech Republic
4 Icelandic Meteorological Office, Reykjavik 108, Iceland
* Correspondence: pavla@hi.is or pavla@lbhi.is or waldhauserova@fzp.czu.cz

Abstract: Particulate matter mass concentrations and size fractions of PM_1, $PM_{2.5}$, PM_4, PM_{10}, and PM_{15} measured in transversal horizontal profile of two dust storms in southwestern Iceland are presented. Images from a camera network were used to estimate the visibility and spatial extent of measured dust events. Numerical simulations were used to calculate the total dust flux from the sources as 180,000 and 280,000 tons for each storm. The mean PM_{15} concentrations inside of the dust plumes varied from 10 to 1600 $\mu g \cdot m^{-3}$ (PM_{10} = 7 to 583 $\mu g \cdot m^{-3}$). The mean PM_1 concentrations were 97–241 $\mu g \cdot m^{-3}$ with a maximum of 261 $\mu g \cdot m^{-3}$ for the first storm. The $PM_1/PM_{2.5}$ ratios of >0.9 and PM_1/PM_{10} ratios of 0.34–0.63 show that suspension of volcanic materials in Iceland causes air pollution with extremely high PM_1 concentrations, similar to polluted urban areas in Europe or Asia. Icelandic volcanic dust consists of a higher proportion of submicron particles compared to crustal dust. Both dust storms occurred in relatively densely inhabited areas of Iceland. First results on size partitioning of Icelandic dust presented here should challenge health authorities to enhance research in relation to dust and shows the need for public dust warning systems.

Keywords: fine dust; Arctic desert; close-to-ultrafine natural particles; volcanic ash

---

1. Introduction

Air pollution from natural sources accounts for a significant part of the total particulate matter pollution on Earth. Deserts, stratovolcanoes, and arable land areas contribute to global air pollution in addition to emissions from industrialized and densely inhibited regions. Desert dust has a remarkable influence on Earth's ecosystems as well as human health. Several studies have shown that suspended desert dust can increase mortality hundreds of kilometers downwind from dust sources [1–6]. Mortality was found to increase by 2%–12% with every 10 $\mu g \cdot m^{-3}$ increase in particulate matter (PM_{10}) concentration. Some of these studies reported increased mortality for the $PM_{2.5}$ dust particulate matter. A decrease of mixing layer height was associated with an increase of daily mortality while the effect of mixing layer thinning on particle toxicity was exacerbated when Saharan dust outbreaks occurred [7]. Positive associations between mass concentrations of larger sizes of particles, such as PM_{15}, were observed for cardiopulmonary and ischemic heart disease causes of death during the long-term studies on air pollution and mortality of the American Cancer Society [8]. The studies on levels and speciation of PM_1 in Europe are, however, scarce [9–11]. The PM_1 fraction has considerable importance in relation to health because of high potential for entering the lungs [12]. Moreover, submicron particles are more likely to travel further distances during the long range transport than larger particles [13].

In spite of a cold and moist climate, Iceland has been identified as the most active and largest Arctic and European desert [14]. Other cold climate and high latitudes regions with considerable dust inputs are Alaska, Greenland, Svalbard, Antarctica, and South America [15–20]. One of the most extreme wind erosion events on Earth was measured in Iceland in 2010 [21]. Annual dust day frequency in Iceland is comparable to the major desert areas of the world [22,23]. Emissions from local dust sources, enhanced by strong winds, affect regional air quality in Iceland, such as in the capital Reykjavik [24]. Particulate Matter (PM_{10}) concentrations during dust events in Reykjavik often exceed the health limit of 50 $\mu g \cdot m^{-3}$ over 24 h [25,26], while PM_{10} concentrations measured during dust events in the vicinity of dust sources (<30 km) exceed the health limit in order of 10–100 times [26–28]. This shows that atmospheric aerosols, mainly dust, can markedly impair air quality in non-polluted Arctic/sub-Arctic regions.

Icelandic dust differs from dust originating from most continental deserts. It is volcanogenic in origin, of basaltic composition, with lower SiO_2 proportions (<50%) and higher Al_2O_3, Fe_2O_3, and CaO contents than crustal dust [23,27–29]. This volcanic dust contains about 80% volcanic glass with numerous large gas bubbles and massive shards. It is extremely angular with sharp-tipped shards and often with curved and concave shard-faces. Fine pipe-vesicular structures of glass, as known from asbestos, can be also found. All these factors suggest that volcanic dust can be easily suspended and have highly negative effects on human health as concluded by Carlsen *et al.* [30].

Studies that provide PM mass concentration measurements during dust storms in Iceland are few and limited in scope, excluding research related to volcanic eruptions [23,24,27,28,31]. None of these studies dealt with the size partitioning of the PM components. Here we presented a study on PM source characteristics of volcanic dust during two dust events from different dust sources in Iceland. The emphasis was given to the fine dust fraction of PM_1. An effort has been made to measure the transverse horizontal profile of dust storms and estimate the spatial extent of such storms in the terms of PM concentrations, dust load computation, and visibility information obtained from cameras.

2. Experiments

2.1. Instrumentation and Measurement Setup

Two dust storms were measured in southwestern Iceland in the summer of 2015. Measurements of both storms began after the dust plume was visible from Reykjavík with transverse horizontal profile measurements through the dust plumes. The source area of the first dust event on 15 June 2015 was Landeyjasandur (Storm 1, Figure 1A), about 100 km from Reykjavík, while the second dust storm on 4 August 2015 originated from the Hagavatn dust source (Storm 2, Figure 1B), about 85 km distance from Reykjavik.

A mobile instrument, aerosol monitor DustTrak DRX 8533EP, was used to measure particulate matter (PM) mass concentrations at several places within the dust plume. The DustTrak instrument provides measurements of mass concentration from 0.001 to 150 $mg \cdot m^{-3}$ with the mass fraction concentrations corresponding to PM_1, $PM_{2.5}$, PM_4, PM_{10}, and the total PM_{15}. Five minute measurements were made at each stop in the dust plume, consisting of 60 five-second sampling periods. The measurement time was short to allow for travel through the dust storm with measurements at as many places as possible. All measurements were calculated as 5-min averages. The TSI DustTraks have given similar results in parallel measurements of PM mass concentration as Beta attenuation instruments (Thermo ESM Andersen FH 62 I-R) [32]. This instrument has been used in Iceland by the Environmental Agency of Iceland (EAI) since 1996 and we have found good relation between PM_{10} concentrations obtained by DustTrak DRX and the Thermo ESM. The error range for absolute values for other PM size ranges could be $\pm$10% [33]. We emphasize that the DustTrak measurements in both storms are complemented by an independent measurement by the Thermo ESM instrument at the EAI in Reykjavik (PM_{10} 30-min maxima of about 300 $\mu g \cdot m^{-3}$ for Storm 1 and about 200 $\mu g \cdot m^{-3}$ for Storm 2). Measurements for Storm 1 with DustTrak at Keldnaholt (Reykjavik) and Thermo ESM at Grensásvegur (Reykjavik), approx. 5 km apart are very similar (PM_{10} 30-min maxima of about 224 $\mu g \cdot m^{-3}$ *vs.* 280 $\mu g \cdot m^{-3}$ at the same time, with the latter being closer to the center of the plume). Similar agreement was found for Storm 2. These numbers indicate that large scale errors sometimes reported for measurements with the TSI DustTraks do not apply here [32,33].

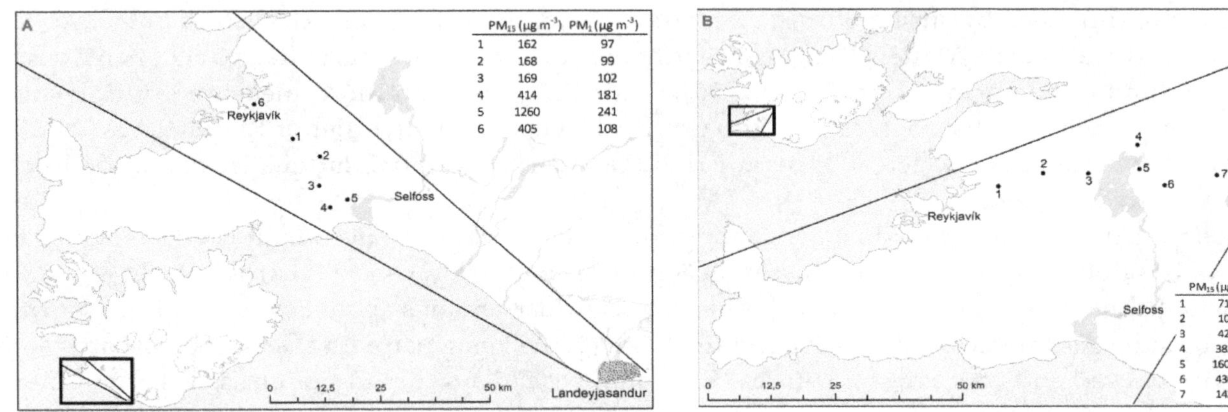

Figure 1. Spatial location of the dust storms. Left figure (**A**, Storm 1) shows the dust storm from Landeyjasandur on 15 June 2015. The measurement locations are marked on the map with the numbers corresponding to the information on the PM_{15} and PM_1 concentrations in the table. Right figure (**B**, Storm 2) shows the dust storm from the Hagavatn on 4 August 2015. The lines on the figures depict the dispersal area of the dust plumes estimated from the images of the Icelandic Road and Coastal Administration web camera network [34].

In this study, we used a unique dataset from a network of active cameras operated by the Icelandic Road and Coastal Administration to find the exact extent and area of the dust plume [34]. A total of 25 cameras were used to estimate the visibility changes during these dust storms. Figure 2 presents the time laps images from the web cameras with corresponding visibility estimations. The houses on the photos are situated at about 1 km from the camera. The mountains in the background are about 3.5 km from the camera. Subsequently, maps of the dust storms were made in ArcMap 10.1.

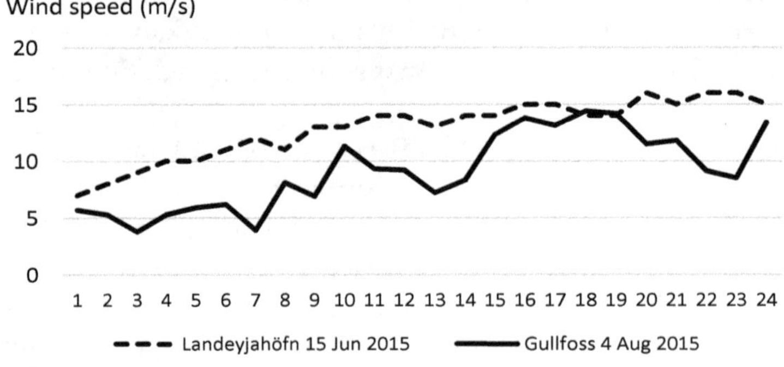

Figure 2. Observations of mean 10 min wind speed from 01-24 UTC at Landeyjahöfn (at the dust source for Storm 1) on 15 June 2015 and at Gullfoss (about 15 km SE of the dust source for Storm 2) on 4 August 2015.

2.2. *Meteorological Conditions and Transport of Dust*

Figure 2 shows the observed 10 min average wind speed at the weather stations Landeyjahöfn (Storm 1), which is close to the dust source, and Gullfoss (Storm 2) which is at a 15 km distance from the source, but does not capture the catabatic wind effects at the source. The wind speed was gradually increasing from about 6 to 16 m$\cdot$ s^{-1} during the Landaeyjasandur dust event (Storm 1). The Hagavatn dust event (Storm 2) occurred with wind speeds from about 4 to 14 m$\cdot$ s^{-1} measured at Gullfoss. A numerical simulation of surface winds during the two events is shown in Figure 3. The simulation was carried out with the numerical model HIRLAM with a horizontal resolution of 5 km. The simulation was initialized at 00UTC the same day, using initial and boundary conditions from the operational suite at the European Centre for Medium range Weather Forecasts. The true winds in the dust source area on 4 August were probably a few m$\cdot$ s^{-1} stronger than observed at Gullfoss. There are substantial horizontal gradients in the wind field close to the dust sources in both cases. In Storm 1 on 15 June, the mountains in S-Iceland generated a corner wind of 6–8 m$\cdot$ s^{-1} greater speed than in the incoming flow. This corner wind extended over a large area over the sea and the SW-coast of Iceland. The dust was lifted where this windstorm blew over the coastal areas. Once in suspension, some of the dust entered a wake area over land with weaker winds. In Storm 2 on 4 August, dust was also generated by locally enhanced winds. There was local acceleration in flow which ran along a major mountain range and had a downslope component. The local acceleration and the winds in general were underestimated by a few m$\cdot$ s^{-1} in the simulation (Figure 3), but the flow pattern and the wind directions are realistic. The orographic flow perturbation is of a smaller scale in Storm 2 on 4 August than in Storm 1 on 15 June, and consequently is harder to reproduce in a numerical simulation.

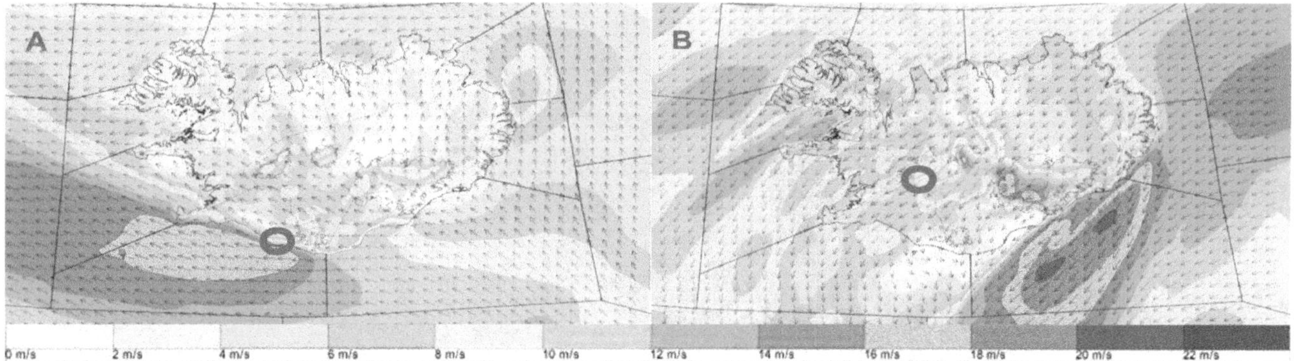

Figure 3. Numerical simulations of 10 m winds (m$\cdot$ s^{-1}) over Iceland at 15 UTC on (**A**) Storm 1 on 15 June 2015 and (**B**) Storm 2 on 4 August 2015. The locations of the dust sources are indicated with a circle.

3. Results and Discussion

3.1. Dust Concentrations and Visibility

The spatial extent and PM concentrations measured during the two dust storms are shown in Figure 1 and Table 1. The mean PM_{15} concentrations inside the dust plume 1 varied from 162 µg$\cdot$ m^{-3} to 1260 µg$\cdot$ m^{-3} (PM_{10} = 158–583 µg$\cdot$ m^{-3}), and from 10 µg$\cdot$ m^{-3} to 1600 µg$\cdot$ m^{-3} (PM_{10} = 7–486 µg$\cdot$ m^{-3}) inside dust plume 2, respectively. These numbers represent rather low concentrations for an ongoing dust storm compared to the long term PM measurements provided by the Environmental Agency of Iceland (EAI) [23,26–28]. This was partly caused by moderate winds not exceeding 16 m$\cdot$ s^{-1} (Figures 2 and 3). The mean PM_1 concentrations were, however, 97–241 µg$\cdot$ m^{-3} during Storm 1 and reached up to 164 µg$\cdot$ m^{-3} during Storm 2. Such high fine dust concentrations have been reported during massive dust storms from lacustrine sediment areas in Iran and during African dust episodes in Barcelona (hourly means 60–70 µg$\cdot$ m^{-3}) [10,35]. The PM_1 maximum of 261 µg$\cdot$ m^{-3} measured during moderate Icelandic dust storms is comparable to the maximum of 495 µg$\cdot$ m^{-3} reported from Iran during a massive dust storm when PM_{10} concentrations exceeded 5000 µg$\cdot$ m^{-3}. Relatively high PM_1 annual means are regularly measured over Greece during African dust outbreaks [36].

Table 1. Particulate matter concentrations PM_{1-15} ($\mu g \cdot m^{-3}$) for both storms. Ratios between different PM values are given.

	PM_1 Average	$PM_{2.5}$ Average	PM_4 Average	PM_{10} Average	Total (PM_{15}) Average	PM_1/PM_{10} Ratio	$PM_{2.5}/PM_{10}$ Ratio	$PM_1/PM_{2.5}$ Ratio	PM_1/PM_4 Ratio	PM_4/PM_{10} Ratio	PM_{10}/PM_{15} Ratio
Storm 1											
1	97	109	130	158	162	0.61	0.69	0.89	0.75	0.82	0.98
2	99	110	130	158	168	0.63	0.70	0.90	0.76	0.82	0.94
3	102	114	137	163	169	0.63	0.70	0.89	0.74	0.84	0.96
4	181	201	248	354	414	0.51	0.57	0.90	0.73	0.70	0.86
5	241	263	322	583	1260	0.41	0.45	0.92	0.75	0.55	0.46
6	108	118	142	224	405	0.48	0.53	0.92	0.76	0.63	0.55
Storm 2											
1	11	12	14	29	71	0.48	0.53	0.92	0.76	0.63	0.55
2	4	4	5	7	10	0.38	0.41	0.92	0.79	0.48	0.41
3	12	13	16	29	42	0.57	0.57	1.00	0.80	0.71	0.70
4	57	61	74	162	383	0.41	0.45	0.92	0.75	0.55	0.69
5	164	174	206	486	1600	0.35	0.38	0.93	0.77	0.46	0.42
6	128	140	177	318	436	0.34	0.36	0.94	0.80	0.42	0.30
7	35	39	48	87	143	0.40	0.44	0.91	0.72	0.56	0.73

The distance of the measurements from the dust sources was <100 km. The source material contains extremely fine particles. Storm 1 originated from the Landeyjarsandur dust hot spot which mostly consists of fine volcanic material from active volcanic systems such as Eyjafjallajökull and the Katla systems [14]. Figure 1B shows the values for Storm 2 which originated from the Hagavatn glacial floodplain. The Hagavatn dust materials are more crystalline in nature compared to most other Icelandic dust sources [29]. The higher PM_1 concentrations during the Storm 1 than Storm 2 can be related to this difference in crystallinity and also to the early suspension in June rather than in August when submicron particles had been already removed.

The camera network from the Icelandic Road and Coastal Administration (IRCA) was used to determine the dispersal area of these two dust storms. Figure 4 depicts a time lapse series of photos from one of these web cameras which is located in Sandskeið near Reykjavík (Point 1 at the Figures 1B and 5B). The visibility was reduced down to 1 km during Storm 2. The camera is located about 3.5 km from the mountain and about 1 km from the house shown in the middle of the photos. The corresponding PM_{10} concentrations to these visibility reductions can be obtained from DustTrak measurements close to this location for the left image only. The closest instrument (EAI) is located about 20 km downwind from this camera. The left picture shows the visibility was >3.5 km corresponding to the PM_{10} of 50 $\mu g \cdot m^{-3}$ by the EAI and 71 $\mu g \cdot m^{-3}$ measured using the DustTrak instrument at the site. The middle photo shows the visibility was about 3.5 km, corresponding to the PM_{10} of 70 $\mu g \cdot m^{-3}$, while the photo on the right shows visibility was about 1 km, corresponding to PM_{10} of 100 $\mu g \cdot m^{-3}$. The PM_{10} concentrations for the middle and right picture at the location of the camera can, therefore, only be retrieved using the visibility–dust formula given by Dagsson-Waldhauserova *et al.* [23]. This formula is based on the long-term observations of PM_{10} and visibility in Iceland. Applying the formula to the visibility estimations from the camera, the PM_{10} concentrations are calculated as <190 $\mu g \cdot m^{-3}$ for the left photo, 370 $\mu g \cdot m^{-3}$ for the middle photo, and >780 $\mu g \cdot m^{-3}$ for the photo on the right.

Impaired visibility was observed at all spots where PM measurements were conducted (Figure 5). These images together with the IRCA camera network allowed us to estimate the spatial extent of the dust plumes as depicted in Figure 1. The total land area affected by Storm 1 was about 2450 km^2 but 4220 km^2 for Storm 2. Numerical simulations and operational radiosoundings at Keflavik revealed the thickness of the well-mixed atmospheric boundary-layer as about 1 km in Strom 1 on 15 June with mean winds of about 15 $m \cdot s^{-1}$. The event lasted for about 8 h, giving a total dust flux from source of about 180,000 tonnes. In Storm 2 on 4 August, the boundary-layer thickness was about 1.3 km and the mean winds were 12 $m \cdot s^{-1}$. This event lasted for about 12 h and the total flux from the source is estimated to be about 280,000 tonnes. Both events can thus be characterized as medium-sized (e.g., [22,37]).

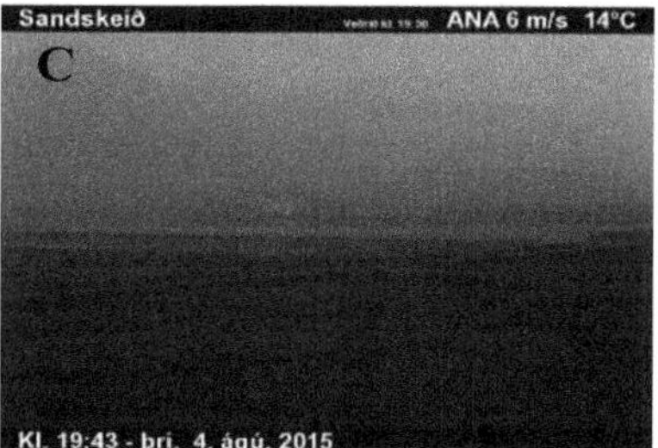

Figure 4. Changes in visibility during Storm 2 on 4 August 2015 when dust was blowing from Hagavatn. Left photo (**A**) shows visibility >3.5 km, corresponding to PM_{10} of 71 $\mu g \cdot m^{-3}$, measured by a DustTrak instrument at the site. The middle photo (**B**) shows visibility of about 3.5 km corresponding to an estimated PM_{10} of 370 $\mu g \cdot m^{-3}$, while the photo on the right (**C**) shows about 1 km visibility, corresponding to an estimated PM_{10} of >780 $\mu g \cdot m^{-3}$ based on the formula from Dagsson-Waldhauserova *et al.* [23]. The images are from a web camera from the Icelandic Road and Coastal Administration.

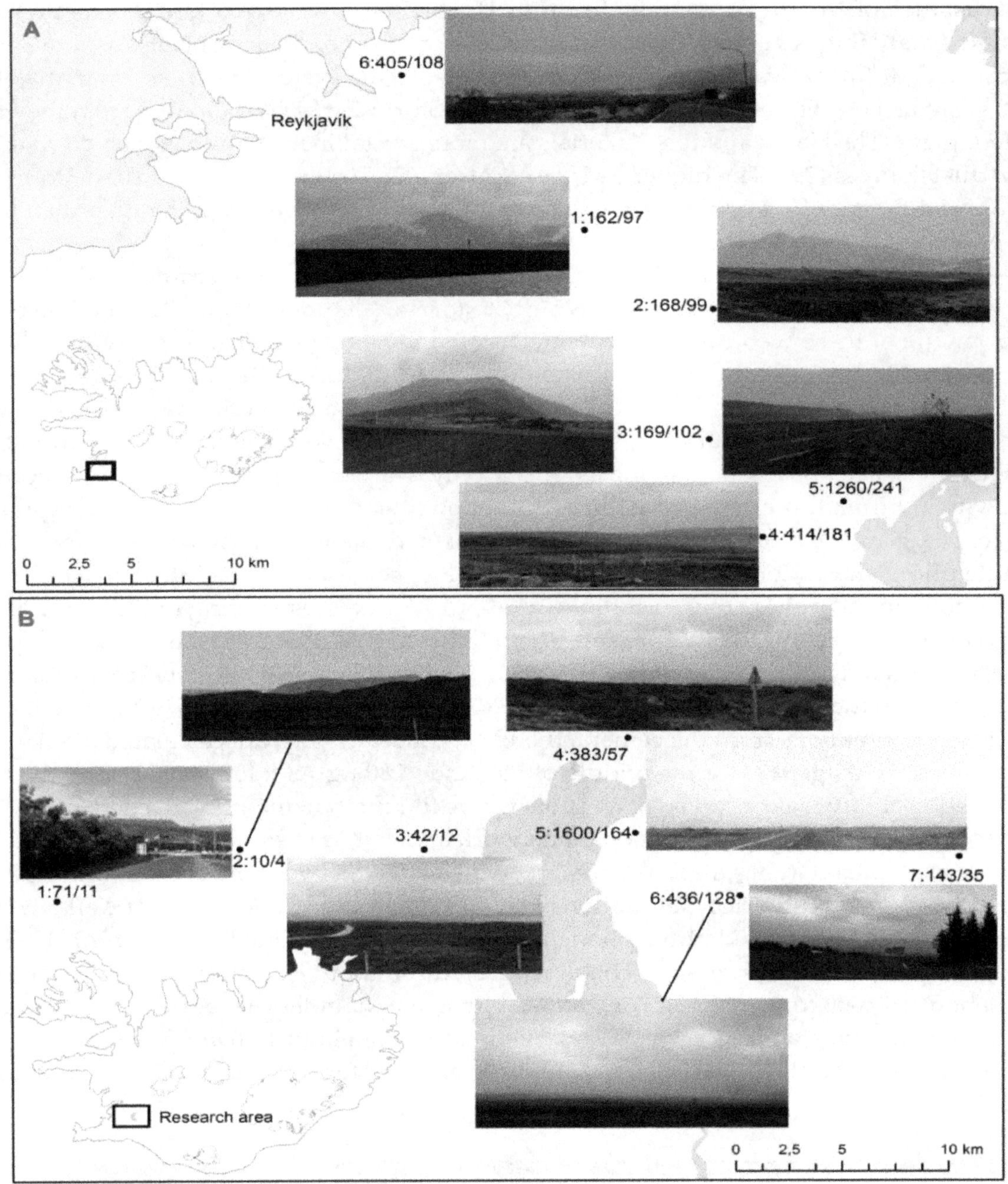

Figure 5. The measurement spots (place number: PM_{15} concentration/PM_1 concentration, $\mu g \cdot m^{-3}$) including a photo from every measurement spot. It can be seen that visibility was more reduced with higher dust concentrations. (**A**) Storm 1 from the Landeyjasandur on 15 June 2015 and (**B**) from Storm 2 on 4 August 2015.

Figure 6 captured how Storm 1 passed relatively densely populated areas such as the town of Hveragerði, population of 2300, and the neighboring town of Selfoss, the eighth largest community in Iceland, with about 6500 inhabitants. The highest PM_{10} concentrations of >500 $\mu g \cdot m^{-3}$ and PM_1 > 200 $\mu g \cdot m^{-3}$ were measured in this area. The margin of the dust plume is very visible on Figure 6A and it can clearly be seen how the visibility changed due to dust in the dust plume. Reduced visibility due to dust in Mosfellsbær (near Reykjavik) is shown on picture 6B. Here the concentrations exceeded 400 $\mu g \cdot m^{-3}$ for PM_{10} and 100 $\mu g \cdot m^{-3}$ for PM_1. The long-term frequency of dust events in Iceland reports about one dust day annually for the capital of Reykjavik [23]. This number of dust storms is highly underestimated judging from our own observations as well as the measurements provided by the EAI. The meteorological stations at the towns of Hveragerði and Selfoss report 3.7 to 6.8 dust days a year.

Figure 6. (**A**) The dust front of Storm 1 approaching to the town Hveragerði, population of 2300, which is near the town Selfoss, the eighth largest community in Iceland, with about 6500 inhabitants. The highest PM concentrations were measured in this area. The margin of the dust plume is very clear. Reduced visibility due to dust in Mosfellsbær (near Reykjavik), overlooking river Leirá, is shown on the right picture (**B**). Here the concentrations exceeded 400 $\mu g \cdot m^{-3}$ for PM_{10} and 100 $\mu g \cdot m^{-3}$ for PM_1.

3.2. Size Partitioning of the PM Components of Icelandic Dust

Mineral dust outbreaks increase both fine and coarse PM concentrations [3,6,10,13,24,38,39]. Table 1 shows the mean PM_{1-15} concentrations at different locations (Figure 1) inside the dust plume. Although the PM_{10} concentrations are moderate (<600 $\mu g \cdot m^{-3}$) for an ongoing dust storm in Iceland, the mean levels of PM_1 are considerably high, such as >97 $\mu g \cdot m^{-3}$ for Storm 1. The proportions of PM_1/PM_{10} are significantly higher for Iceland than for any other dust events we have found in the literature. The PM_1/PM_{10} ratio ranged from 0.41 to 0.63 for Storm 1, and 0.34 to 0.57 for Storm 2, respectively (Table 1). Perez *et al.* [9] reported the PM_1/PM_{10} ratios were relatively stable during African dust outbreaks, where the mean annual average decreased from 0.48 to 0.42 for dust days. Arizona dust outbreaks also had even more stable PM_1/PM_{10} ratios of between 0.17 and 0.22 (0.18 on average) [38]. Claiborn *et al.* [39] reported no increase in PM_1 during dust storms within the USA. In Iceland, the more severe the dust event was the lower PM_1/PM_{10} ratio was observed. The same trend was reported from Iran where the PM_1/PM_{10} ratio decreased from 0.4 to 0.05 during the dust storms with the mean ratio of 0.14. Generally, the PM_1/PM_{10} ratio <0.4 is attributed to the summer season with high dust suspension as summarized from 22 studies on size-segregated particulate matter ratios [40]. These comparisons showed that Icelandic volcanic dust is extremely fine compared to the crustal dust. Such high proportions (>60% of PM_1 in PM_{10}) as obtained during Storm 1 have been reported for urban air pollution, but not for natural dust. The PM_1 proportion of 57%–60% in PM_{10} was found, for example, at four sites in Austria [41] while the PM_1/PM_{10} ratio between 0.45 and 0.74 was found at a polluted urban site of Taipei in Taiwan [42].

Table 1 shows that the $PM_1/PM_{2.5}$ ratio ranged from 0.89 to 0.94, thereby confirming that most of the fine dust particles were of submicron size. This is contrary to what has been reported on such ratios during dust events elsewhere. The $PM_1/PM_{2.5}$ ratio was 0.49 during dust events in the USA [39] while the $PM_1/PM_{2.5}$ was ranging from 0.05 to 0.8, with a corresponding mean value 0.55, during dust storms in Iran [10]. Values such as 90% of $PM_1/PM_{2.5}$ were reported from urban sites or cities such Graz in Austria [41]. High $PM_1/PM_{2.5}$ and PM_1/PM_{10} ratios imply that $PM_{2.5}$ or PM_{10} mainly consist of submicron particles that have a greater health impact than larger particles [6,10,12]. Detailed mineralogical and geochemical analyses of Icelandic dust revealed fine pipe-vesicular structures of volcanic glasses, as known from asbestos and high content of heavy metals [27,28]. Such structured submicron particles can likely have even more destructive effects on human and animal health, as reported by Carlsen *et al.* [30]. Figures 1 and 6 illustrate that the dust plumes with high PM concentrations are passing the most densely inhibited areas in Iceland. The frequency of such events is up to 135 dust days annually in Iceland with many crossing populated areas [23]. Currently, no warnings for the general public are issued.

About 90% of the $PM_{2.5}$ particles were attributed to submicron particle fractions, and the ratios of $PM_{2.5}/PM_{10}$ were similar to the PM_1/PM_{10} (Table 1). The mean $PM_{2.5}/PM_{10}$ ratio was 0.61 for Storm 1 and 0.44 for Storm 2. This is similar to what was found during dust outbreaks in Spain [9], but considerably higher than the 0.3 ratio reported from the USA [39]. The $PM_{2.5}/PM_{10}$ ratios in Iran ranged from 0.1 to 0.5 with the mean of 0.23 [10]. The lower values were related to the high dust season. The high proportion of fine materials in Iceland is attributed to the small grain size at the dust sources, which is a result of glacial action producing fine-grained materials. These materials are further sorted by glacio-fluvial processes at the glacial margins and in glacial rivers [14]. This unceasing glacial and glacio-fluvial action ensures continuous re-supply of the fine grained materials, in contrast to larger aeolian bodies on the continents.

The high ratios of PM_{10}/PM_{15} shown in Table 1 suggest a low proportion of particles >10 μm. However, <50% proportion of PM_{10} on PM_{15} in many cases suggest the presence of larger particles. We did not find any relation between the PM_{10} concentration and PM_{10}/PM_{15} ratio. There is, however, limited information on this size range of PM in the literature [8].

Our results show that Icelandic dust contains fine submicron particles, as was reported by Dagsson-Waldhauserova *et al.* [27]. Suspended dust measured at the Mælifellssandur glacial floodplain resulted in the high proportion of close-to-ultrafine particles, such as 0.3–0.37 μm. Generation of such fine particles is associated with mechanical processes of glaciers and fluvial processes [27]. However, even the finest lacustrine sediments produced from the most active dust hot spots of the world, such as the Bodele Depression in the Sahara or the Sistan Basin in Iran do not consist of such high amounts of submicron particles as observed in Icelandic dust [3,10,13,43,44].

4. Conclusions

The study of two dust events shows the usefulness of combining photos obtained by surveillance cameras and portable dust measurement instruments for identifying the extent, magnitude, and grain size characteristics of single dust storms in Iceland. It shows that common dust storms are of several hundred thousand tons of magnitude from relatively well-defined main dust sources. The *in situ* measurements of two moderate Icelandic dust storms in 2015 show that aeolian dust can be very fine. The study highlights that suspended volcanic dust in Iceland can have extremely high PM_1 concentrations that are comparable to urban pollution in Europe or Asia. The $PM_1/PM_{2.5}$ ratios are generally low during dust storms outside of Iceland, much lower than >0.9 and PM_1/PM_{10} ratios of 0.34–0.63 found in our study. The extremely high proportions of submicron particles are predicted to travel long distances. Moreover, such submicron particles pose considerable health risks because of their high potential for entering the lungs. Icelandic volcanic glass often has fine pipe-vesicular structures, known from asbestos, and has a high content of heavy metals. The two dust events with high PM concentrations reported here passed the most densely inhabited areas of Iceland and influenced an area of 2450 km^2 during Storm 1 and 4220 km^2 in Storm 2. The mean frequency of dust events in Iceland is about 135 dust days annually, however, health risk warnings for the general public are not being issued. The data provided stresses the need for such a warning system and is an important step towards its development. In light of the small size of the dust reported here, in addition to the high frequency of the dust events, it is vital to step up integrated dust and health research in Iceland. Furthermore, dust has influence on weather and climate in general, and the fine fraction of the Icelandic dust has bearings for weather forecast and climate change predictions.

Acknowledgments: This research was funded by the Icelandic Research Fund (Rannis) Grant No. 152248-051 and the Recruitment Fund of the University of Iceland. We acknowledge the Nordic Center of Excellence (NCoE) Top Research Initiative "Cryosphere-atmosphere interactions in a changing Arctic climate" (CRAICC).

We acknowledge that the Icelandic Road and Costal administration allowed us to use images from their web-camera network to use in this research.

Author Contributions: A.M. and O.A. conceived and designed the experiments; A.M. performed the experiments; A.M., H.O., and P.D-W. analyzed the data and contributed reagents/materials/analysis tools; P.D-W., A.M., H.O., and O.A. wrote the paper.

Abbreviations

The following abbreviations are used in this manuscript:

PM	Particulate matter
EAI	Environmental Agency of Iceland
IRCA	Icelandic road and coastal administration

References

1. Perez, L.; Tobias, A.; Querol, X.; Künzli, N.; Pey, J.; Alastuey, A. Coarse particles from Saharan dust and daily mortality. *Epidemiology* **2008**, *19*, 800–807. [CrossRef] [PubMed]
2. Karanasiou, A.; Moreno, N.; Moreno, T.; Viana, M.; de Leeuw, F.; Querol, X. Health effects from Sahara dust episodes in Europe: Literature review and research gaps. *Environ. Int.* **2012**, *47*, 107–114. [CrossRef] [PubMed]
3. Mallone, S.; Stafoggia, M.; Faustini, A.; Gobbi, G.P.; Marconi, A.; Forastiere, F. Saharan dust and associations between particulate matter and daily mortality in Rome, Italy. *Environ. Health Perspect.* **2011**, *119*, 1409–1414. [CrossRef] [PubMed]
4. Stafoggia, M.; Zauli-Sajani, S.; Pey, J.; Samoli, E.; Alessandrini, E.; Basagaña, X.; Cernigliaro, A.; Chiusolo, M.; Demaria, M.; Díaz, J.; *et al.* Desert dust outbreaks in southern Europe: Contribution to daily PM_{10} concentrations and short-term associations with mortality and hospital admissions. *Environ. Health Perspect.* **2015**, *124*, 4. [CrossRef] [PubMed]
5. Hyewon, L.; Ho, K.; Yasushi, H.; Youn-Hee, L.; Seungmuk, Y. Effect of Asian dust storms on daily mortality in seven metropolitan cities of Korea. *Atmos. Environ.* **2013**, *79*, 510–517. [CrossRef]
6. The World Health Organization. *Health Effects of Particulate Matter, Policy Implications for Countries in Eastern Europe, Caucasus and Central Asia*; WHO Regional Publications European Series: Copenhagen, Denmark, 2013; p. 15.
7. Pandolfi, M.; Tobias, A.; Alastuey, A.; Sunyer, J.; Schwartz, J.; Lorente, J.; Pey, J.; Querol, X. Effect of atmospheric mixing layer depth variations on urban air quality and daily mortality during Saharan dust outbreaks. *Sci. Total Environ.* **2014**, *494–495*, 283–289. [CrossRef] [PubMed]
8. Krewski, D.; Jerrett, M.; Burnett, R.T.; Ma, R.; Hughes, E.; Shi, Y.; Turner, M.C.; Pope, C.A.; Thurston, G.; Calle, E.E.; *et al.* Extended follow-up and spatial analysis of the American Cancer Society study linking particulate air pollution and mortality. *Res. Rep. Health Eff. Inst.* **2009**, *140*, 5–114. [PubMed]
9. Perez, N.; Pey, J.; Querol, X.; Alastuey, A.; Lopez, J.M.; Viana, M. Partitioning of major and trace components in PM_{10}, $PM_{2.5}$ and PM_1 at an urban site in Southern Europe. *Atmos. Environ.* **2008**, *42*, 1677–1691. [CrossRef]
10. Shahsavani, A.; Naddafi, K.; Jafarzade Haghighifard, N.; Mesdaghinia, A.; Yunesian, M.; Nabizadeh, R.; Arahami, M.; Sowlat, M.; Yarahmadi, M.; Saki, H.; *et al.* The evaluation of PM10, PM2.5, and PM1 concentrations during the Middle Eastern Dust (MED) events in Ahvaz, Iran, from April through September 2010. *J. Arid Environ.* **2012**, *77*, 72–83. [CrossRef]
11. Carbone, C.; Decesari, S.; Paglione, M.; Giulianelli, L.; Rinaldi, M.; Marinoni, A.; Cristofanelli, P.; Didiodato, A.; Bonasoni, P.; Fuzzi, S.; *et al.* 3-year chemical composition of free tropospheric PM1 at the Mt. Cimone GAW global station—South Europe—2165 m a.s.l. *Atmos. Environ.* **2014**, *87*, 218–227. [CrossRef]
12. Colls, J.; Tiwary, A. *Air Pollution Measurement, Modelling and Mitigation*, 3rd ed.; Routledge: London, UK; New York, NY, USA, 2010.
13. Mahowald, N.M.; Albani, S.; Kok, J.F.; Engelstaeder, S.; Scanza, R.; Ward, S.; Flanner, M.G. The size distribution of desert dust aerosols and its impact on the Earth system. *Aeol. Res.* **2014**, *15*, 53–71. [CrossRef]

14. Arnalds, O.; Dagsson-Waldhauserova, P.; Olafsson, H. The Icelandic volcanic aeolian environment: Processes and impacts—A review. *Aeol. Res.* **2016**, *20*, 176–195. [CrossRef]
15. Nickling, W.G. Eolian sediment transport during dust storms: Slims River Valley, Yukon Territory. *Can. J. Earth Sci.* **1978**, *15*, 1069–1084. [CrossRef]
16. Dornbrack, A.; Stachlewska, I.S.; Ritter, C.; Neuber, R. Aerosol distribution around Svalbard during intense easterly winds. *Atmos. Chem. Phys.* **2010**, *10*, 1473–1490. [CrossRef]
17. Lancaster, N.; Nickling, W.G.; Gillies, J.A. Sand transport by wind on complex surfaces: Field studies in the McMurdo Dry Valleys, Antarctica. *J. Geophys. Res.* **2010**, *115*. [CrossRef]
18. Crusius, J.; Schroth, A.W.; Gasso, S.; Moy, C.M.; Levy, R.C.; Gatica, M. Glacial flour dust storms in the Gulf of Alaska: Hydrologic and meteorological controls and their importance as a source of bioavailable iron. *Geophys. Res. Lett.* **2011**, *38*. [CrossRef]
19. Bullard, J.E. Contemporary glacigenic inputs to the dust cycle. *Earth Surf. Proc. Land.* **2013**, *38*, 71–89. [CrossRef]
20. Lamy, F.; Gersonde, R.; Winckler, G.; Esper, O.; Jaeschke, A.; Kuhn, G.; Ullermann, J.; Martinez-Garcia, A.; Lambert, F.; Kilian, R. Increased dust deposition in the Pacific Southern Ocean during glacial periods. *Science* **2014**, *343*, 403–407. [CrossRef] [PubMed]
21. Arnalds, O.; Thorarinsdottir, E.F.; Thorsson, J.; Dagsson-Waldhauserova, P.; Agustsdottir, A.M. An extreme wind erosion event of the fresh Eyjafjallajokull 2010 volcanic ash. *Nat. Sci. Rep.* **2013**, *3*, 1257. [CrossRef] [PubMed]
22. Dagsson-Waldhauserova, P.; Arnalds, O.; Olafsson, H. Long-term frequency and characteristics of dust storm events in Northeastern Iceland (1949–2011). *Atmos. Environ.* **2013**, *77*, 117–127. [CrossRef]
23. Dagsson-Waldhauserova, P.; Arnalds, O.; Olafsson, H. Long-term variability of dust events in Iceland (1949–2011). *Atmos. Chem. Phys.* **2014**, *14*, 13411–13422. [CrossRef]
24. Thorsteinsson, T.; Gisladottir, G.; Bullard, J.; McTainsh, G. Dust storm contributions to airborne particulate matter in Reykjavík, Iceland. *Atmos. Environ.* **2011**, *45*, 5924–5933. [CrossRef]
25. The World Health Organization. Air Quality Guidelines Global Update 2005. In Proceedings of the Report on a Working Group Meeting, Bonn, Germany, 18–20 October 2005.
26. The Environment Agency of Iceland. *Air Quality in Iceland*. Available online: http://ust.is/the-environment-agency-of-iceland/ (accessed on 14 February 2016).
27. Dagsson-Waldhauserova, P.; Arnalds, O.; Olafsson, H.; Skrabalova, L.; Sigurdardottir, G.; Branis, M.; Hladil, J.; Skala, R.; Navratil, T.; Chadimova, L.; *et al.* Physical properties of suspended dust during moist and low-wind conditions in Iceland. *Icel. Agric. Sci.* **2014**, *27*, 25–39.
28. Dagsson-Waldhauserova, P.; Arnalds, O.; Olafsson, H.; Hladil, J.; Skala, R.; Navratil, T.; Chadimova, L.; Meinander, O. Snow-dust storm: A case study from Iceland, 7 March 2013. *Aeol. Res.* **2015**, *16*, 69–74. [CrossRef]
29. Baratoux, D.; Mangold, N.; Arnalds, O.; Bardintzeff, J.M.; Platevoet, B.; Gregorie, M.; Pinet, P. Volcanic sands of Iceland—Diverse origins of Aeolian sand deposits revealed at Dyngjusandur and Lambahraun. *Earth Surf. Proc. Land* **2011**, *36*, 1789–1808. [CrossRef]
30. Carlsen, H.K.; Gislason, T.; Forsberg, B.; Meister, K.; Thorsteinsson, T.; Johannsson, T.; Finnbjornsdottir, R.; Oudin, A. Emergency hospital visits in association with volcanic ash, dust storms and other sources of ambient particles: A time-series study in Reykjavík, Iceland. *Int. J. Environ. Res. Public Health* **2015**, *12*, 4047–4059. [CrossRef] [PubMed]
31. Blechschmidt, A.-M.; Kristjansson, J.E.; Olafsson, H.; Burkhart, J.F.; Hodnebrog, Ø. Aircraft-based observations and high-resolution simulations of an Icelandic dust storm. *Atmos. Chem. Phys.* **2012**, *12*, 7949–7984. [CrossRef]
32. Branis, M.; Hovorka, J. Performance of a photometer DustTrak in various indoor and outdoor environments. In Proceedings of the Abstracts of the 2005 Evaluations and Assessment Conference (EAC 2005), Ghent, Belgium, 28 September–10 October 2005; p. 535.
33. Wallace, L.A.; Wheeler, A.J.; Kearney, J.; Van Ryswyk, K.; You, H.; Kulka, R.H.; Rasmussen, P.E.; Brook, J.R.; Xu, X. Validation of continuous particle monitors for personal, indoor, and outdoor exposures. *J. Expo. Sci. Environ. Epidemiol.* **2011**, *21*, 49–64. [CrossRef] [PubMed]
34. The Icelandic Road and Coastal Administration. *Road, Web Cameras*. Available online: http://www.road.is/travel-info/web-cams/ (accessed on 4 August 2015).

35. Pey, J.; Rodríguez, S.; Querol, X.; Alastuey, A.; Moreno, T.; Putaud, J.P.; Van Dingenen, R. Variations of urban aerosols in the western Mediterranean. *Atmos. Environ.* **2008**, *42*, 9052–9062. [CrossRef]
36. Theodosi, C.; Grivas, G.; Zarmpas, P.; Chaloulakou, A.; Mihalopoulos, N. Mass and chemical composition of size-segregated aerosols (PM_1, $PM_{2.5}$, PM_{10}) over Athens, Greece: Local versus regional sources. *Atmos. Chem. Phys.* **2011**, *11*, 11895–11911. [CrossRef]
37. Arnalds, O.; Olafsson, H.; Dagsson-Waldhauserova, P. Quantification of iron rich volcanogenic dust emissions and deposition over ocean from Icelandic dust sources. *Biogeosciences* **2014**, *11*, 6623–6632. [CrossRef]
38. Lundgren, D.A.; Hlaing, D.N.; Rich, T.A.; Marple, V.A. PM10/PM2.5/PM1.0 data from a trichotomous sampler. *Aerosol Sci. Tech.* **1996**, *25*, 353–357. [CrossRef]
39. Claiborn, C.S.; Finn, D.; Larson, T.; Koenig, J. Windblown dust contributes to high $PM_{2.5}$ concentrations. *J. Air Waste Manag. Assoc.* **2000**, *50*, 1440–1445. [CrossRef] [PubMed]
40. Speranza, A.; Caggiano, R.; Margiotta, S.; Summa, V.; Trippett, S. A clustering approach based on triangular diagram to study the seasonal variability of simultaneous measurements of PM10, PM2.5 and PM1 mass concentration ratios. *Arab. J. Geosci.* **2016**, *9*, 132. [CrossRef]
41. Gomišček, B.; Hauck, H.; Stopper, S.; Preining, O. Spatial and temporal variation of PM1, PM2.5, PM10 and particle number concentration during the AUPHEP-project. *Atmos. Environ.* **2004**, *38*, 3917–3934. [CrossRef]
42. Li, C.-S.; Lin, C.-H. $PM_1/PM_{2.5}/PM_{10}$ Characteristics in the urban atmosphere of Taipei. *Aerosol Sci. Technol.* **2002**, *36*, 469–473. [CrossRef]
43. Todd, M.C.; Washington, R.; Martins, J.V.; Dubovik, O.; Lizcano, G.; M'Bainayel, S.; Engelstaedter, S. Mineral dust emission from the Bodélé Depression, northern Chad, during BoDEx 2005. *J. Geophys. Res.* **2007**, *112*. [CrossRef]
44. Kaskaoutis, D.G.; Rashki, A.; Houssos, E.E.; Goto, D.; Nastos, P.T. Extremely high aerosol loading over Arabian sea during June 2008: The specific role of the atmospheric dynamics and Sistan dust storms. *Atmos. Environ.* **2014**, *94*, 374–384. [CrossRef]

9

Composition and Sources of Particulate Matter Measured near Houston, TX: Anthropogenic-Biogenic Interactions

Jeffrey K. Bean [1], Cameron B. Faxon [1], Yu Jun Leong [2], Henry William Wallace [2], Basak Karakurt Cevik [2], Stephanie Ortiz [3], Manjula R. Canagaratna [4], Sascha Usenko [3], Rebecca J. Sheesley [3], Robert J. Griffin [2] and Lea Hildebrandt Ruiz [1,*]

1 McKetta Department of Chemical Engineering, The University of Texas at Austin, Austin, TX 78712, USA; jbean15@utexas.edu (J.K.B.); cfaxon1@gmail.com (C.B.F.)
2 Department of Civil and Environmental Engineering, Rice University, Houston, TX 77005, USA; yu.jun.leong@rice.edu (Y.J.L.); wwallaceiv@gmail.com (H.W.W.); bk4@rice.edu (B.K.C.); rob.griffin@rice.edu (R.J.G.)
3 Department of Environmental Science, Baylor University, Waco, TX 76798, USA; Stephanie_Ortiz1@baylor.edu (S.O.); sascha_usenko@baylor.edu (S.U.); rebecca_sheesley@baylor.edu (R.J.S.)
4 Aerodyne Research Inc., Billerica, MA 01821, USA; mrcana@aerodyne.com
* Correspondence: lhr@che.utexas.edu

Abstract: Particulate matter was measured in Conroe, Texas (~60 km north of downtown Houston, Texas) during the September 2013 DISCOVER-AQ campaign to determine the sources of particulate matter in the region. The measurement site is influenced by high biogenic emission rates as well as transport of anthropogenic pollutants from the Houston metropolitan area and is therefore an ideal location to study anthropogenic-biogenic interactions. Data from an Aerosol Chemical Speciation Monitor (ACSM) suggest that on average 64 percent of non-refractory PM_1 was organic material, including a high fraction (27%–41%) of organic nitrates. There was little diurnal variation in the concentrations of ammonium sulfate; however, concentrations of organic and organic nitrate aerosol were consistently higher at night than during the day. Potential explanations for the higher organic aerosol loadings at night include changing boundary layer height, increased partitioning to the particle phase at lower temperatures, and differences between daytime and nighttime chemical processes such as nitrate radical chemistry. Positive matrix factorization was applied to the organic aerosol mass spectra measured by the ACSM and three factors were resolved—two factors representing oxygenated organic aerosol and one factor representing hydrocarbon-like organic aerosol. The factors suggest that the measured aerosol was well mixed and highly processed, consistent with the distance from the site to major aerosol sources, as well as the high photochemical activity.

Keywords: organic aerosol; ambient measurements; diurnal variation; positive matrix factorization

---

1. Introduction

Air quality in the United States has received increased attention in recent years as regulations tighten and cities strive to reduce concentrations of airborne pollutants. Ozone and atmospheric particulate matter (PM) are two pollutants that have received increased attention as health effects become clearer [1]. Particulate matter is linked to a range of respiratory and cardiovascular diseases [2]. High ozone levels can also lead to respiratory problems [3]—especially in more sensitive groups such as children, the elderly, and those with asthma. Many regions struggle to meet compliance with the

National Ambient Air Quality Standard (NAAQS) [4] for ozone and PM set by the U.S. Environmental Protection Agency (EPA).

The U.S. EPA recently lowered the annual NAAQS for $PM_{2.5}$ (particulate matter with diameter below 2.5 μm) from 15 to 12 $\mu g \cdot m^{-3}$ [5]. This new annual standard brings numerous additional metropolitan regions including Houston, TX to near non-attainment for $PM_{2.5}$. This underlines the importance of understanding the composition and sources of $PM_{2.5}$ in these areas. The EPA has also announced that the NAAQS for ozone will be lowered from 75 to 70 ppb [6]—a level that will require action for many metropolitan regions. Houston is an important area for air quality research as the fourth largest U.S. city and one that struggles to meet air quality standards. As a major center for the energy and chemical industry, Houston must continuously inspect, analyze, and improve its air quality in order to stay below the NAAQS and improve the health of its inhabitants. Regional photo-chemical models are used to inform policy makers, but these models must be validated with ambient measurements. Measurements can also be used for source apportionment of air pollution.

Recognizing the importance of ambient measurements, several large-scale ambient measurement campaigns have been conducted in Texas [7]. The biggest campaign was the Texas Air Quality Study in 2000 (TexAQS 2000). A key discovery of this campaign was the important role of highly reactive volatile organic compounds (HRVOCs) in ozone production [8]. The Gulf Coast Aerosol Research and Characterization Study (GC-ARCH), a companion study to TexAQS, was focused on spatial and temporal variability in PM, as well as understanding its formation and transformation in southeast Texas [9]. The TexAQS 2000 campaign was followed up with TexAQS II in 2005–2006, a key finding of which was the magnitude of background concentrations of pollutants in Texas, which adds to the complexity of understanding and improving air quality. The 2009 Study of Houston Atmospheric Radical Precursors (SHARP) campaign uncovered the previously underestimated role of nitrous acid (HONO) in Texas air [10]. Since 2010, many smaller-scale studies in Texas have added to our understanding of the complex effects of oil and gas activity on air quality [11–13]. The amount of effort that has been applied towards understanding air quality in Texas highlights the importance of this research in meeting NAAQS and improving human health.

Previous studies have found that a large fraction of particulate matter in Houston is organic aerosol (OA) [9,14]. Sources of OA in Houston include primary organic aerosol (POA) and secondary organic aerosol (SOA) [15] from urban anthropogenic activity, the petrochemical industry, and fires, as well as SOA from biogenic volatile organic compounds (VOCs) [9,14,16]. Understanding the sources and formation of OA is therefore very complex, and significant uncertainties remain. Conroe, TX, the location of the measurements reported here, is located ~60 km north of the urban center of Houston. The measurement site is in an area that is influenced by anthropogenic emissions from Houston that have been diluted and atmospherically processed since emission. The area is subject to high biogenic emission rates and is located near the start of the piney woods that extend through the US Southeast—a big difference from the grassy prairies that extend west and south throughout Texas. This ecosystem transition near Conroe makes it an interesting place to explore the effects of the ecosystems on observed PM. The interaction of biogenic VOCs and anthropogenic oxidants is very important as it helps explain why radiocarbon analysis in places like the U.S. Southeast show that biogenic (modern) carbon constitutes more than half of SOA, yet SOA correlates with anthropogenic tracers like CO [17]. Recent work [18–20] has begun to explore these interactions and their implications for air quality in places with high levels of biogenic VOCs.

Here we report measurements taken as part of Deriving Information on Surface Conditions from Column and Vertically Resolved Observations Relevant to Air Quality (DISCOVER-AQ) [21] during the period of 24 August–1 October 2013. A main purpose of this most recent large-scale ambient measurement campaign, which was organized through NASA, was to improve the interpretation of ground-level pollutant concentrations from satellite data by taking simultaneous measurements from space, by plane and on the ground. This manuscript focuses on measurements taken at a ground site in Conroe, TX, where various instruments were deployed. The focus of this work is the composition and size distribution of PM_1 (particulate matter with diameter below 1 μm), which was measured with an Aerosol Chemical Speciation Monitor (ACSM) and a Scanning Electrical Mobility Spectrometer (SEMS). The purpose of these measurements was to better characterize the sources and processes which influence the concentrations of PM in this area. An improved understanding of Houston PM is essential in formulating ways to decrease concentrations and more generally manage the air quality in this region.

2. Experimental

2.1. Site Description

The data were obtained at an air quality monitoring ground site in Conroe, TX (30.350278° N, 95.425000° W) situated next to the Lone Star Executive Airport in Montgomery county. The site is located approximately 60 km NNW from the Houston, TX urban center and approximately 125 km NW of the nearest coastline. The area surrounding Conroe, TX is primarily affected by pollution in the outflow of air from Houston, which hosts significant energy and petrochemical industries in addition to a large urban population. The regional atmospheric chemistry is also influenced by marine air from the Gulf of Mexico. The site itself is located in the middle of a field adjacent to the airport, with a gravel parking lot nearby and bordered by trees approximately 200 m to the North. The Conroe region is where the ecosystem transforms from prairie and marsh to piney woods, which then extend north and east through much of the Southeastern United States.

2.2. Instrumentation and Data Analysis

A permanent Texas Commission on Environmental Quality (TCEQ) ambient measurement station exists at this site and provided continuous meteorological data for the duration of the campaign [22]. Measured parameters included wind speed, wind direction, solar irradiance, temperature, and relative humidity. The site also housed NO_x and O_3 monitors, as well as a Tapered Element Oscillating Microbalance (TEOM) for measurements of $PM_{2.5}$ mass concentrations. During DISCOVER-AQ a temporary ground site (an air-conditioned trailer) was set up adjacent to the permanent station. This temporary site housed an NO_2 monitor (Model AS32M from Environnement) which utilizes cavity attenuated phase shift spectroscopy (CAPS) to provide a direct absorption measurement of nitrogen dioxide [23]. NO_x was measured using a chemiluminescence NO_x monitor (Teledyne Model 200E), and O_3 was measured by direct UV absorption (Teledyne, 400E). An Aerosol Chemical Speciation Monitor (ACSM, Aerodyne Research) [24] was used to measure the mass concentrations of non-refractory species in PM_1 including sulfate, nitrate, ammonium, and organics. A Scanning Electrical Mobility System (SEMS, Brechtel Manufacturing) was used to characterize particle size distributions and mass concentrations of PM_1. A High Resolution Time-of-Flight Chemical Ionization Mass Spectrometer (HR-ToF-CIMS, Aerodyne Research) [25–28] was employed to measure gas-phase species. All sample lines extended out the trailer and to a vertical level of approximately 10 feet. Teflon® tubing (1/4 inch) was used to sample all gas-phase compounds and copper tubing (1/2 inch) was used for particle-phase instruments.

Filter measurements of $PM_{2.5}$ were taken on site as described in Section 2.2.4. During approximately 61 h of the campaign the University of Houston-Rice University mobile air quality laboratory (MAQL) was parked at the measurement site. The instrumentation on the MAQL included a suite of photochemical trace gas instrumentation, a photoacoustic spectrometer for measurement of particle-phase polycyclic aromatic hydrocarbons, and a High-Resolution Time-of-Flight Aerosol Mass Spectrometer (HR-ToF-AMS, Aerodyne Research). The HR-ToF-AMS measures the same PM_1 species as the ACSM but in a size-resolved manner based on the vacuum aerodynamic diameter, and the time of flight mass spectrometer enables measurements at much higher mass and time resolution. The co-location of the HR-ToF-AMS and the ACSM enables comparison of PM_1 measurements.

2.2.1. Aerosol Chemical Speciation Monitor

Data analysis and instrument operation were performed in IGOR Pro (WaveMetrics) using the "ACSM Local" software package. The ACSM was set to scan between *m/z* 12 and 159 with a dwell time of 0.5 s, resulting in a scan time of 80 s. The instrument was set to alternate between sampling mode and filter mode, where the filter sample is used to characterize the gas-phase background. This results in a cycle time of 160 s. Further averaging over 25 min intervals was performed in the post-analysis of the data (see Appendix A). The vaporizer temperature was set at 600 °C (as is standard) for fast vaporization of ammonium sulfate. The ACSM measures only non-refractory (NR) PM_1, *i.e.*, compounds that flash vaporize at the heater temperature of 600 °C. Quantification of aerosol concentrations measured by the ACSM is complicated by incomplete transmission of larger particles through the aerodynamic lens and particle bounce at the vaporizer. The ACSM collection efficiency (CE) for these data was estimated to be 0.5, which resulted in good agreement with ancillary measurements (Figure A1). Additional details on instrument calibration, data preparation, and adjustments to the standard fragmentation table [29] are provided in Appendix A. The HR-ToF-AMS operates similarly to the ACSM but at higher mass and time resolution due to its time of flight mass spectrometer (as opposed to the quadrupole mass spectrometer used by the ACSM). Details on the HR-ToF-AMS operation and data collected during the DISCOVER-AQ campaign will be presented in a forth coming publication [30].

The ACSM provides two main measures of PM_1: bulk composition (concentrations of organics, nitrate, sulfate, and ammonium) and the total mass spectrum from which the organic mass spectrum can be derived. The organic mass spectrum can be used to characterize the extent of oxidation of the measured organic aerosol. The organic mass at *m/z* 44 mostly correspond to the CO_2^+ ion [31] and can therefore be used as a semi-empirical measure of the extent of oxidation in the system. Aiken *et al.* [31] showed that f_{44}, the fraction of the total organic signal due to mass at *m/z* 44, can be used to estimate the oxygen to carbon ratio (O:C) in the organic aerosol. The correlation between O:C and f_{44} was recently updated to [32]:

$$(\mathrm{O:C})_{\mathrm{f44}} = 4.31 \times f_{44} + 0.079 \tag{1}$$

Aiken *et al.* [31] also found a significant correlation between the ratio of organic mass to organic carbon (OM:OC) and O:C. This relationship was found to be applicable to field data as well as laboratory data and is described by:

$$(\mathrm{OM:OC}) = 1.29 \times \mathrm{O:C} + 1.17 \tag{2}$$

Thus, the observed f_{44} can be used to estimate O:C and OM:OC of the organic aerosol measured at Conroe. These estimates can be compared with values from the HR-ToF-AMS, which directly computes O:C and OM:OC from elemental analysis of the high resolution measurements (see Appendix B).

2.2.2. Positive Matrix Factorization

Positive Matrix Factorization (PMF) was applied to the organic aerosol mass spectra measured by the ACSM [33]. The PMF2 algorithm (version 4.2) by P. Paatero was used to solve the bilinear unmixing problem as represented and described below. PMF has proven useful in the analysis of ambient organic aerosol data, and details of the mathematical model, its application, output evaluation, and factor interpretation have been described elsewhere [34–38]. A key assumption is that the measured dataset can be separated into a number of constant components (here, ACSM mass spectra) contributing varying concentrations over time. The problem is represented in matrix form by:

$$X = GF + E \tag{3}$$

where X is an m × n matrix of the measured data with m rows of average mass spectra (number of time periods = m) and n columns of time series of each *m/z* sampled (number of *m/z* sampled and fit = n). F is a p × n matrix with p factor profiles (constant mass spectra), G is an m × p matrix with the corresponding factor contributions, and E is the m × n matrix of residuals. G and F are fit to minimize the sum of the squared and uncertainty-scaled residuals [33]. The number of factors is chosen by determining when added factors fail in explaining additional dataset variability.

The ACSM dataset was prepared for PMF analysis by first selecting only organic fragments below *m/z* 100, as higher *m/z* fragments exhibited very low concentrations and added significant error to the analysis. Peaks with a signal to noise ratio below 2 were downweighted by a factor of 2. The peaks, which are calculated from *m/z* 44 (*m/z* 16–18.44), were downweighted to remove the extra influence of *m/z* 44 on PMF solutions. The PMF2 algorithm was used in exploration mode with fpeak set from −1 to 1 by 0.2 increments.

2.2.3. High Resolution Time of Flight Chemical Ionization Mass Spectrometer

The HR-ToF-CIMS was set to alternate between positive (hydronium-water clusters) and negative (iodide-water clusters) chemical ionization in half hour intervals. Hydronium-water cluster ionization is more sensitive than iodide-water cluster ionization to less oxidized compounds such as early oxidation products from terpenes and isoprene. For both cases ultra-high purity N_2 was first passed through water, then across a methyl iodide permeation tube, and then ionized as it passed through a radioactive source of Po-210. The increased humidity helped dampen the effects of the changing RH in the sample gas. Ionized compounds were pulsed in a "V" shape through a time-of-flight region during measurement to obtain a mass spectrum. Some data obtained through iodide ionization have been described previously [39] and here we focus on data from water cluster ionization.

Data from the HR-ToF-CIMS were analyzed in Igor Pro (Wavemetrics) using Tofware, the software provided with the instrument. The data were first mass calibrated based on HR-ToF-CIMS reagent ions and other known ions. The baseline was subtracted and the average peak shape was found so it could be used for high resolution analysis, through which multiple ions can be identified at any given integer mass to charge ratio (*m/z*). Analyte ion concentrations were then normalized by dividing by the reagent ion concentrations, the sum of H_3O^+, $H_3O^+(H_2O)$ and $H_3O^+(H_2O)_2$,and then multiplying by the average sum of the three reagent signals (to maintain the units of ion counts s^{-1}).

2.2.4. Filter Measurements

A high volume $PM_{2.5}$ sampler (Tisch Environmental, Cleves, OH, USA; 226 lpm), on loan from the US EPA, was used to collect daily samples. $PM_{2.5}$ samples were collected over 23.5 h (6 a.m. to 5:30 a.m.). Sample media consisted of quartz fiber filters (QFF) which were baked at 550 °C for 12 h in individual foil packets prior to sampling. QFF were stored in freezers (−10 °C) pre- and post-sampling. $PM_{2.5}$ was collected on 102 mm diameter QFF (Pall Corporation, Port Washington, NY, USA), and samplers were calibrated prior to field deployment. Field blanks were collected throughout the campaign for each type of sampler and handled in the same manner as ambient samples. The QFF were analyzed for organic and elemental carbon (OCEC) using a thermal-optical method (NIOSH-5040) on Baylor University's thermo-optical transmission (TOT) carbon analyzer (Sunset Laboratories, Tigard, OR, USA) [40]. Sample aliquots were also sent to the Desert Research Institute (DRI-Nevada) for inorganic ion analysis (sulfate, nitrate, ammonium, chloride and potassium).

Radiocarbon abundance (^{14}C) was analyzed on filter samples in order to determine the contributions of contemporary and fossil emissions to Houston's ground-based carbonaceous PM. Contemporary sources include biomass burning and biogenic emissions, and they include the presence of ^{14}C. Fossil sources include combustion and non-combustion emission sources with depleted ^{14}C. Ambient $PM_{2.5}$ filter subsamples were taken for analysis to give ~60 µg of total organic carbon [41] for measurement of the ^{14}C signal on the accelerator mass spectrometer. Subsamples were acidified over hydrochloric acid using a desiccator for 12 h to remove carbonate, and dried in a muffler oven at 60 °C for one hour. ^{14}C abundance measurements were performed at the National Oceanic

Sciences Accelerator Mass Spectrometry (NOSAMS) facility at Woods Hole Oceanographic Institute (Woods Hole, MA, USA).

In order to apportion total organic carbon (TOC) based on ^{14}C abundance, $\Delta^{14}C$ end members are chosen based on the sampling region and used in the following equation:

$$\Delta^{14}C_{sample} = \Delta^{14}C_{contemporary} \times f_{contemporary} + \Delta^{14}C_{fossil} \times (1 - f_{contemporary}) \tag{4}$$

The contemporary end member used for this study was 67.5‰, an average of the 2010 biomass burning end member ($\Delta^{14}C$ = 107.5‰) corresponding to wood smoke and the 2010 biogenic end member ($\Delta^{14}C$ = 28‰) corresponding to primary and secondary biogenic emissions, meat cooking and combustion of grass, prunings and agricultural waste [42,43]. The fossil fuel end member was −1000‰ [44]. Results from NOSAMS are reported as % contemporary, with contribution from fossil carbon equaling 1-$f_{contemporary}$.

2.2.5. Diurnal Patterns: Analysis of Statistical Significance and Patterns

We conducted one-way analysis of variance (ANOVA) tests for organics, sulfate, nitrate and each PMF factor as dependent variables with time of day as the independent variable. ANOVA tests determine whether there are statistically significant differences in the mean values of the dependent variables [45]. While ANOVA tests determine statistical significance of variation by time of day, they cannot quantify or characterize the diurnal cycle. Thus, we also conducted harmonic analysis [45,46] to characterize the diurnal cycle. In brief, the general harmonic function is given by:

$$y_t = \overline{y} + C_k cos\,(2\pi t/n - \varnothing_k) \tag{5}$$

where t is the time (1–24 in our diurnal analysis), $\overline{y}$ is the mean of the time series (e.g., y_t is the mean value of f_{44} during hour t, $\overline{y}$ is the mean value for the whole campaign), C_k is the amplitude of the kth harmonic, n is the period (n = 24 here) and φ is the phase. Using only the first harmonic, we can estimate the amplitude by [45,46]:

$$C_1 = \left[A_1^2 + B_1^2\right]^{1/2} \tag{6}$$

where

$$A_1 = 2/n \times \sum y_t cos\,(2\pi t/n) \tag{7a}$$

$$B_1 = 2/n \times \sum y_t sin\,(2\pi t/n) \tag{7b}$$

The phase is then given by:

$$\varphi_1 = tan^{-1}\,(B_1/A_1) \pm \pi \qquad \text{if } A_1 < 0 \tag{8a}$$

$$\varphi_1 = tan^{-1}\,(B_1/A_1) \qquad \text{if } A_1 > 0 \tag{8b}$$

$$\varphi_1 = \pi/2 \qquad \text{if } A_1 = 0 \tag{8c}$$

The portion of the variance explained by the first harmonic, analogous to a correlation coefficient (R^2) commonly computed in regression analysis, is given by:

$$V_1 = C_1^2/2s^2 \tag{9}$$

where s is the standard deviation of the n values. The phase simply describes to what extent the observed cycle is offset from a standard cosine curve. The amplitude describes the magnitude of the diurnal cycle.

3. Results

This work combines PM measurements from several different instruments. Measurements from different instruments generally agreed well as discussed in Appendix B.

3.1. Bulk Concentrations and Diurnal Cycle

Figure 1 shows a time series of particle size distributions (top), a time series of bulk concentrations measured by the ACSM (bottom), and the campaign-average bulk concentration (right). The ACSM nitrate measurements (sum of NO^+ and $NO_2{}^+$ fragments) can be attributed to inorganic nitrate and/or -ONO_2 functional groups on organic nitrates. Measurements indicate that the nitrate measured by the ACSM in Conroe is mostly organic. One indication of this is the NO^+:$NO_2{}^+$ ratio in ACSM measurements, which is estimated from the fragmentation table-corrected unit mass resolution data. In the ACSM used for this study a ratio of 2.6–3.9 has been measured for ammonium nitrate. For organic nitrate this ratio has varied but has always been greater than 5 for this instrument. The average NO^+:$NO_2{}^+$ ratio for this data set is 13.4, which is consistent with organic nitrate (and inconsistent with ammonium nitrate). Filter measurements also indicate that nitrate measured by the ACSM was mostly organic (see Appendix B): filter measurements of inorganic nitrate are significantly lower than ACSM measurements of total nitrate (Figure A4C). Further, the molar ratio of NH_4:SO_4 indicates that on average there was no excess NH_4 as required for the formation of ammonium nitrate.

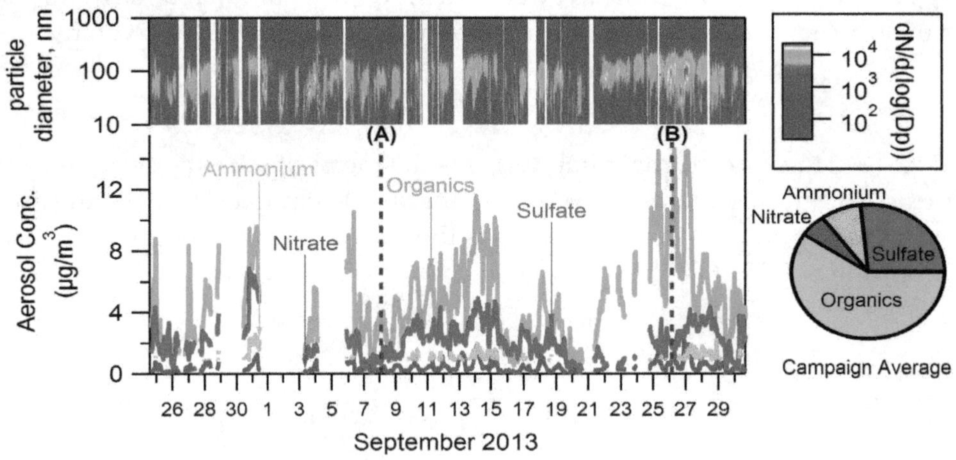

Figure 1. Campaign measurements from ACSM, SEMS.

As seen in Figure 1, a large portion of PM_1 measured in Conroe was organic (64% on average, including nitrate). PM composition from filter measurements agreed with ACSM measurements—71% of PM from the filter samples was OM, with most of the remainder being ammonium and sulfate (as well as 1.7% EC). Sulfate is a significant part of PM_1 in Conroe. Ammonium concentrations were often below the detection limit of the ACSM but when it was measured, the aerosol had an average ammonium/sulfate molar ratio of 2. Figure 1 also shows the PM_1 number distributions from the SEMS. Nucleation events are not easily identified and do not seem to play a major role in PM concentrations in this area during this time period.

Consistent diurnal profiles were seen for both organics and nitrate in PM_1 measurements. Figure 2 shows the average (median) diurnal variation of organics, nitrate, sulfate and total PM_1 measured by the SEMS; the error bars indicate the 25th and 75th percentiles. ANOVA reveals statistically significant variation by time of day for organic, nitrate, and total PM_1 concentrations ($p < 10^{-16}$), but no statistically significant variation by time of day for sulfate concentrations ($p = 0.65$). Harmonic analysis suggests that the phase (between 0 and 2π) for these diurnal trends is 0.4, 0.7, and 0.5 for organics, nitrate, and (SEMS) PM_1 respectively-indicating that concentrations of these species increase and decrease at approximately the same time. The first harmonic explains 78%, 88% and 87% of the variance for organics, nitrate, and (SEMS) PM_1, respectively. The analysis further reveals that the amplitude-to-mean ratio of the nitrate diurnal is 0.55, compared to the amplitude-to-mean ratio of the organics and PM_1 diurnal profiles which are 0.29 and 0.20, respectively.

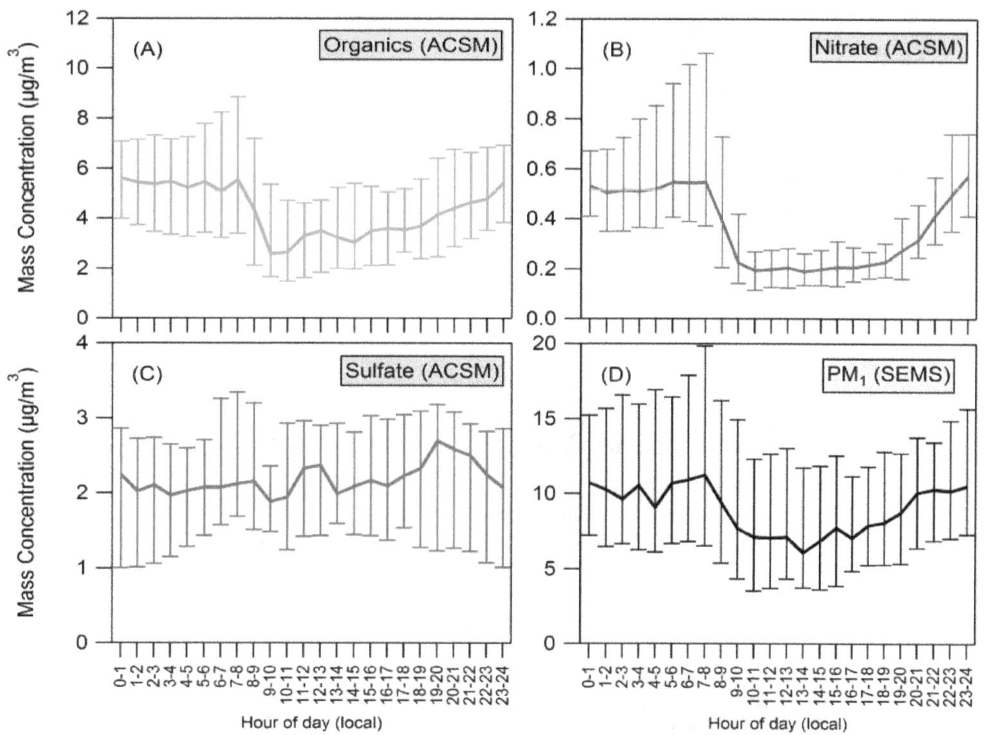

Figure 2. (**A**–**C**) Diurnal plots for organics, nitrate, and sulfate measured by the ACSM; (**D**) PM_1 measured by the SEMS. Median values are plotted, with error bars showing the 25th and 75th percentiles.

3.2. Positive Matrix Factorization

Various PMF solutions (obtained by varying the number of factors and other PMF settings, See Section 2.2.2) were examined and evaluated with respect to mathematical diagnostics and ancillary data (not included in the PMF analysis, e.g., ACSM-sulfate). The three-factor solution was found to best represent these data. Possible solutions of up to 8 factors were considered but factor splitting was observed and no additional information was obtained from the use of more than three factors.

The mass spectra and diurnal cycles of the three factors are shown in Figure 3. Two of the factors resemble oxygenated organic aerosol (OOA), the other factor resembles fresher organic aerosol. We name the more oxidized OAA factor (f_{43} = 4.4%, f_{44} = 22.7%) MO-OOA (more oxidized OOA) and the less oxidized OOA factor (f_{43} = 14.8%, f_{44} = 7.6%) LO-OOA (less oxidized OOA). The third factor has mass spectral signatures representative of hydrocarbon like organic aerosol (HOA) and biomass burning organic aerosol (BBOA), but we refer to the third factor (f_{43} = 4.6%, f_{44} = 3.2%) as HOA for simplicity.

The time series of MO-OOA showed a correlation with the time series of sulfate measured by the ACSM (R^2 = 0.46), whereas LO-OOA did not (R^2 = 0.10). Thus, MO-OOA correlated with a low-volatility inorganic component (sulfate). LO-OOA and HOA showed correlation with NO_x (R^2 = 0.35 and 0.34, respectively), a proxy for fresh anthropogenic emissions, while MO-OOA did not (R^2 = 0.06). We also examined correlations of the factor profiles with factor profiles identified in previous work [47]. The MO-OOA profile correlated most strongly with previously identified MO-OOA (R^2 = 0.92), the LO-OOA profile correlated most strongly with previously identified LO-OOA

(R^2 = 0.92), and the HOA correlated most strongly to previously identified HOA (R^2 = 0.67) and BBOA (R^2 = 0.74).

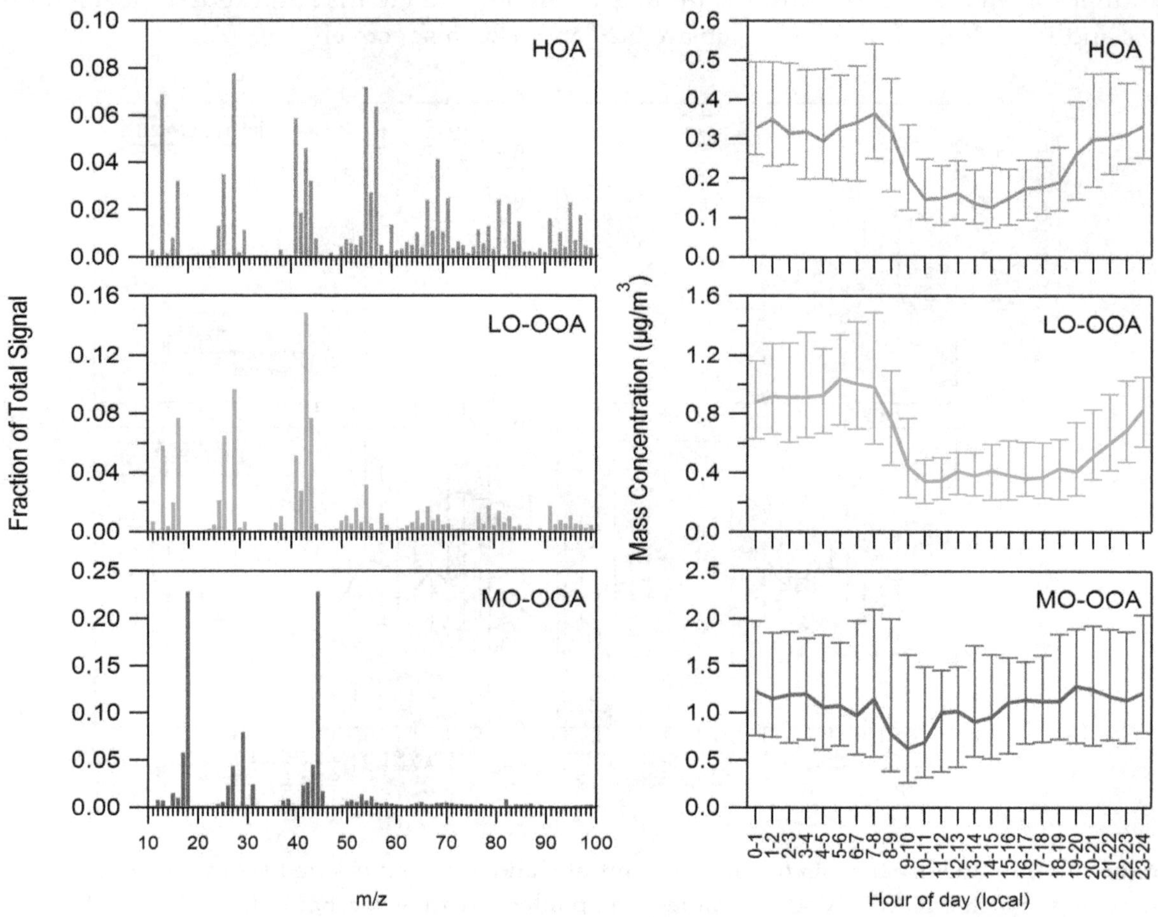

Figure 3. Mass spectra (**left**) and diurnal profiles (**right**) of PMF factors.

Figure 3 (right panel) shows the diurnal cycle of the three PMF factors. According to ANOVA, all three factors exhibited statistically significant variation by time of day ($p < 10^{-16}$ for HOA and LO-OOA, $p = 6 \times 10^{-8}$ for MO-OOA). LO-OOA and HOA exhibited a clear pattern with higher concentrations at night, the same pattern exhibited by total OA (see Section 3.1). MO-OOA did not show this clear pattern, presumably because during the afternoon some LO-OOA and HOA is converted to the MO-OOA, which is more highly oxidized. MO-OOA can also form directly from oxidized VOCs. Harmonic analysis suggests that the diurnal cycle of LO-OOA has an amplitude-to-mean ratio of 0.53 and phase of 0.8 and can explain 84% of the variance; the diurnal cycle of HOA has an amplitude-to-mean ratio of 0.41, phase of 0.5 and can explain 79% of the variance. These two PMF factors (LO-OOA and HOA) hence have diurnal cycles of similar phase, which is also similar to the phase of the diurnal cycle of total OA (Section 3.1).

Figure 4 shows time series of the factors in terms of fraction of total organics (the sum of all 3 factors). The 12 days before 6 September were included in PMF calculations but excluded from Figure 4 to facilitate viewing of radiocarbon results. HOA can constitute 30% or more of OA on days when overall PM concentrations are low (7 September, 16–21 September). However, fresh emissions represented by HOA constitute a smaller fraction (less than 20%) of PM on high concentrations days, such as 10–15 September and 25–28 September. On these higher concentration days a larger fraction of the increased PM levels are due to MO-OOA (and LO-OOA to a lesser extent), consistent with atmospheric conditions which transport highly processed OA or allow existing OA to become highly oxidized. The results of radiocarbon analysis (see Section 2.2.4) are also shown in Figure 4.

Fossil carbon constituted as little as 10% of carbon in OA during the low concentration period from 21–23 September but was approximately 30% of carbon during the high concentration period from 25–28 September, suggesting that a higher fraction of OA originates from fossil sources of carbon on higher concentrations days.

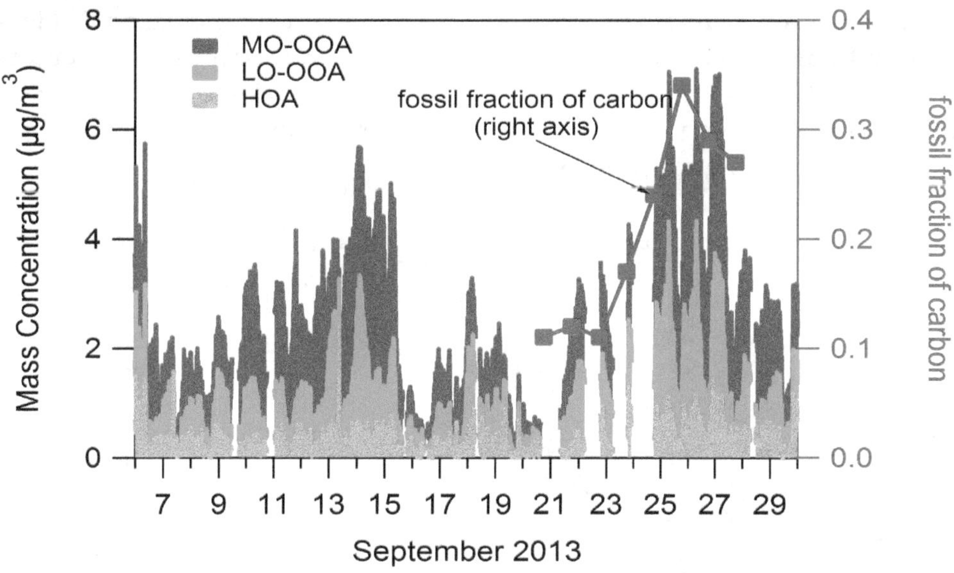

Figure 4. Time series of PMF factors in terms of fractional concentrations.

4. Discussion

4.1. Composition of PM and Source Regions

Because very little inorganic nitrate appears to be present (see Section 3.1), we assume that all nitrate measured by the ACSM is organic in order to estimate the organic nitrate contribution to organic aerosol. For an assumed MW range of 200–300 $g \cdot mol^{-1}$ [48] organic nitrates constitute 27%–41% of organic aerosol. If nitrate was overestimated by up to 60% (Figure A3) then organic nitrates would still constitute 18%–27% of OA. Either estimate would suggest that organic nitrates play a larger role in Conroe than has been measured in other areas. Using the same estimate for molecular weight of organic nitrates, Xu *et al.* [19] estimated that organic nitrates constitute 5%–16% of OA during the summer in Alabama and Georgia locations. Mylones *et al.* [49] assumed an average molecular weight of 150 $g \cdot mol^{-1}$ for organic nitrates and calculated that they are 13% of OA in Los Angeles. Studies in Houston using the same methods as Mylones *et al.* have observed an organic nitrate fraction similar to the one observed in Los Angeles [50,51]. O'Brien *et al.* [52] estimated organic nitrates constituted 17% of OA in Los Angeles in 1975. The prominence of organic nitrates in OA highlights the importance of anthropogenic emissions for this region as nitrate formation requires anthropogenic NO_x or NO_3 in addition to VOCs.

Concentrations of PM_1 vary both diurnally and over the course of several days. Increases and decreases in PM_1 concentrations in the timeframe of days and weeks are often associated with changes in regional air flow. Figure 5 shows 72-h back trajectories calculated using the Hybrid Single Particle Lagrangian Integrated Trajectory Model (HYSPLIT) [53]. Figure 5A,B shows the trajectory for characteristic lower and higher concentration days, respectively. HYSPLIT uses archived meteorological data to compute the back trajectory of a particle or parcel of air which arrives at a location at a specified time. The trajectory ensemble method is used, in which grid points are offset by small amounts to produce multiple potential trajectories as shown in Figure 5. Back trajectories for times indicated by vertical dashed lines "A" and "B" in Figure 1 are shown in Figure 5A,B, respectively. The air source of a characteristic high concentration day is slower moving continental air (Figure 5B) while a low concentration day is supplied with quickly moving oceanic air with significant vertical mixing (Figure 5A). PMF results (Section 3.2) suggest that high concentration days are the result of increased levels of OOA but not HOA. The 7-day period of radiocarbon results (Figure 4) shows higher portions of fossil carbon in OA during times of increased concentrations. The fact that fossil carbon increases along with MO-OOA and LO-OOA while HOA does not suggests that oxidized anthropogenic emissions are a larger contributor during this time of increased PM levels, consistent with transport of pollutants from the Houston metropolitan and/or industrial centers. On average, 87% of the measured PM_1 organics was due to OOA, which is representative of organic aerosol that has been processed in the atmosphere, highlighting the importance of atmospheric processing on controlling fine PM concentrations in Conroe.

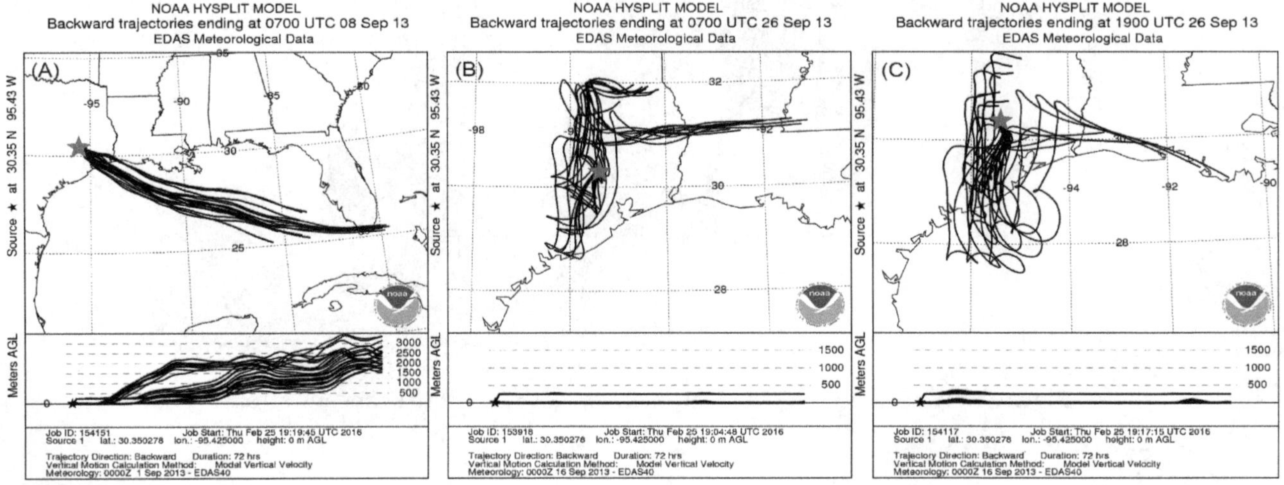

Figure 5. HYSPLIT 72-h back trajectories showing the differences between lower (**A**) and higher (**B**) concentration days. The low and high trajectories correspond to 2:00, 8 September and 2:00 26 September, respectively. Seventy-two-hour back trajectories are also shown for 14:00 26 September (**C**), so (**B**,**C**) show a typical 12 h difference.

While the source of air mass can explain variation in OA over the course of days and weeks, it does not adequately explain the consistent diurnal variation that was observed. HYSPLIT back-trajectories (Figure 5B,C) show that there are often only small differences between day and night air sources. During the measurement campaign, the average nighttime (0:00–6:00) winds were more easterly (average 48°) and daytime (12:00–18:00) winds were more southerly (average 137°). Daytime winds were typically stronger (average 6.3 miles/h) than nighttime winds (average 2.7 miles/h). Despite these differences between day and night Figure 6 shows that there is significant variation in wind speed and direction during the day and night. This variation suggests that regional air flow is not a main factor in the observed diurnal cycle. In addition, the higher nighttime concentrations were observed when wind was predominantly from the east, which would likely be a cleaner air mass than the daytime, southeastern winds which pass through Houston. Thus, for Conroe the source of the air mass appears to play a large role in multi-day and weekly high and low concentration trends but has significantly less influence on the daily trends in OA levels.

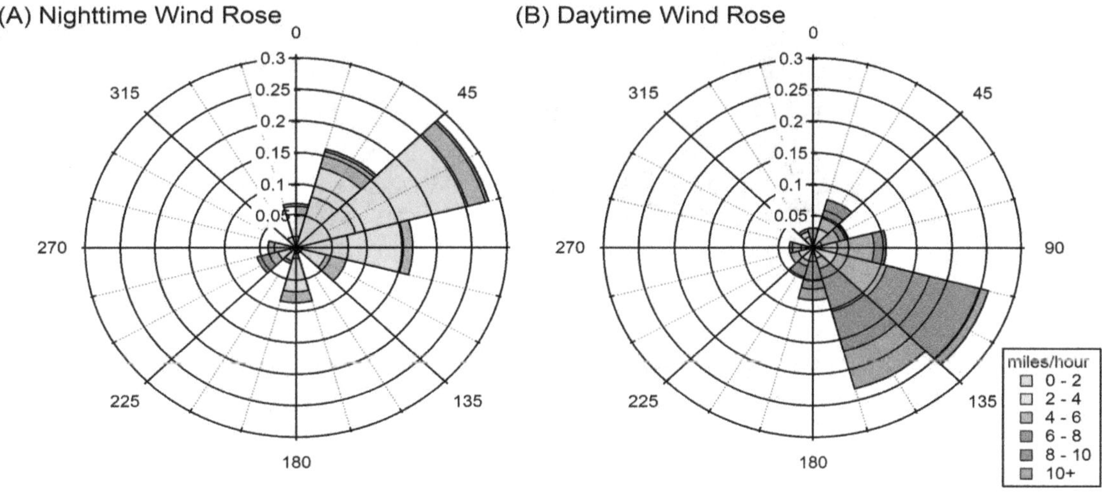

Figure 6. Wind rose plots for (**A**) night (0:00–6:00) and (**B**) day (12:00–18:00) averaged over the course of the campaign.

4.2. Influences on Diurnal Cycle

The strong diurnal cycle for organics and nitrate but lack of diurnal cycle for sulfate is consistent with a more regional source of sulfate and a more local source of organics and organic nitrates. The higher amplitude-to-mean ratio of nitrate indicates that the diurnal trend is especially prominent for nitrate and suggests that organic nitrates play a large role in the observed diurnal trends. The more pronounced diurnal changes in organic nitrates could be attributed to evaporation with increasing daytime temperatures or nighttime growth due to NO_3 chemistry. NO_3, formed from the reaction of NO_2 and O_3, is considered a night-time oxidant because it photolyzes quickly during the day.

The shape of the diurnal profile shows highest concentrations at night and quick decreases in concentration during daylight hours. Photo-oxidation of organics predominantly decreases their vapor pressure and can result in overall increases of organic particulate matter during the day. The diurnal cycle of the organic aerosol O:C ratio (Figure 7) suggests that organic aerosol is more oxygenated during daylight hours, as expected with increasing photochemical activity. PMF results also support this, as daytime decreases are seen in HOA and LO-OOA, potentially indicating conversion to MO-OOA. However, total organic concentrations also decrease during this time, indicating that photochemical activity is not the main factor affecting concentrations of OA.

Some meteorological factors may play a significant role in diurnal trends. Temperature and boundary layer height (BLH) effects on concentrations of PM are explored in Figure 8. Temperature increases when the sun rises, increasing the saturation vapor pressure of particle-phase compounds which causes the higher vapor-pressure species to evaporate. According to absorptive partitioning theory [54,55], the gas-particle partitioning of an organic species depends on its vapor pressure and the concentration of organic material already in the condensed phase. The fraction of a compound *i* in the particle phase (Y_i) is given by [55]:

$$Y_i = \left(1 + \frac{C_i^*}{C_{OA}}\right)^{-1} \tag{10}$$

where C_i^* is a function of the vapor pressure, If C_i^* is known at one temperature then it can be predicted at a second temperature if ΔH_{vap} is known using the Clausius-Clapeyron equation [48]:

$$C^*\left(T_2\right) = C^*\left(T_1\right) exp\left(\frac{\Delta H_{vap}}{R}\left(\frac{1}{T_1} - \frac{1}{T_2}\right)\right) \tag{11}$$

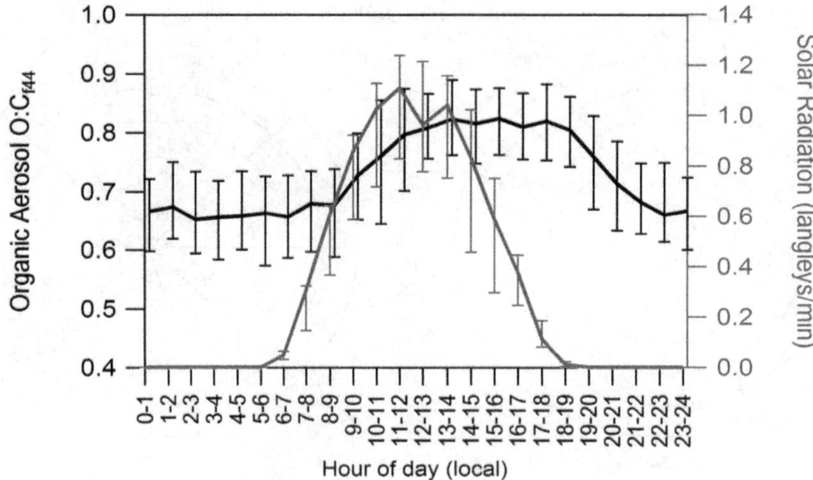

Figure 7. Diurnal cycle of organic aerosol oxygen to carbon ratio (O:C_{f44}) and solar radiation.

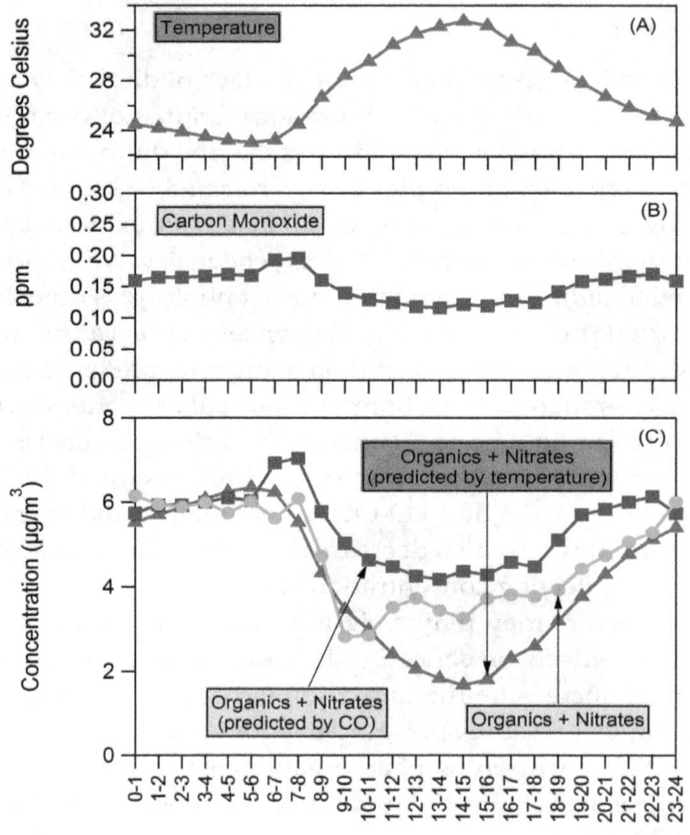

Figure 8. Diurnal profiles of (**A**) temperature and (**B**) carbon monoxide. (**C**) shows the diurnal profile of OA as well as the diurnal profile of OA predicted by temperature and boundary layer height changes. Starting concentrations in predictions are the average of measured concentrations between 0:00 and 6:00.

In a simplified yet illustrative calculation, we assume that $\Delta H_{vap} = 40\ kJ \cdot mol^{-1}$ [55] and assume an initial set of C_i^* values of those used by Murphy and Pandis [56] for high-NO_x terpene SOA. We assume this set of C_i^* values applies to the average OA concentration (including both the organics and nitrate measured by the ACSM) measured from 0:00 to 6:00 (the time period when concentrations and temperatures were stable) at the average temperature from 0:00 to 6:00 and then calculate the expected

OA concentrations, based on observed temperature changes, for the entire day as seen in Figure 8C. Though these assumptions greatly simplify evaporation behavior of the organic aerosol, it is illustrative to see that the resulting predicted OA concentrations match the observed trend. Temperature is likely to be partially responsible for the observed diurnal cycle.

The effect of BLH on atmospheric mixing may also play a role, and a similar estimate was performed to illustrate the effect this might have. Carbon monoxide (CO) measurements from the TCEQ Jones Forest site (~10 miles southwest of Conroe site) were used for an estimate of mixing effects. The average CO and OA concentrations between 0:00 and 6:00 were used as a baseline, and then this baseline was diluted or concentrated based on the CO concentrations as seen in Figure 8C. The increase in predicted OA from 6–8 a.m. is most likely a reflection of traffic conditions, but otherwise the CO-predicted OA concentrations match the trend of measurements and show the effect that BLH and mixing may have had on daytime concentrations as the air mass was diluted. Tucker *et al.* [57] observed that BLH effects on pollutant concentrations in the Houston region are complicated and depend on many factors including the location of source air and turbulence levels. The lack of significant diurnal variation of sulfate could suggest that the effect of BLH on observed concentrations is lower than suggested by our analysis, and/or that BLH and daytime oxidation of SO_2 leading to sulfate had opposite effects. It is also consistent with a more regional source of sulfate and similar concentrations above and below the boundary layer. Daytime oxidation of organics is also expected to increase concentrations of OA and partially offset the changes due to temperature and BLH. Nonetheless, the shape of the organic aerosol diurnal variation is consistent with changes in either BLH or temperature, and both are potential influences on the observed diurnal trend.

In order to further explore this diurnal trend we considered $PM_{2.5}$ (TEOM) measurements at four Houston area sites operated by the TCEQ [22]. Figure 9 shows diurnal cycles of $PM_{2.5}$ measurements averaged over the month of September and January taken at Conroe (Figure 9A, same site as our location), Kingwood (Figure 9B, midway between Conroe and downtown Houston), Clinton (Figure 9C, downtown Houston location), and Fayette County (Figure 9D, a rural Texas location). Figure 9 shows that the diurnal trend observed in Conroe is not specific to that area; similar patterns are observed in all three of the other areas, and use of ANOVA reveals statistically significant variation by time of day at all four of these monitoring sites. A similar trend is seen in the winter at all locations as also shown in Figure 9. This indicates that colder temperatures and shorter daylight hours do not eliminate the trend, though the decrease in PM concentration starts later in the day.

Variation in the hourly averaged $PM_{2.5}$ concentrations measured at the Conroe location are reasonably well described with first order harmonic analysis, which can explain 54% of the variation in TEOM readings. Only 15%, 14% and 6% of the variation is explained by first-order harmonic analysis for Clinton, Kingwood, and Fayette County, respectively. We estimate the magnitude of the diurnal cycle as $(\mathrm{Avg}_{\mathrm{high}} - \mathrm{Avg}_{\mathrm{low}})/\mathrm{Mean}$, where $\mathrm{Avg}_{\mathrm{high}}$ and $\mathrm{Avg}_{\mathrm{low}}$ are calculated as the mean of the six highest and lowest concentrations in the diurnal trend, respectively. The magnitude of diurnal variation using this method is 0.57 for Conroe and 0.45, 0.39, and 0.39 for Kingwood, Clinton, and Fayette County, respectively, suggesting that $PM_{2.5}$ concentrations measured at Conroe exhibited the strongest diurnal cycle. It is notable that neither Clinton, which has the most anthropogenic influence, nor Fayette County, which has the least anthropogenic influence, has the most pronounced diurnal profile. Distance from the coast can affect diurnal temperature patterns, with coastal areas having milder temperature swings. However, TCEQ data show that the Conroe and Fayette County sites have nearly identical diurnal temperature profiles, suggesting that more than temperature is needed to describe the observed diurnal trend in organic aerosol concentrations. The strong diurnal cycle in Conroe may in part be due to the interaction of anthropogenic oxidants with biogenic hydrocarbons. Vegetation type may play a large role in this as the Conroe area is where the ecosystem transforms from prairie and marsh to the piney woods of the US Southeast, which are known to have higher biogenic emissions [58]. Xu *et al.* [19] saw a similar diurnal pattern for a less oxidized OA factor in PMF analysis from measurements in Alabama and Georgia, places that have a piney woods ecosystem.

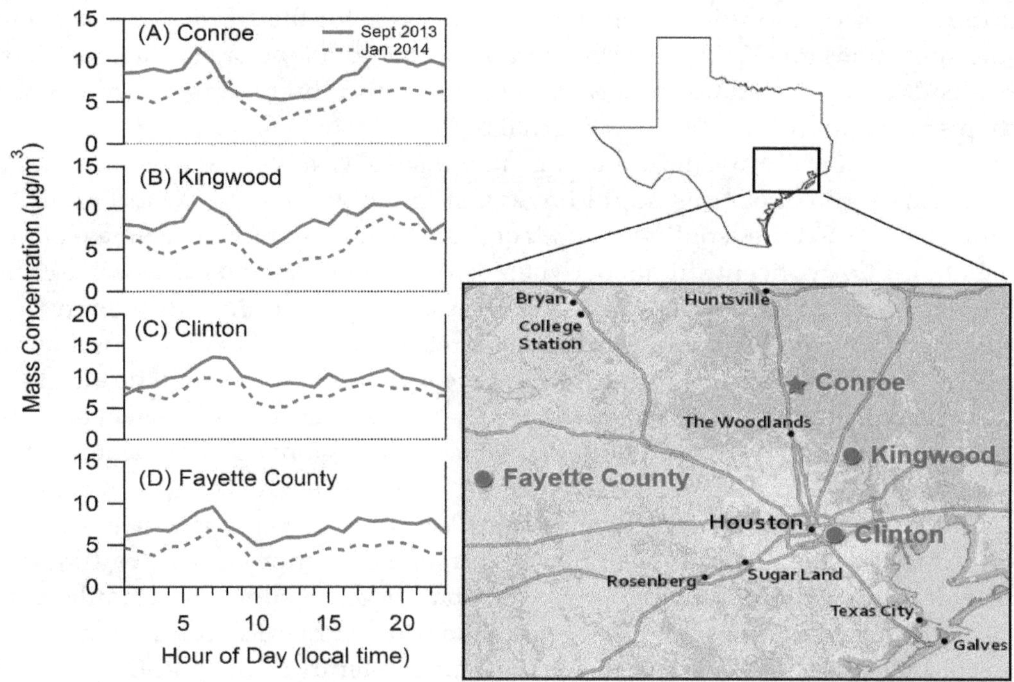

Figure 9. Diurnal plots for Conroe and neighboring areas based on tapered element oscillating microbalance (TEOM) measurements of $PM_{2.5}$ by the Texas Commission on Environmental Quality (TCEQ). (**A**) Conroe, (**B**) Kingwood, (**C**) Clinton, (**D**) Fayette County.

Observations from the HR-ToF-CIMS support the hypothesis that biogenic VOCs are an important contributor to the diurnal cycle seen in OA in Conroe. Recent work [19,20] has suggested that monoterpene reactions with the NO_3 radical at night are a significant source of SOA. Gas-phase organic nitrates observed during DISCOVER-AQ, which likely formed from monoterpenes (C_{10}), also exhibited a diurnal trend of elevated concentrations at night. Data from the HR-ToF-CIMS in Figure 10 show increased levels of monoterpene organic nitrates at night, the time when monoterpene and NO_3 concentrations are typically highest [59]. Increasing concentrations are also seen in the time just after sunrise when monoterpene concentrations are still high and NO concentrations are increasing. Lee *et al.* [59] observed a similar increase in gas phase concentrations of biogenic organic nitrates in the hours following sunrise in the SOAS campaign.

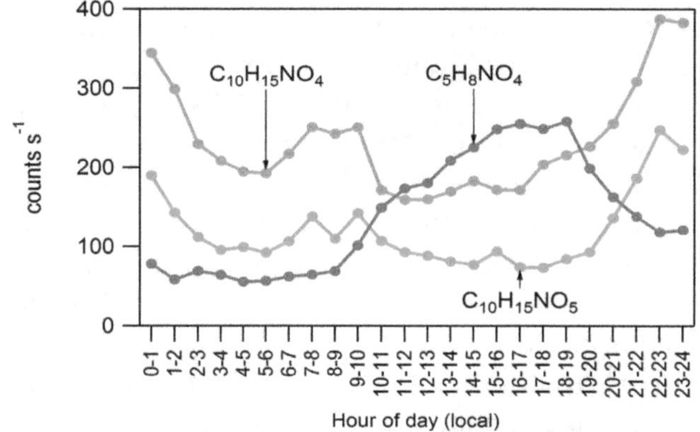

Figure 10. Diurnal plot of gas-phase organic nitrates which appear to have formed from biogenic compounds. These HR-ToF-CIMS data are taken from the high-concentration period of 11–14 September.

The SIMPOL.1 model [60] was used to estimate changes in volatility from the oxidation of VOCs. The SIMPOL.1 model predicts vapor pressure based on molecular functional groups and in this case was used to predict changes in vapor pressure to biogenic VOCs due to the addition of nitrate and hydroxyl functional groups. Using Equation (10) and conditions in Conroe we find that α-pinene or β-pinene which have been oxidized to $C_{10}H_{16}NO_4$ would partition less than 1% to the particle phase. $C_{10}H_{16}NO_5$ would partition 11%–25%, and $C_{10}H_{16}NO_6$ would partition 95%–98% to the particle phase. Thus, $C_{10}H_{16}NO_6$ and compounds that are more oxidized are not likely found at high concentrations in the gas phase but may be important components of the particle phase as was observed by Lee *et al.* [59]. Most oxidized gas-phase hydrocarbons observed with the HR-ToF-CIMS have a diurnal cycle with elevated daytime concentrations due to photochemistry similar to $C_5H_8NO_4$ shown on Figure 10. $C_5H_8NO_4$ is likely an isoprene hydroxyl-nitrate that is formed through an isoprene peroxy-radical and NO, a similar mechanism to the early morning formation of monoterpene nitrates but different from the night-time monoterpene nitrate formation mechanism through reaction with NO_3. Isoprene nitrates could be partially responsible for the diurnal trend in OA in the region. Though they mostly form during the day when isoprene concentrations are highest, according to the SIMPOL.1 model a compound such as $C_5H_8NO_6$ would partition 15% to the particle phase during the day but nearly double that (28%) during the night, which would increase total organic nitrate concentrations.

5. Conclusions

Measurements were taken in Conroe, TX during 24 August–1 October 2013, as part of DISCOVER-AQ. Organic aerosol (OA) was a major component of the measured particulate matter, constituting 64% of PM_1, and up to 41% of the measured OA in the region was organic nitrates. Through PMF the OA was divided into three factors: Two factors were classified as OOA—one more oxidized (MO-OOA) and one less (LO-OOA). A third factor, named HOA, had similarities to hydrocarbon-like OA and biomass burning OA. The LO-OOA and the HOA displayed diurnal cycles in which concentrations increased in the evening and decreased in the morning. This pattern was also seen in the bulk ACSM measurements of organics and nitrate. Night-time chemistry between biogenic compounds (isoprene, terpenes) and anthropogenic oxidants (O_3, NO_3) appears to contribute to this variation. Temperature and changes in boundary layer height also appear to contribute to the trend.

Understanding diurnal and multi-day trends in PM levels is crucial as regions continue to strive to achieve lower PM levels. Both the anthropogenic and biogenic drivers that cause concentrations to fluctuate need to be understood and correctly modeled for policy-makers to make informed decisions about regulations. Decreasing PM formation can be especially challenging in locations such as Conroe, TX where organic aerosol formation appears to be strongly influenced by the interaction of biogenic hydrocarbons and anthropogenic oxidants.

Acknowledgments: This work was financed in part through six grants from the Texas Commission on Environmental Quality (TCEQ), administered by The University of Texas through the Air Quality Research Program (Projects 12-012, 12-032, 13-022, 14-009, 14-024 and 14-029). The contents, findings opinions and conclusions are the work of the authors and do not necessarily represent findings, opinions or conclusions of the TCEQ. The authors would like to thank the US EPA for the loan of the PM2.5 high volume sampler. The authors gratefully acknowledge the NOAA Air Resources Laboratory (ARL) for the provision of the HYSPLIT transport and dispersion model and/or READY website (http://www.ready.noaa.gov) used in this publication.

Author Contributions: Jeffrey Bean, Cameron Faxon, and Lea Hildebrandt Ruiz planned, prepared and conducted the ambient measurements. Jeffrey Bean analyzed the ambient ACSM and SEMS data presented in this work. Under the guidance of Robert Griffin, Yu Jun Leong, Basak Karakurt Cevik, and H. William Wallace operated the HR-ToF-AMS and analyzed the subsequent data. Under the guidance of Rebecca Sheesley and Sascha Usenko, Stephanie Ortiz analyzed the filter samples, which were collected by Jeffrey Bean and Cameron Faxon. Under the guidance of Lea Hildebrandt Ruiz and Manjula Canagaratna, Jeffrey Bean conducted and interpreted PMF. The manuscript was written by Jeffrey Bean with the guidance of Lea Hildebrandt Ruiz.

Appendix A: ACSM Calibration and Data Preparation

A1. ACSM Calibration

The nitrate ionization efficiency (IE) of the ACSM, as well as the relative ionization efficiencies (RIEs) of sulfate and ammonium were measured four times between 24 August and 30 September (the time period during which ACSM data were acquired) using dried ammonium nitrate and ammonium sulfate particles with a diameter of 300 nm. The ratio of IE to the MS airbeam (AB) was constant for these calibrations (within noise), so the average IE/AB value of 3.29×10^{-11} Hz^{-1} was used for the whole campaign, and the IE was determined at any point by multiplying IE/AB by the current AB. The RIE of ammonium measured during the IE calibrations ranged from 4.57 to 5.82, and the measured RIE of sulfate ranged from 0.49 to 0.67. The variation in the values appeared random; therefore the average values of 5.02 and 0.57 were used for the entire campaign for ammonium and sulfate, respectively. The flow rate in the ACSM was 100 $cm^3 \cdot min^{-1}$. Lens alignment and flow calibrations were performed at the beginning of the campaign.

A2. Adjustments to the Standard Fragmentation Table

The collected data were analyzed using a standard AMS fragmentation table and batch table [29], with a few modifications: The fragmentation patterns of air at *m/z* 44 (CO_2^+), *m/z* 29 ($N^{15}N^+$) and *m/z* 16 (O^+) were evaluated using filter data that were collected continuously throughout the campaign. $N^{15}N^+$ and CO_2^+ were calculated as constant fractions of the N_2^+ signal at m/z 28; the calculated fractions were 7.3×10^{-3} and 1.2×10^{-3} for $N^{15}N^+$ and CO_2^+, respectively. O^+ was calculated as a constant fraction of N^+; the calculated ratio was 0.48. The correction for CO_2^+ from air using the N_2^+ signal was calculated by averaging the filter measurements throughout the campaign when particle-phase organics were below 1 $\mu g \cdot m^{-3}$ in order to avoid interference of organics being interpreted as CO_2^+ from air. The correction for $N^{15}N^+$ was calculated as an average of all filter data throughout the campaign.

A3. Data Averaging

For bulk composition analysis (organics, sulfate, ammonium, nitrate), every 10 ACSM data points were averaged, resulting in a time resolution of approximately 25 min (including 12.5 min of averaged sample and 12.5 min of averaged filter data), and 1475 data points throughout the campaign. (ACSM measurements were taken 24 August–30 September 2013). The following detection limits were then calculated considering the 12.5 min sample averaging time [24]: 0.440 $\mu g \cdot m^{-3}$ (ammonium), 0.229 $\mu g \cdot m^{-3}$ (organics), 0.037 $\mu g \cdot m^{-3}$ (sulfate), 0.017 $\mu g \cdot m^{-3}$ (nitrate). Application of the detection limits resulted in removal of 68% of the ammonium data, no removal of sulfate data, and removal of 0.3% and 0.7% of organics and nitrate data, respectively. A 42-h period (4–5 September) in which the airbeam was abnormally high and a 68-h period (1–3 September) during which the vaporizer temperature was set to 700 °C were also removed. The f_{44} data were cleaned as follows: first, every five data points were averaged. Then datapoints for which $f_{44} < 0$ or $f_{44} > 1$ were removed since these are not physically possible. (7 data points were below zero, and 1 data point was above 1). Then, every four data points were averaged again for an overall time resolution of approximately 50 min. Then data were removed for which the signal of organics at *m/z* 44 (*i.e.*, $f_{44} \times$ org) was below the detection limit of organics for the 25 min averaging time. This resulted in removal of 17% of the final averaged data.

Appendix B: Comparison of Co-Located Instruments

Measurements across different instruments generally agreed throughout the campaign. In Figure A1 we show the comparison between PM_1 mass concentrations measured by the ACSM (corrected for CE) and by the SEMS. The volume concentration from the SEMS was converted to mass concentration using the density 1.77 $g \cdot cm^{-3}$ for ammonium and sulfate and 1.4 $g \cdot cm^{-3}$ for organics and nitrate [61]. On average the SEMS measured higher PM_1 mass (slope = 1.35), which could be due to uncertainties in the density estimate and the SEMS measurement including refractory compounds which are not measured by the ACSM.

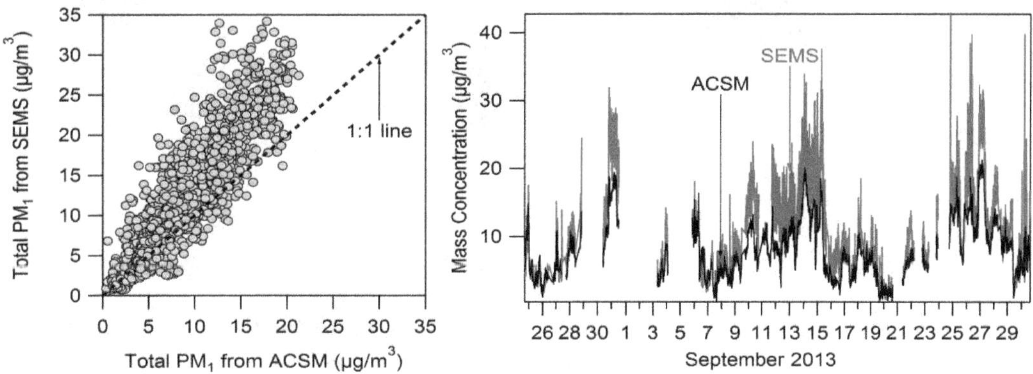

Figure A1. Comparison of PM_1 mass measured by the SEMS and ACSM throughout the campaign.

Figure A2 compares O:C and OM:OC measurements and estimations by the ACSM and HR-ToF-AMS when the HR-ToF-AMS was at the Conroe site (see Section 2.2.1). There is relatively good agreement in f_{44} and total organic aerosol mass (Figures A2A and A3A); however, the O:C calculated from measured f_{44} using Equation (1) (O:C$_{f44}$) is significantly higher than the O:C calculated from elemental analysis of the high resolution HR-ToF-AMS spectra (Figure A2B). Despite this difference the calculated OM:OC is similar to the OM:OC from elemental analysis of the co-located HR-ToF-AMS (Figure A2C). This is important as OM:OC is used to convert filter measurements of organic carbon to organic mass (as described below).

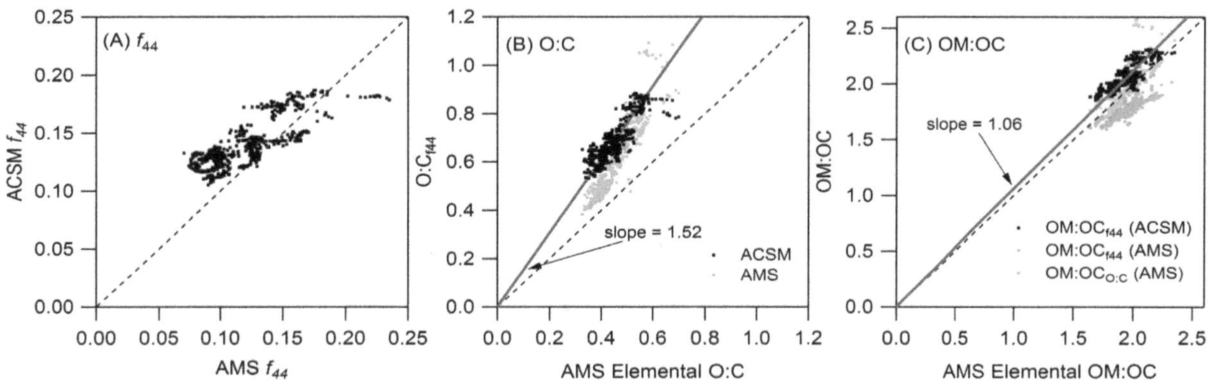

Figure A2. Comparison of (**A**) f_{44}; (**B**) O:C and (**C**) OM:OC between the HR-ToF-AMS and ACSM.

ACSM PM_1 measurements are also compared with $PM_{2.5}$ measurements from filter samples (see Section 2.2.4) and the TCEQ-operated TEOM (see Figure A4). The filter measurements of OC are converted to organic mass using the calculated OM:OC ratio described in Section 2.2.1. In general, measurements from the filters are consistent with those from the ACSM, suggesting that the majority of the mass in $PM_{2.5}$ is found in particles with a diameter below 1 μm (Figure A4A,B,D). The total concentrations from the ACSM and filter measurements are also consistent with the TCEQ-operated TEOM (Figure A4D). The total filter measurement also includes elemental carbon (about 7% of total carbon), which is not measured by the ACSM.

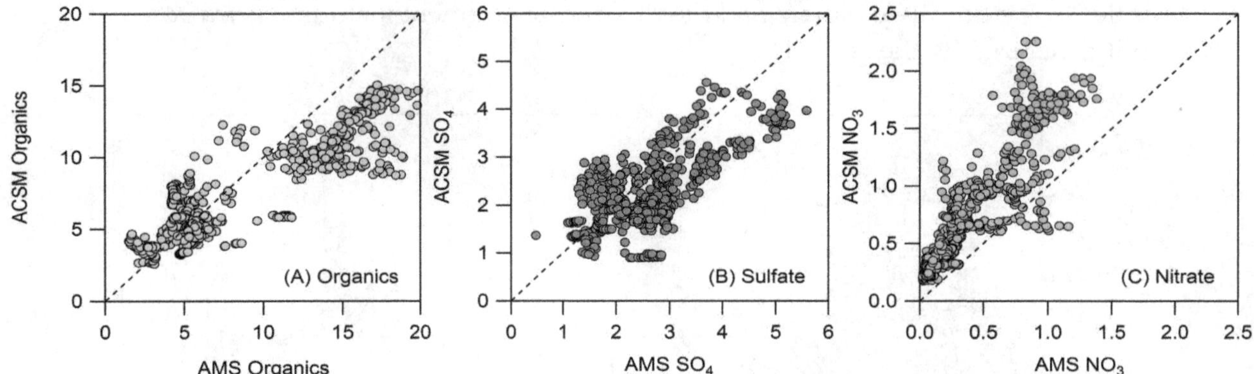

Figure A3. ACSM and HR-ToF-AMS comparison of **(A)** organics, **(B)** sulfate, and **(C)** nitrate. HR-ToF-AMS measurements are high resolution while ACSM measurements are unit mass resolution.

Figure A3 shows that, with the exception of nitrate, speciated measurements between the HR-ToF-AMS (high resolution) and ACSM (unit mass resolution) were reasonably consistent during the times when the HR-ToF-AMS was co-located (61 h over the course of the campaign). Nitrate was measured 60% higher by the ACSM, on average, than by the HR-ToF-AMS. Unit mass resolution measurements of nitrate from the ACSM rely on the standard fragmentation table to estimate the split of *m/z* 30 between nitrate (NO^+) and organics (mostly CH_2O^+). Conditions with high levels of CH_2O^+ can result in over-prediction of nitrate by the ACSM. The HR-ToF-AMS directly measures NO^+ and CH_2O^+. In this campaign, the ACSM measurements of nitrate were, on average, 60% higher than the high resolution measurements of nitrate by the HR-ToF-AMS; ACSM measurements were only 20% higher than unit mass resolution measurements by the HR-ToF-AMS, which rely on the same fragmentation table as the ACSM.

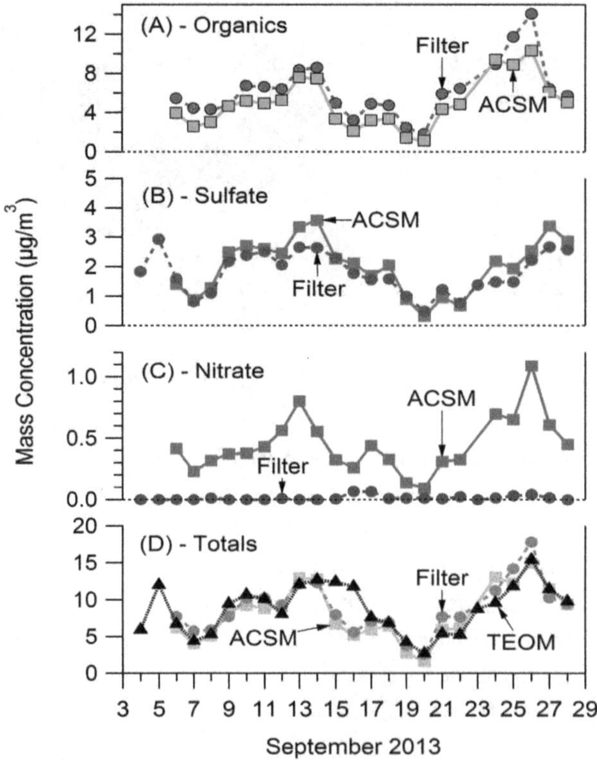

Figure A4. Instrument comparisons between the PM_1 ACSM, $PM_{2.5}$ filter, and $PM_{2.5}$ TEOM measurements for. **(A)** Organics, **(B)** Sulfate, **(C)** Nitrate, **(D)** Totals.

References

1. Lim, S.S.; Vos, T.; Flaxman, A.D.; Danaei, G.; Shibuya, K.; Adair-Rohani, H.; Amann, M.; Anderson, H.R.; Andrews, K.G.; Aryee, M.; *et al.* A comparative risk assessment of burden of disease and injury attributable to 67 risk factors and risk factor clusters in 21 regions, 1990–2010: A systematic analysis for the Global Burden of Disease Study 2010. *Lancet* **2012**, *380*, 2224–2260. [CrossRef]
2. Dockery, D.; Pope, C.A.; Xiping, X.; Spengler, J.D.; Ware, J.H.; Fay, M.E.; Ferris, B.G.; Speizer, F.E. An association between air pollution and mortality in six U.S. cities. *N. Engl. J. Med.* **1993**, *329*, 1753–1759. [CrossRef] [PubMed]
3. Tong, D.Q.; Yu, S.; Kan, H. Ozone exposure and mortality. *N. Engl. J. Med.* **2009**, *360*, 2788–2789. [PubMed]
4. NAAQS Table. Available online: https://www.epa.gov/criteria-air-pollutants/naaqs-table (accessed on 26 February 2016).
5. United States Environmental Protection Agency. National Ambient Air Quality Standards for Particulate Matter. *Fed. Reg.* **2013**, *78*, 3085–3287.
6. United States Environmental Protection Agency. National ambient air quality standards for ozone. *Fed. Reg.* **2015**, *80*, 75233–75411.
7. Allen, D.T.; McDonald-Buller, E.C.; McGaughey, G.R. *State of the Science of Air Quality in Texas: Scientific Findings from the Air Quality Research Program (AQRP) and Recommendations for Future Research*; Air Quality Research Program: Austin, TX, USA, 2016.
8. Allen, D.; Estes, M.; Smith, J.; Jeffries, H. *Accelerated Science Evaluation of Ozone Formation in the Houston-Galveston Area*; University of Texas: Austin, TX, USA, 2001.
9. Allen, D.T.; Fraser, M. An overview of the gulf coast aerosol research and characterization study: The Houston fine particulate matter supersite. *J. Air Waste Manag. Assoc.* **2006**, *56*, 456–466. [CrossRef] [PubMed]
10. Olaguer, E.; Kolb, C.; Lefer, B.; Rappenglueck, B.; Zhang, R.; Pinto, J. Overview of the SHARP campaign: Motivation, design, and major outcomes. *J. Geophys. Res. Atmos.* **2014**, *119*, 2597–2610. [CrossRef]
11. Allen, D.T.; Torres, V.M.; Thomas, J.; Sullivan, D.W.; Harrison, M.; Hendler, A.; Herndon, S.C.; Kolb, C.E.; Fraser, M.P.; Hill, A.D.; *et al.* Measurements of methane emissions at natural gas production sites in the United States. *Proc. Natl. Acad. Sci. USA* **2013**, *110*, 17768–17773. [CrossRef] [PubMed]
12. Pasci, A.; Kimura, Y.; McGaughey, G.; McDonald-Buller, E.; Allen, D.T. Regional ozone impacts of increased natural gas use in the Texas power sector and development in the Eagle Ford shale. *Environ. Sci. Technol.* **2015**, *49*, 3966–3973.
13. Zavala-Araiza, D.; Sullivan, D.W.; Allen, D.T. Atmospheric hydrocarbon emissions and concentrations in the Barnett Shale natural gas production region. *Environ. Sci. Technol.* **2014**, *48*, 5314–5321. [CrossRef] [PubMed]
14. Bahreini, R.; Ervens, B.; Middlebrook, A.M.; Warneke, C.; de Gouw, J.A.; DeCarlo, P.F.; Jimenez, J.L.; Brock, C.A.; Neuman, J.A.; Ryerson, T.B.; *et al.* Organic aerosol formation in urban and industrial plumes near Houston and Dallas, Texas. *J. Geophys. Res. Atmos.* **2009**, *114*, 1–17. [CrossRef]
15. Murphy, B.N.; Donahue, N.M.; Robinson, A.L.; Pandis, S.N. A naming convention for atmospheric organic aerosol. *Atmos. Chem. Phys.* **2014**, *14*, 5825–5839. [CrossRef]
16. Parrish, D.D.; Allen, D.T.; Bates, T.S.; Estes, M.; Fehsenfeld, F.C.; Feingold, G.; Ferrare, R.; Hardesty, R.M.; Meagher, J.F.; Nielsen-Gammon, J.W.; *et al.* Overview of the second texas air quality study (TexAQS II) and the Gulf of Mexico atmospheric composition and climate study (GoMACCS). *J. Geophys. Res. Atmos.* **2009**, *114*, 1–28. [CrossRef]
17. Weber, R.; Sullivan, A.; Peltier, R.; Russell, A.; Yan, B.; Zheng, M.; de Gouw, J.; Warneke, C.; Brock, C.; Holloway, J.; *et al.* A study of secondary organic aerosol formation in the anthropogenic-influenced southeastern United States. *J. Geophys. Res. Atmos.* **2007**, *112*. [CrossRef]
18. Xu, L.; Guo, H.; Boyd, C.M.; Bougiatioti, A.; Cerully, K.M.; Hite, J.R.; Isaacman-Vanwertz, G.; Kreisberg, N.M.; Olson, K.; Koss, A.; *et al.* Effects of anthropogenic emissions on aerosol formation from isoprene and monoterpenes in the southeastern United States. *Proc. Natl. Acad. Sci. USA* **2015**, *112*, 37–42. [CrossRef] [PubMed]
19. Xu, L.; Suresh, S.; Guo, H.; Weber, R.J.; Ng, N.L. Aerosol characterization over the southeastern United States using high-resolution aerosol mass spectrometry: Spatial and seasonal variation of aerosol composition and sources with a focus on organic nitrates. *Atmos. Chem. Phys.* **2015**, *15*, 7307–7336. [CrossRef]

20. Boyd, C.M.; Sanchez, J.; Xu, L.; Eugene, A.J.; Nah, T.; Tuet, W.Y.; Guzman, M.I.; Ng, N.L. Secondary organic aerosol formation from the β-pinene+NO3 system: Effect of humidity and peroxy radical fate. *Atmos. Chem. Phys.* **2015**, *15*, 7497–7522. [CrossRef]
21. DISCOVER-AQ Home. Available online: http://discover-aq.larc.nasa.gov/ (accessed on 26 February 2016).
22. Texas Air Monitoring Information System (TAMIS) Web Interface. Available online: http://www17.tceq.texas.gov/tamis/ (accessed on 26 February 2016).
23. Kebabian, P.L.; Wood, E.C.; Herndon, S.C.; Freedman, A. A Practical Alternative to Detection of Nitrogen Dioxide: Cavity Attenuated Phase Shift Spectroscopy. *Environ. Sci. Technol.* **2008**, *42*, 6040–6045. [CrossRef] [PubMed]
24. Ng, N.L.; Herndon, S.C.; Trimborn, A.; Canagaratna, M.R.; Croteau, P.L.; Onasch, T.B.; Sueper, D.; Worsnop, D.R.; Zhang, Q.; Sun, Y.L.; *et al.* An Aerosol Chemical Speciation Monitor (ACSM) for routine monitoring of the composition and mass concentrations of ambient aerosol. *Aerosol Sci. Technol.* **2011**, *45*, 780–794. [CrossRef]
25. Bertram, T.H.; Kimmel, J.R.; Crisp, T.A.; Ryder, O.S.; Yatavelli, R.L.N.; Thornton, J.A.; Cubison, M.J.; Gonin, M.; Worsnop, D.R. A field-deployable, chemical ionization time-of-flight mass spectrometer. *Atmos. Meas. Tech.* **2011**, *4*, 1471–1479. [CrossRef]
26. Yatavelli, R.L.N.; Lopez-Hilfiker, F.; Wargo, J.D.; Kimmel, J.R.; Cubison, M.J.; Bertram, T.H.; Jimenez, J.L.; Gonin, M.; Worsnop, D.R.; Thornton, J.A. A chemical ionization high-resolution time-of-flight mass spectrometer coupled to a Micro Orifice Volatilization Impactor (MOVI-HRToF-CIMS) for analysis of gas and particle-phase organic species. *Aerosol Sci. Technol.* **2012**, *46*, 1313–1327. [CrossRef]
27. Lee, B.H.; Lopez-Hilfiker, F.D.; Mohr, C.; Kurtén, T.; Worsnop, D.R.; Thornton, J.A. An iodide-adduct high-resolution time-of-flight chemical-ionization mass spectrometer: Application to atmospheric inorganic and organic compounds. *Environ. Sci. Technol.* **2014**, *48*, 6309–6317. [CrossRef] [PubMed]
28. Aljawhary, D.; Lee, A.K.Y.; Abbatt, J.P.D. High-resolution chemical ionization mass spectrometry (ToF-CIMS): Application to study SOA composition and processing. *Atmos. Meas. Tech.* **2013**, *6*, 3211–3224. [CrossRef]
29. Allan, J.D.; Delia, A.E.; Coe, H.; Bower, K.N.; Alfarra, M.R.; Jimenez, J.L.; Middlebrook, A.M.; Drewnick, F.; Onasch, T.B.; Canagaratna, M.R.; *et al.* A generalised method for the extraction of chemically resolved mass spectra from Aerodyne aerosol mass spectrometer data. *J. Aerosol Sci.* **2004**, *35*, 909–922. [CrossRef]
30. Leong, Y.J.; Sanchez, N.P.; Wallace, H.W.; Karakurt Cevik, B.; Hernandez, C.S.; Han, Y.; Choi, Y.; Flynn, J.H.; Massoli, P.; Floerchinger, C.; *et al.* Overview of Surface Measurements and Spatial Characterization of Submicron Particulate Matter during the DISCOVER-AQ 2013 Campaign in Houston, TX, USA, **2016**. in preparation.
31. Aiken, A.C.; Decarlo, P.F.; Kroll, J.H.; Worsnop, D.R.; Huffman, J.A.; Docherty, K.S.; Ulbrich, I.M.; Mohr, C.; Kimmel, J.R.; Sueper, D.; *et al.* O/C and OM/OC Ratios of primary, secondary, and ambient organic aerosols with high-resolution time-of-flight aerosol mass spectrometry. *Environ. Sci. Technol.* **2008**, *42*, 4478–4485. [CrossRef] [PubMed]
32. Canagaratna, M.R.; Jimenez, J.L.; Kroll, J.H.; Chen, Q.; Kessler, S.H.; Massoli, P.; Hildebrandt Ruiz, L.; Fortner, E.; Williams, L.R.; Wilson, K.R.; *et al.* Elemental ratio measurements of organic compounds using aerosol mass spectrometry: Characterization, improved calibration, and implications. *Atmos. Chem. Phys.* **2015**, *15*, 253–272. [CrossRef]
33. Paatero, P.; Tapper, U. Positive matrix factorization: A nonnegative factor model with optimal utilization of error estimates of data values. *Environmetrics* **1994**, *5*, 111–126. [CrossRef]
34. Hildebrandt, L.; Engelhart, G.J.; Mohr, C.; Kostenidou, E.; Lanz, V.A.; Bougiatioti, A.; DeCarlo, P.F.; Prevot, A.S.H.; Baltensperger, U.; Mihalopoulos, N.; *et al.* Aged organic aerosol in the eastern Mediterranean: The finokalia aerosol measurement experiment–2008. *Atmos. Chem. Phys.* **2010**, *10*, 4167–4186. [CrossRef]
35. Hildebrandt, L.; Kostenidou, E.; Lanz, V.A.; Prevot, A.S.H.; Baltensperger, U.; Mihalopoulos, N.; Donahue, N.M.; Pandis, S.N. Sources and atmospheric processing of organic aerosol in the Mediterranean: Insights from aerosol mass spectrometer factor analysis. *Atmos. Chem. Phys.* **2011**, *11*, 12499–12515. [CrossRef]
36. Lanz, V.A.; Alfarra, M.R.; Baltensperger, U.; Buchmann, B.; Hueglin, C.; Prevot, A.S.H. Source apportionment of submicron organic aerosols at an urban site by factor analytical modelling of aerosol mass spectra. *Atmos. Chem. Phys.* **2007**, *7*, 1503–1522. [CrossRef]

37. Lanz, V.A.; Prévôt, A.S.H.; Alfarra, M.R.; Weimer, S.; Mohr, C.; DeCarlo, P.F.; Gianini, M.F.D.; Hueglin, C.; Schneider, J.; Favez, O.; *et al.* Characterization of aerosol chemical composition with aerosol mass spectrometry in Central Europe: An overview. *Atmos. Chem. Phys.* **2010**, *10*, 10453–10471. [CrossRef]
38. Ulbrich, I.M.; Canagaratna, M.R.; Zhang, Q.; Worsnop, D.R.; Jimenez, J.L. Interpretation of organic components from Positive Matrix Factorization of aerosol mass spectrometric data. *Atmos. Chem. Phys.* **2009**, *9*, 2891–2918. [CrossRef]
39. Faxon, C.; Bean, J.; Hildebrandt Ruiz, L. Inland concentrations of ClNO2 in Southeast Texas suggest chlorine chemistry significantly contributes to atmospheric reactivity. *Atmosphere* **2015**, *6*, 1487–1506. [CrossRef]
40. Birch, M.; Cary, R. Elemental carbon-based method for monitoring occupational exposures to particulate diesel exhaust. *Aerosol Sci. Technol.* **1996**, *25*, 221–241. [CrossRef]
41. Zaveri, R.A.; Shaw, W.J.; Cziczo, D.J.; Schmid, B.; Ferrare, R.A.; Alexander, M.L.; Alexandrov, M.; Alvarez, R.J.; Arnott, W.P.; Atkinson, D.B.; *et al.* Overview of the 2010 Carbonaceous Aerosols and Radiative Effects Study (CARES). *Atmos. Chem. Phys.* **2012**, *12*, 7647–7687. [CrossRef]
42. Zotter, P.; El-Haddad, I.; Zhang, Y.; Hayes, P.L.; Zhang, X.; Lin, Y.; Wacker, L.; Schnelle-Kreis, J.; Abbaszade, G.; Zimmermann, R.; *et al.* Diurnal cycle of fossil and nonfossil carbon using radiocarbon analyses during CalNex. *J. Geophys. Res. Atmos.* **2014**, *119*, 6818–6835. [CrossRef]
43. Barrett, T.E.; Robinson, E.M.; Usenko, S.; Sheesley, R.J. Source contributions to wintertime elemental and organic carbon in the western arctic based on radiocarbon and tracer apportionment. *Environ. Sci. Technol.* **2015**, *49*, 11631–11639. [CrossRef] [PubMed]
44. Gustafsson, Ö.; Kruså, M.; Zencak, Z.; Sheesley, R.J.; Granat, L.; Engström, E.; Praveen, P.S.; Rao, P.S.P.; Leck, C.; Rodhe, H. Brown clouds over South Asia: Biomass or fossil fuel combustion? *Science* **2009**, *323*, 495–498. [CrossRef] [PubMed]
45. Atkinson-Palombo, C.M.; Miller, J.A.; Balling, R.C., Jr. Quantifying the ozone "weekend effect" at various locations in Phoenix, Arizona. *Atmos. Environ.* **2006**, *40*, 7644–7658. [CrossRef]
46. Wilks, D.S. *Statistical Methods in the Atmospheric Science*; Academic Press: San Diego, CA, USA, 1995.
47. Ng, N.L.; Canagaratna, M.R.; Jimenez, J.L.; Zhang, Q.; Ulbrich, I.M.; Worsnop, D.R. Real-time methods for estimating organic component mass concentrations from aerosol mass spectrometer data. *Environ. Sci. Technol.* **2011**, *45*, 910–916. [CrossRef] [PubMed]
48. Rollins, A.W.; Browne, E.C.; Min, K.-E.; Pusede, S.E.; Wooldridge, P.J.; Gentner, D.R.; Goldstein, A.H.; Liu, S.; Day, D.A.; Russell, L.M.; *et al.* Evidence for NOx control over nighttime SOA formation. *Science* **2012**, *337*, 1210–1212. [CrossRef] [PubMed]
49. Mylonas, D.T.; Allen, D.T.; Ehrmanf, S.H.; Pratsins, S.E. The sources and size distributions of organonitrates in Los Angeles aerosol. *Atmos. Environ.* **1991**, *25A*, 2855–2861. [CrossRef]
50. Garnes, L.A.; Allen, D.T. Size distributions of organonitrates in ambient aerosol collected in Houston, Texas. *Aerosol Sci. Technol.* **2002**, *36*, 983–992. [CrossRef]
51. Laurent, J.-P.; Allen, D.T. Size distributions of organic functional groups in ambient aerosol collected in Houston, Texas. *Aerosol Sci. Technol.* **2004**, *38*, 60–67. [CrossRef]
52. O'Brien, R.J.; Holmes, J.R.; Bockian, A.H. Formation of photochemical aerosol from hydrocarbons. chemical reactivity and products. *Environ. Sci. Technol.* **1975**, *9*, 568–576. [CrossRef]
53. Stein, A.F.; Draxler, R.R.; Rolph, G.D.; Stunder, B.J.B.; Cohen, M.D.; Ngan, F. NOAA'S HYSPLIT atmospheric transport and dispersion modeling system. *Am. Meteorol. Soc.* **2015**, *96*, 2059–2077. [CrossRef]
54. Pankow, J.F. An absorption model of gas/particle partitioning of organic compounds in the atmosphere. *Atmos. Environ.* **1994**, *28*, 185–188. [CrossRef]
55. Donahue, N.M.; Robinson, A.L.; Stanier, C.O.; Pandis, S.N. Coupled partitioning, dilution, and chemical aging of semivolatile organics. *Environ. Sci. Technol.* **2006**, *40*, 2635–2643. [CrossRef] [PubMed]
56. Murphy, B.N.; Pandis, S.N. Exploring summertime organic aerosol formation in the eastern United States using a regional-scale budget approach and ambient measurements. *J. Geophys. Res.* **2010**, *115*, D24216. [CrossRef]
57. Tucker, S.C.; Banta, R.M.; Langford, A.O.; Senff, C.J.; Brewer, W.A.; Williams, E.J.; Lerner, B.M.; Osthoff, H.D.; Hardesty, R.M. Relationships of coastal nocturnal boundary layer winds and turbulence to Houston ozone concentrations during TexAQS 2006. *J. Geophys. Res.* **2010**, *115*, 1–17. [CrossRef]

58. Vizuete, W.; Junquera, V.; McDonald-Buller, E.; McGaughey, G.; Yarwood, G.; Allen, D. Effects of temperature and land use on predictions of biogenic emissions in Eastern Texas, USA. *Atmos. Environ.* **2002**, *36*, 3321–3337. [CrossRef]
59. Lee, B.H.; Mohr, C.; Lopez-Hilfiker, F.D.; Lutz, A.; Hallquist, M.; Lee, L.; Romer, P.; Cohen, R.C.; Iyer, S.; Kurtén, T.; *et al.* Highly functionalized organic nitrates in the southeast United States: Contribution to secondary organic aerosol and reactive nitrogen budgets. *Proc. Natl. Acad. Sci. USA* **2016**, *113*, 1516–1521. [CrossRef] [PubMed]
60. Pankow, J.F.; Asher, W.E. SIMPOL.1: A simple group contribution method for predicting vapor pressures and enthalpies of vaporization of multifunctional organic compounds. *Atmos. Chem. Phys.* **2008**, *8*, 2773–2796. [CrossRef]
61. Ng, N.L.; Chhabra, P.S.; Chan, A.W.H.; Surratt, J.D.; Kroll, J.H.; Kwan, A.J.; McCabe, D.C.; Wennberg, P.O.; Sorooshian, A.; Murphy, S.M.; *et al.* Effect of NOx level on secondary organic aerosol (SOA) formation from the photooxidation of terpenes. *Atmos. Chem. Phys. Discuss.* **2007**, *7*, 10131–10177. [CrossRef]

10

Comparison of Land-Use Regression Modeling with Dispersion and Chemistry Transport Modeling to Assign Air Pollution Concentrations within the Ruhr Area

Frauke Hennig [1,*], Dorothea Sugiri [1], Lilian Tzivian [1], Kateryna Fuks [1], Susanne Moebus [2], Karl-Heinz Jöckel [2], Danielle Vienneau [3,4], Thomas A.J. Kuhlbusch [5], Kees de Hoogh [3,4], Michael Memmesheimer [6], Hermann Jakobs [6], Ulrich Quass [5] and Barbara Hoffmann [1,7]

1. Working group of Environmental Epidemiology of Cardiovascular Aging and Allergies, IUF-Leibniz Research Institute for Environmental Medicine, Auf'm Hennekamp 50, Düsseldorf 40225, Germany; Dorothee.Sugiri@IUF-Duesseldorf.de (D.S.); Liliana.Tzivian@IUF-Duesseldorf.de (L.T.); Kateryna.Fuks@IUF-Duesseldorf.de (K.F.); B.Hoffmann@uni-Duesseldorf.de (B.H.)
2. Institute for Medical Informatics, Biometry and Epidemiology, University Hospital, University Duisburg-Essen, Essen 45141, Germany; Susanne.Moebus@uk-essen.de (S.M.); k-h.joeckel@uk-essen.de (K.-H.J.)
3. Department of Epidemiology and Public Health, Swiss Tropical and Public Health Institute, Socinstr. 57, Basel 4051, Switzerland; danielle.vienneau@unibas.ch (D.V.); c.dehoogh@unibas.ch (K.H.)
4. University of Basel, Basel 4003, Switzerland
5. IUTA e.V., Air Quality & Sustainable Nanotechnology Unit, Duisburg, Germany; tky@iuta.de (T.K.); quass@iuta.de (U.Q.)
6. Rhenish Institute for Environmental Research (RIU), Aachenerstr. 209, 50931 Köln, Germany; mm@eurad.uni-koeln.de (M.M.); Hermann.Jakobs@eurad.uni-koeln.de (H.J.)
7. Heinrich Heine University of Düsseldorf, Medical Faculty, Deanery of Medicine, Moorenstraße 5, Düsseldorf 40225, Germany

* Correspondence: Frauke.hennig@iuf-duesseldorf.de

Abstract: Two commonly used models to assess air pollution concentration for investigating health effects of air pollution in epidemiological studies are Land Use Regression (LUR) models and Dispersion and Chemistry Transport Models (DCTM). Both modeling approaches have been applied in the Ruhr area, Germany, a location where multiple cohort studies are being conducted. Application of these different modelling approaches leads to differences in exposure estimation and interpretation due to the specific characteristics of each model. We aimed to compare both model approaches by means of their respective aims, modeling characteristics, validation, temporal and spatial resolution, and agreement of residential exposure estimation, referring to the air pollutants $PM_{2.5}$, PM_{10}, and NO_2. Residential exposure referred to air pollution exposure at residences of participants of the Heinz Nixdorf Recall Study, located in the Ruhr area. The point-specific ESCAPE (European Study of Cohorts on Air Pollution Effects)-LUR aims to temporally estimate stable long-term exposure to local, mostly traffic-related air pollution with respect to very small-scale spatial variations ($\leqslant$100 m). In contrast, the EURAD (European Air Pollution Dispersion)-CTM aims to estimate a time-varying average air pollutant concentration in a small area (*i.e.*, 1 km^2), taking into account a range of major sources, e.g., traffic, industry, meteorological conditions, and transport. Overall agreement between EURAD-CTM and ESCAPE-LUR was weak to moderate on a residential basis. Restricting EURAD-CTM to sources of local traffic only, respective agreement was good. The possibility of combining the strengths of both applications will be the next step to enhance exposure assessment.

Keywords: air pollution; Land use regression; chemistry-transport dispersion-model

1. Introduction

A large number of epidemiological studies have shown associations between short-and/or long-term exposure to outdoor air pollution and adverse health effects [1]. Traditionally, adverse health effects of air pollution have been divided into effects of short-term variations in air pollution concentrations, mainly influenced by meteorology, and effects of long-term exposure to air pollution, where contrasts rely on spatial variation of air pollution concentrations. Early approaches on assessing exposure to air pollution used average air pollution concentrations of the nearest monitoring station as a surrogate of personal exposure, assuming homogeneity among air pollution concentrations within the area surrounding the monitoring station, or even within the whole city [2]. Considering short-term health effects in ecological time-series studies on air pollution and mortality, it seems reasonable to assume such a spatially-uniform temporal elevation or reduction in air pollution concentration because they are dependent on the underlying meteorological conditions. When considering long-term health effects on an individual basis, however, the spatial and spatio-temporal variations are of great importance given that outdoor air pollution concentrations vary on a small spatial scale, e.g., within 100 m of a busy road [3]. More recent epidemiological studies have, thus, approached such small-scale intra-urban variation of air pollution concentrations by using different types of models, such as Land Use Regression (LUR) models, Dispersion Models (DM), chemistry Transport Model Models (CTM), a combination of DM+CTM (DCTM), hybrid models, or other alternatives [4,5].

The LUR method, first developed by Briggs *et al.* [6] in the Small Area Variations In Air quality and Health (SAVIAH study), uses linear (least squared) regression models to predict monitoring air pollution data with Geographic Information System (GIS)-based data reflecting pollutant conditions. Compared to other approaches, LUR models were built to predict temporally-stable long-term air pollution concentrations applicable to the smallest spatial scale (point-specific), e.g., home residences.

DMs are in general mathematical simulation models to estimate air pollution concentrations by means of numerical descriptions of deterministic (physical, chemical, and fluid dynamical) processes of the dispersion of air pollutants in the ambient atmosphere, and typically include data on emissions, meteorological conditions, and topography [3].

CTMs model the variability in space and time of chemical concentrations in the atmosphere, using three-dimensional numerical models to simulate processes of emission, transport, chemical transformation, diffusion and deposition, using emissions, meteorological information, and land use as input. Most often DMs and CTMs (DCTM) are combined in practice, resulting in spatio-temporal estimations. Usually DMs and CTMs estimate air pollution concentrations on a coarser spatial scale compared to the point-specific LUR, e.g., a grid of 1 or 5 km^2.

LUR models were developed to estimate exposure concentration at the finest spatial resolution and have been increasingly used in epidemiological studies due to their relatively low cost and easy implementation, developed either on the basis of purpose-designed monitoring campaigns or routine monitoring measurements and appropriate geographic predictors of sources [7]. In contrast, DCTMs have been developed for air quality, *i.e.*, prediction, regulation and management, putting high demands on data requirements, costs and the complexity of modeling [6].

So far, only a few studies compared the performance of LUR and dispersion modeling for estimating exposure to nitrogen dioxide (NO_2). While some studies suggested that LUR models explained small-scale variations in air pollution concentrations as well or even better than various dispersion models [8–10], Beelen *et al.* [11] showed that the dispersion models performed better than LUR models regarding monitored and modeled concentrations on several validation sites. Most recently, de Hoogh *et al.* [12] investigated agreement between LUR and DM modeling approaches aiming to estimate residential exposure to NO_2 and particulate matter (PM) with an aerodynamic diameter $\leqslant 10$ μm and $\leqslant 2.5$ μm (PM_{10}, $PM_{2.5}$) within the European Study of Cohorts for Air Pollution Effects (ESCAPE). Comparisons across 4–13 cohorts, including the Heinz Nixdorf Recall (Risk Factors, Evaluation of Coronary calcium and Lifestyle) (HNR) study, located in the Ruhr area in Germany, yielded moderate to good correlations between LUR and DM (or DCTM) for NO_2 (0.39–0.90) and for

PM_{10} and $PM_{2.5}$ (0.23–0.81). However, single correlation coefficients for the HNR study were below 0.4 for all three pollutants [12], raising the question of comparability of the two different exposure modelling approaches. So far, most studies on the comparison of different modeling strategies focused on the residential agreement of estimated exposure concentrations, disregarding the potential reasons for the disagreement between different modelling approaches, as well as respective strengths and limitations. Although all exposure metrics are equally used as a surrogate of personal exposure in epidemiological studies, exposure modeling is strongly influenced by the spatial and temporal variation of exposure and exposure sources [5]. Furthermore, aims, application, input data but also the complexity of models might differ, yielding not only different exposure estimates but consequently different health effect estimates in terms of magnitude and/or statistical significance [5,13].

In the Ruhr area in Germany, the location of multiple epidemiological studies, e.g., the Heinz Nixdorf Recall study, air pollution concentrations have been modeled with a LUR model as part of the European Study of Cohorts for Air Pollution Effects (ESCAPE-LUR), as well as with a European Air Quality and Dispersion Model which is a DCTM (EURAD-CTM) as part of several research projects investigating health effects of residential air pollution exposure. In this article, we aim to compare the ESCAPE-LUR model and the EURAD-CTM model focusing on their respective strengths and limitations. To do so, we compare model approaches by means of their respective aim, application characteristics, validation, temporal, and spatial resolution and by means of residential agreement. In addition, we evaluated the agreement of modeled air pollution concentrations by EURAD-CTM and measured air pollution concentrations at ESCAPE-LUR monitoring sites for overlapping time windows. Air pollutants of interest are $PM_{2.5}$, PM_{10}, and NO_2.

2. Methods

2.1. Study Area

The Heinz Nixdorf Recall (Risk Factors, Evaluation of Coronary calcium and Lifestyle) (HNR) study area covers a region of approximately 600 km^2 and is located in the highly urbanized Ruhr Area in the west of Germany, including the cities of Mülheim, Essen, and Bochum. In addition to that the HNR study area is located within N3, one of the smallest sequential nests developed for the air pollution modelling purposes of EURAD-CTM. We used locations (x,y) (Gauss–Krüger coordinates) of 4809 residences, located within the HNR study area. According to the Ruhr Regional Association, land use in the area can be roughly divided into agricultural (~40%), built-up (~40%), and forest and other regions (e.g., water) (~20%) [14]. The population density of the Ruhr area is about 2100 inhabitants per 1 km^2, and in terms of traffic density the area is one of the densest in the whole of Europe (Figure 1). As an urban area, almost one fifth of the working population is occupied in the industrial sector. Among many industrial areas, the majority of steel and coal industry is located in Duisburg, in the west of the Ruhr area, including the biggest steelwork in Europe. Furthermore, Europe's largest inland harbor is located directly west of the study area in Duisburg. Intensive shipping takes place on the Rhine, which flows through Duisburg from south to north.

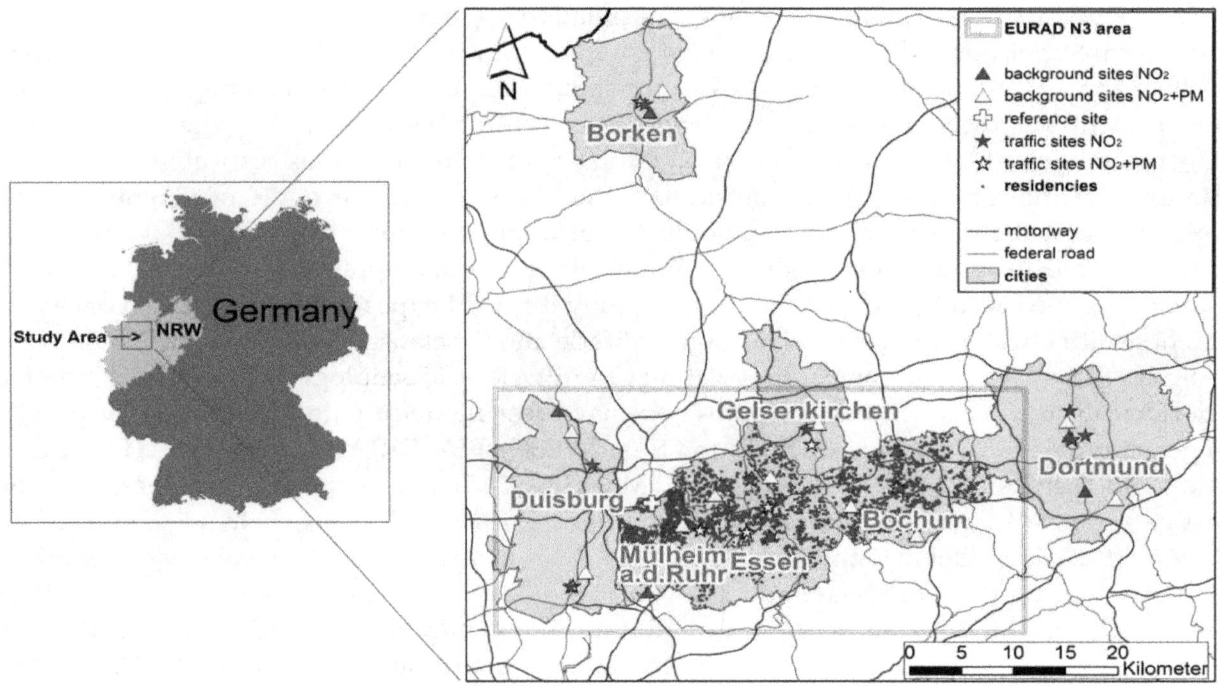

Figure 1. Study area, residences, and monitoring sites.

2.2. *Exposure Assesment*

2.2.1. EURAD-CTM

The EURAD-CTM model [15] is a validated time dependent three-dimensional chemistry transport model [16–19] developed to predict daily concentrations of air pollutants on a horizontal grid resolution of 1 km^2 (Table 1). The EURAD-CTM model system is a multi-layer, multi-grid model system for the simulation of transport, chemical transformation, and deposition of tropospheric constituents [20], and consists of five major parts (Figure S1): (1) the meteorological driver version 3 (MM5V3) [21]; (2) two pre-processors for preparation of meteorological fields and observational data; (3) the EURAD Emission Model EEM [22], and (4) the Chemistry Transport Model (CTM); including (5) a model for aerosol dynamics in Europe (MADE) [16,18,23,24]. An additional procedure includes data assimilation on an hourly basis, using routine measurement data of monitoring sites in North Rhine-Westphalia (NRW) provided by the local environmental agency: State Agency for Nature, Environment, and Consumer Protection (LANUV-NRW) [25–27] (intermitted 3d-var) (Figure S1). EURAD-CTM calculations are performed using a one-way nesting scheme to take long-range transport into account. Nested grid domains ranged from a European scale (N0: 125 km), to central Europe (N1: 25 km), to NRW (N2: 5 km) in Germany, to the Ruhr area (N3: 1 km), while the vertical resolution is the same for all model domains (40 m) ([18,20]). In addition to long-range transport, the formation of atmospheric gases and PM is also included in the model, *i.e.*, the formation of secondary particles in the atmosphere from primary emitted gaseous pollutants from NO_2, sulfur dioxide (SO_2), ammonia (NH_3), and Volatile Organic Compounds (VOC) during the transport [19]. Long-range transport and formation of secondary particles in the atmosphere can contribute considerably to the particle mass concentration in NRW and the Ruhr area, e.g., more than 50% [28]. The EURAD-CTM is driven by emissions due to anthropogenic and biogenic sources [29]. Anthropogenic emissions are taken from officially-available databases as EMEP-grid [30] for Europe and from the LANUV-NRW. The EURAD-CTM emission input is further structured with respect to different source categories according to the Selected Nomenclature for Sources of Air Pollution (SNAP-97) [31], including traffic, industry, and other source categories.

Table 1. Characteristics of the ESCAPE-LUR and EURAD-CTM approaches to estimate air pollution concentrations.

-	Land use regression (ESCAPE-LUR)	European Air Quality and Dispersion Chemistry Transport Model (EURAD-CTM)
Model Type	Linear regression model, to predict annual averages derived from selected monitored concentrations with land use data	Mesoscale chemistry transport model involving emissions, transport, diffusion, chemical transformation, wet and dry deposition, and sedimentation of gases and aerosols
Aim & Application	Estimation of long-term traffic-related air pollution for population-based exposure studies and epidemiological health outcome analyses	1) Air pollution modeling (forecasts, episode analysis, trend analysis, reduction scenarios) and Chemical data assimilation studies for Europe, Central Europe and several German States; 2) Exposure estimation in population-related exposure studies
Model Input	1) Data: • Annual mean AP concentration (for details see Table S1); • Land use density in 100, 300, 500, 1000, and 5000 m buffers: ○ Industry ○ Seaport ○ urban green ○ semi-natural ○ forested areas ○ number of inhabitants • Traffic data in 25, 50, 100, 300, 500, and 1000 m buffers: ○ distance ○ (heavy) traffic intensity on the nearest road and nearest major road ○ (heavy) traffic load on all roads and major roads)	1) Data: • Model area projection topography • Land use • Meteorological initial and boundary values • Anthropogenic emission data (according to the Selected Nomenclature for Sources of Air Pollution (SNAP-97)) • Chemical initial and boundary values, • Long-range transport, • Photolysis frequencies. 2) Procedures (Figure S3): • Mesoscale meteorological model (MM5) driven by global meteorological fields provided by NCEP (http://www.ncep.noaa.gov/), • EPC, anthropogenic and non-anthropogenic emission modules (EEM-A, EEM-B), • Aerosol dynamics module (MADE), • Data assimilation [a]
Modelled Air Pollutants	**$PM_{2.5}$, PM_{10}, NO_2** (additional pollutants: $PM_{2.5}$ absorbance, PM coarse, NO, NOx)	**$PM_{2.5}$, PM_{10}, NO_2** (additional pollutants: PM_1, O_3, SO_2, CO, PNC, NH_4, NO_3, SO_4, BC, EC)
Temporal Resolution (Output)	Yearly mean concentration (October16, 2008 until October 15, 2009)	Any temporal resolution > day within October 2000 until December 2003 and January 2006 until December 2008 is possible; e.g., 7-,14-, 21-,28-,91-,182-, and 365-day mean concentration
Model Validation	a) Goodness of fit (cf. Table S2): $PM_{2.5}$ (R^2 = 0.85), PM_{10} (R^2 = 0.66), NO_2 (R^2 = 0.88) b) Leave-one-out cross-validation: $PM_{2.5}$ (R^2 = 0.74), PM_{10} (R^2 = 0.59), NO_2 (R^2 = 0.82)	Validation for daily mean concentration in N3 area with routine measurements (mean bias, correlation); year: a) Before data assimilation: PM_{10} (−6.5, 0.45); 2006 NO_2 (4.0, 0.39); 2007 b) After data assimilation PM_{10} (−0.9, 0.93); 2006 NO_2 (0.6, 0.95); 2007
Spatial Resolution	Point-specific	1 km × 1 km grid
Additional Features	1) XRF-Model for air pollutant constituents 2) Back-extrapolating back in time and for specific time windows	Source-specific air pollutant concentrations (only local traffic (TRA), only local industry (IND))

[a] only for PM_{10} and NO_2 for the considered time period.

Output of the EURAD-CTM calculations consists of chemical compounds, such as atmospheric particle mass, number density, and particle size distribution, as well as concentration of atmospheric gases, photo oxidants, and a set of volatile organic compounds on an hourly basis for each grid. EURAD-CTM estimates of PM_{10} and NO_2 concentrations are assimilated using measurements from all available routine monitoring sites within the region of interest. For the Ruhr area there exists a maximum of ten monitoring sites, including different air pollution data bases [25]. Using ArcView 9.2, location of residences were assigned to a 1 × 1 km^2-grid and then matched to the corresponding grid-based air pollutant concentration, allowing both short-term (daily mean concentrations) and long-term (annual mean concentrations) assignment of exposure. The basis of daily mean concentration allows us to calculate exposure for any temporal resolution with a minimum of one day. Model runs for the EURAD-CTM within N3 were done for the examination periods of the HNR study (2000–2003 and 2006–2008). Thus, we are able to assign exposure concentrations of yearly-mean concentrations for the years 2001, 2002, 2003, 2006, 2007, and 2008 and personalized exposure concentrations of 1-, 7-, 28-, 91-, 182-, and 365-day mean concentrations prior to the date of examination.

As an add-on feature it was possible to model source-specific Air Pollution (AP) concentration with EURAD-CTM [28]. Briefly, within EURAD-CTM we estimated AP concentration suppressing local sources within the smallest grid domain (N3), such as traffic and industry by setting to them to zero (AP_{noTRA} or AP_{noIND} respectively). We then calculated local traffic-specific or industry-specific AP by taking the difference $AP_{TRA} = AP - AP_{noTRA}$ or $AP_{IND} = AP - AP_{noIND}$, respectively. In earlier studies, we applied this method to compare the health effects of PM, emitted from local traffic and local industrial sources within the Ruhr area on levels of highly-sensitive C-reactive protein, a marker of systemic inflammation [32].

2.2.2. ESCAPE-LUR

LUR models were developed to estimate temporally-stable spatial-variant concentrations of long-term exposure to traffic-related air pollutants as part of the ESCAPE study (Table 1). Following the definition of LUR describes a standardized model building procedure developed within the ESCAPE study, here the ESCAPE-LUR. The ESCAPE-LUR defines a linear prediction model for an air pollutant concentration, including annual mean air pollution concentrations as a dependent variable and geographic data on traffic, industry, and population density as potential predictors (independent variables). Predictor data were collected in a Geographical Information System (GIS), based on CORINE 2000 definitions [33]. The procedure of model development was standardized within the ESCAPE study and included a forward selection of predictors based on the incremental improvement in R^2 [34–36]. A predictor was added if addition of the predictor yielded an improvement of R^2 by more than 1%, if the coefficient conformed to the pre-specified direction, and if the direction of previously selected predictors did not change. In addition, predictors with a *p*-value > 0.1 were removed, while predictors with a variance inflation factor (VIF) > 3 and Cook's Distance (Cook's D) >1 were further investigated. To avoid extrapolation, estimated concentrations were truncated at the highest observed value. Annual air pollution concentrations were based on a measurement campaign in the study area of interest, including three periods of a 14-day measurement to cover all seasons (cold, warm, and one intermediate temperature season) from October 2008 until October 2009. The reason for the choice of 14-days was the settings design of the ESCAPE-LUR measurement campaign, which was conducted with discontinuous particle measurement devices (Harvard impactors). Measurements were conducted at 20–40 monitoring sites, placed at locations which were characteristic of traffic and background pollutant concentrations to measure PM (at 20 sites) and NO_2 (at 40 sites) (Figure 1, Table S1). One additional background reference site was chosen to measure PM and NO_2 continuously during a complete year (starting in October 2008) so that all discontinuous site-specific measurements could be adjusted to derive a long-term annual average. Measurement data from the reference site was only used for adjustment and not for ESCAPE-LUR model development. A separate LUR model was developed for each air pollutant and validated via Leave-One-Out Cross Validation (LOOCV), excluding one monitoring site at a time. Other choices of model validation are possible, e.g., hold-out cross validation, which has recently been proposed to perform better [37]. However, in this manuscript we hold onto the ESCAPE-LUR.

Since ESCAPE included two cohorts located within NRW, namely the HNR study and the Study on the influence of air pollution on lung function, inflammation, and aging (SALIA), the ESCAPE-LUR measurement campaign was combined for both studies and ranged from the urban Ruhr area to the more rural city of Borken (Figure 1) [34,36]. ESCAPE-LUR for $PM_{2.5}$ included heavy traffic load (1 km buffer), industry (5 km buffer), population density (1 km buffer), and the x-coordinate of the location of interest as predictors with an explained variance of $R^2 = 0.85$ (LOOCV-$R^2 = 0.74$) (Table S2) [34]. ESCAPE-LUR for PM_{10} included heavy traffic load (1 km buffer) and population density (1 km buffer) with an explained variance of $R^2 = 0.66$ (LOOVC-$R^2 = 0.59$) (Table S1) [34], ESCAPE-LUR for NO_2

included industry (5 km buffer), population density (100 m buffer), inland or seaport (5 km buffer) and traffic load (100 m buffer) with an explained variance of R^2 = 0.88 (LOOVC-R^2 = 0.82) (Table S1) [36]. (Heavy) traffic load referred to total (heavy-duty) traffic load of all roads in a buffer (sum of (traffic intensity × length of all segments)), industry referred to industrial, commercial, and transport units in a certain buffer; inland or seaport referred to the respective area within a buffer and population density to the number of inhabitants in a certain buffer. Uncertainty was evaluated as residuum's mean squared error in the LOOCV-approach, which was 0.61 for $PM_{2.5}$, 1.44 for PM_{10}, and 3.19 for NO_2.

Based on the coordinates of residence, located within the study area, annual mean concentrations were estimated using the ESCAPE-LUR prediction models and the relevant GIS predictors. In order to estimate AP concentration back in time, LUR modeling offers the method of back-extrapolation using a ratio or absolute difference method. Briefly, routine monitoring data should be available in order to account for differences of AP concentrations back in time [38]. Within the ESCAPE study, back-extrapolated AP estimations referred to a two year average (± 365 days of the examination day) in order to avoid any time-specific outliers. An additional feature offered by ESCAPE-LUR is the possibility to estimate exposure concentration as an average per month or trimester, e.g., before pregnancy, which might be of interest when investigating birth cohorts.

2.3. *Statistical Analysis*

Conducted statistical analysis referred to air pollutants $PM_{2.5}$, PM_{10}, and NO_2, estimated using the EURAD-CTM and the ESCAPE-LUR model. First, we described EURAD-CTM grid-based concentrations for the whole HNR study area for the years 2001–2003 and 2006–2008 by mean and standard deviation (mean ± SD) as well as minimum and maximum (Min, Max). Secondly, we described residence-based exposures derived from the EURAD-CTM and from the ESCAPE-LUR by mean ± SD (Min, Max) and Person's correlation coefficients for the most closely matched annual time-window: year 2008 for EURAD-CTM *vs.* annual mean ESCAPE-LUR (*i.e.*, based on measurements from October 2008 until October 2009). Considered air pollutants were $PM_{2.5}$, PM_{10}, and the gas NO_2. In addition, we calculated Spearman's correlation coefficient between 14-day mean air pollution concentrations measured at ESCAPE measurement sites (traffic and background) and 14-day mean air pollution concentrations calculated by EURAD-CTM for the grid cells that included an ESCAPE measurement site within the time period of October 2008–December 2008.

To evaluate an overall agreement between routinely measured air pollution concentrations, we compared annual mean concentrations of three routine monitoring stations provided by LANUV, located within the Ruhr area, and thus within EURAD specific grid cells (gc), with annual estimated air pollution concentrations estimated by EURAD-CTM and ESCAPE-LUR. Details of routine measurement stations are given in Table S3. Referred monitoring sites are the above mentioned reference site in Mülheim-Styrum (STYR) (gc: 679), an additional background site, located in Essen-Vogelheim (EVOG) (gc: 942), and one traffic site, located at a highly trafficked road in Essen (VESN) (gc: 690). For the comparison with the EURAD-CTM we considered annual mean concentrations from January 1, 2008 until December 31, 2008, while for the comparison with the ESCAPE-LUR we considered annual means from October 16, 2008 until October 15, 2009 in order to match the time window of the ESCAPE measurement campaign. Annual mean concentrations modeled by the ESCAPE-LUR referred to the location (coordinate points) of monitoring sites. In addition to that we calculated Pearson's correlation coefficients between daily measurements of LANUV monitoring sites and daily estimations by EURAD-CTM for the year 2008.

With regard to different temporal resolution, we compared EURAD-CTM air pollution concentration estimates to measured air pollution concentrations on a monthly basis to yearly mean concentrations (2006, 2007, and 2008) estimated by EURAD-CTM in two of the above mentioned grid cells (679 and 690). In contrast we visualized time-dependent measurements of the two corresponding routine monitoring sites (STYR and VESN) on a monthly basis as well as the temporally stable air pollution concentration estimated by ESCAPE-LUR for the specific locations of routine monitoring sites. For ESCAPE-LUR values we used the original, not back-extrapolated values, since during the study period of 2006–2008, no substantial changes of long-term air pollutant concentrations were observed at the routine monitoring sites, therefore not having a meaningful influence on the back-extrapolated values.

With respect to the additional feature of source-specific estimation of air pollution concentrations, we further investigated the correlation of traffic-specific and industry-specific EURAD-CTM (EURAD-CTM_{TRA} and EURAD-CTM_{IND}, respectively) and ESCAPE-LUR concentrations at residence as well as at locations of specific ESCAPE measurement sites.

Statistical analysis were carried out with the statistical software R version 3.1.3 (2015-03-09) [39].

3. Results and Discussion

3.1. Comparison of Residence-Based EURAD-CTM and ESCAPE-LUR

Residence-based air pollution concentrations (for 4809 residences within the HNR study area) estimated by EURAD-CTM as yearly-mean air pollution concentrations for the years 2001–2003 (not including 2000 since modeling did not start before October 2000), 2006–2008 and estimated yearly mean air pollution concentrations by ESCAPE-LUR as well as back-extrapolated ESCAPE-LUR air pollution concentration estimates are presented in Table 2 for $PM_{2.5}$, PM_{10}, and NO_2 and visualized in Figure 2 for the year 2008 (EURAD-CTM) and October 2008–October 2009 (ESCAPE-LUR), respectively.

Table 2. Description of residence-based air pollutant exposure estimates $PM_{2.5}$, PM_{10}, and NO_2 from EURAD-CTM and ESCAPE-LUR for 4809 residences within the HNR study area.

–	$PM_{2.5}$ Mean ± SD (Min, Max)	PM_{10} Mean ± SD (Min, Max)	NO_2 Mean ± SD (Min, Max)
	EURAD-CTM (μg/m³)		
2001 year-mean	16.6 ± 1.5 (14.0, 21.6)	21.2 ± 2.9 (17.0, 30.1)	42.2 ± 4.2 (28.2, 55.4)
2002 year-mean	16.8 ± 1.4 (14.3, 21.2)	20.4 ± 1.9 (16.7, 27.0)	39.3 ± 3.8 (27.5, 50.2)
2003 year-mean	18.2 ± 1.4 (15.5, 22.7)	22.4 ± 3.3 (17.8, 32.4)	42.7 ± 4.1 (30.1, 56.1)
2006 year-mean	16.2 ± 1.3 (13.9, 21.2)	21.0 ± 3.7 (16.5, 34.2)	40.0 ± 4.8 (27.1, 57.2)
2007 year-mean	15.7 ± 1.3 (13.4, 20.3)	19.8 ± 2.9 (15.7, 30.8)	37.7 ± 4.5 (26, 53.7)
2008 year-mean	14.6 ± 1.1 (12.5, 19.0)	18.0 ± 2.3 (14.9, 25.1)	37.5 ± 3.9 (26.3, 47.9)
	ESCAPE-LUR (μg/m³)		
back-extrapolated (2-year averages)	–	30.3 ± 2.1 (25.5, 38.7)	30.5 ± 5.0 (19.3, 62.0)
Year 2008–2009	18.4 ± 1.0 (16.0, 21.4)	27.7 ± 1.8 (23.9, 34.7)	30.1 ± 4.9 (19.8, 62.4)
	Difference (μg/m³)		
ΔESCAPE-LUR (2008–09) EURAD-CTM (2008)	3.7 ± 1.3 (−0.7, 7.0)	9.8 ± 2.4 (0.9, 16.5)	−7.4 ± 4.9 (−26.8, 18.9)

On a residential basis, estimated $PM_{2.5}$ and PM_{10} concentrations revealed a consistent decline since 2006 (Table 2). Considering the back-extrapolated ESCAPE-LUR and ESCAPE-LUR, we also observed a decline over time. Observed declines are accounted for by ongoing nation- and state-wide air quality regulations.

Comparing EURAD-CTM (2008) and ESCAPE-LUR (2008–09), however, we saw that the overall mean of the ESCAPE-LUR was considerably higher compared to the overall yearly-mean of EURAD-CTM ($\Delta PM_{2.5} = 3.7 \pm 1.3\ \mu g/m^3$ and ΔPM_{10} $9.8 \pm 2.4\ \mu g/m^3$, respectively). Ranges for $PM_{2.5}$ estimated by EURAD-CTM were slightly smaller than estimated by ESCAPE-LUR (5.4 *vs.* 6.5 $\mu g/m^3$), while ranges for PM_{10} were more similar for both models (10.8 *vs.* 10.0 $\mu g/m^3$). Smaller ranges of air pollution concentrations from EURAD-CTM are not unexpected due to the smoothing pattern within 1 km^2.

Explanations for the difference in mean concentrations for PM might be a consequence of the finer spatial resolution of the ESCAPE-LUR, since high exposure peaks in a very close proximity to busy roads are better captured with this model than with the EURAD-CTM, especially considering that residences are usually located close to the roads and not randomly distributed across a certain area.

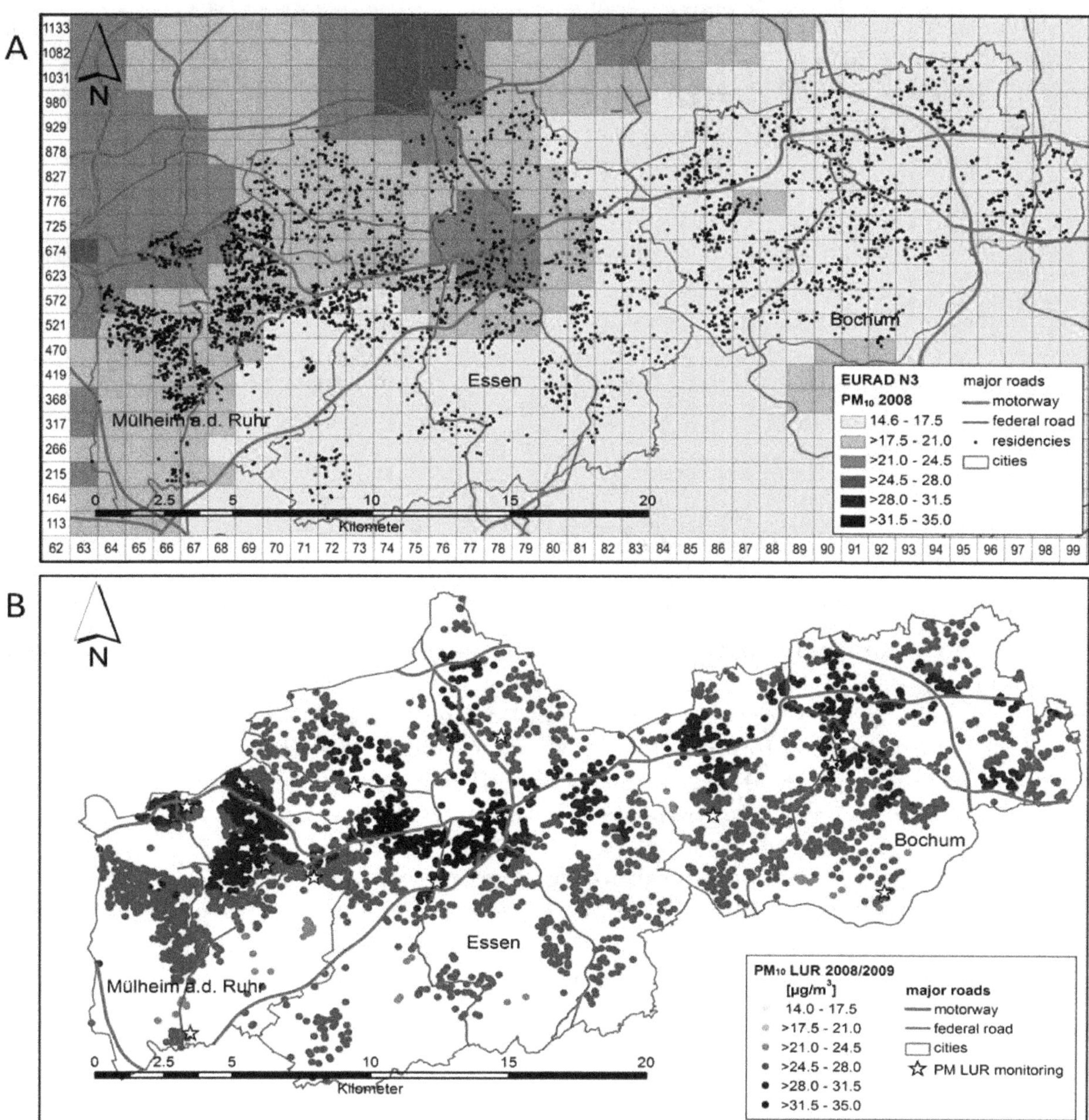

Figure 2. Spatial distribution of EURAD-CTM (1 km^2, yearly mean 2008, (**A**)) and ESCAPE-LUR (yearly mean October 2008–October 2009, (**B**)) at 4809 residences within the HNR study area for PM_{10}.

Pearson's correlation coefficients between models were rather weak for both, $PM_{2.5}$ and PM_{10}, with 0.33. This rather weak correlation has been reported earlier [12] and is not unexpected due to the different spatial resolution but also due to the different spatial distribution of PM concentrations for the two modelling approaches within the study area (Figure 2 and Figure S2): while we observed a west-to-east gradient for EURAD-CTM with higher concentrations in the west, estimated concentrations of ESCAPE-LUR revealed only a slight west-to-east gradient, which was prominently overlapped by an additional decreasing north-to-south and local hot spots, e.g., in Essen at a motorway intersection. In our study area the decreasing west-to-east gradient mirrors the distribution of industrial locations, e.g., metallurgical-industry and Europe's largest inland harbour in Duisburg, located to the west of the study area (Figure 1), as well as transported emissions from other countries in the west of study area, e.g., the Netherlands or Great Britain. The decreasing north-to-south gradient on the other hand is consistent with the population density and the location of major arterial roads within our study area [32].

NO_2 concentrations estimated by EURAD-CTM showed an overall decrease between 2001 with 42.2 μg/m^3 and 2008 with 37.7 μg/m^3, while a change between the ESCAPE-LUR and the back-extrapolated ESCAPE-LUR was not observed. Yet, in contrast to PM, temporally-stable NO_2 concentrations estimated by ESCAPE-LUR were systematically lower than estimated by EURAD-CTM ($\Delta NO_2 = -7.4 \pm 4.9$ μg/m^3). One explanation for this difference could be a misrepresentation of industrial sources within the ESCAPE modeling approach: "industry" referred to industrial, commercial and transport units in a certain buffer, giving no information of the emission of such sources. Ranges of concentrations, however, were twice as big for the ESCAPE-LUR compared to the EURAD-CTM (42.4 *vs.* 21.9 μg/m^3), probably driven by greater small-scale variations due to point-specific estimates and the consideration of traffic load within a buffer of 100 m. Unlike spatial gradients for $PM_{2.5}$ and PM_{10}, we observed a more pronounced northwest-to-southeast-gradient for EURAD-CTM for NO_2, while the distribution of NO_2 by ESCAPE-LUR did not reveal a clear gradient, but local hot spots near major roads or motorway intersections (Figure S2). Similar to PM, correlation between EURAD-CTM NO_2 and ESCAPE-LUR NO_2 was rather weak with a correlation coefficient of 0.4.

3.2. Comparison of Estimated and Measured Air Pollution Concentrations

3.2.1. Comparison between 14-Day Mean ESCAPE-LUR Measurements and EURAD-CTM Estimates

In order to evaluate EURAD-CTM estimates we compared estimated 14-day mean AP concentrations by EURAD-CTM to available 14-day measurements taken during the ESCAPE measurement campaign. Descriptive statistics and correlation coefficients of these 14-day mean measured air pollution concentrations at ESCAPE measurement sites (background, traffic (*cf.* Table S2), and both) and the respective 14-day mean air pollution concentrations estimated by EURAD-CTM in the corresponding grid cells are shown in Table 3 for air pollutants $PM_{2.5}$, PM_{10}, and NO_2.

Table 3. Description of 14-day mean measured air pollution concentrations at ESCAPE measurement sites (background and/or traffic) and 14-day mean air pollution concentration estimations of EURAD-CTM in the corresponding grid cells for $PM_{2.5}$, PM_{10}, and NO_2.

Background	**ESCAPE Site (μg/m^3)**	**EURAD-CTM (μg/m^3)**	**Spearman Correlation Coefficient (*r*)**
	Mean ± SD	**Mean ± SD**	
$PM_{2.5}$ (N = 9)	17.78 ± 2.40	19.80 ± 5.80	0.34
PM_{10} (N = 9)	26.12 ± 4.70	23.29 ± 5.98	0.93
NO_2 (N = 16)	37.85 ± 6.21	50.82 ± 10.07	0.34
		traffic	
$PM_{2.5}$ (N = 6)	19.75 ± 3.75	21.78 ± 6.96	0.43
PM_{10} (N = 6)	29.26 ± 4.95	26.97 ± 7.68	0.37
NO_2 (N = 13)	50.43 ± 9.83	58.04 ± 10.33	0.60
		Background + traffic	
$PM_{2.5}$ (N = 15)	18.57 ± 3.05	20.59 ± 6.13	0.45
PM_{10} (N = 15)	27.37 ± 4.89	24.77 ± 6.71	0.77
NO_2 (N = 29)	43.49 ± 10.13	54.06 ± 10.65	0.55

Overall, 14-day mean EURAD-CTM estimates for $PM_{2.5}$ are slightly higher than mean of 14 daily measurements at the ESCAPE sites, while EURAD-CTM estimates for PM_{10} are slightly lower and EURAD-CTM estimates for NO_2 are considerably higher, especially regarding the ESCAPE background site (Table 3).

The highest correlation coefficient (r) was observed for PM_{10} between EURAD-CTM and ESCAPE background sites ($r = 0.93$), while the lowest correlation was observed for PM_{10} between EURAD-CTM and ESCAPE traffic sites ($r = 0.37$). This finding is not unexpected, regarding the aim, input, and construction of the two modeling approaches (Table 1): the EURAD-CTM aims to assess an average concentration in a 1 km^2 grid cell, taking into account long-range transport rather than locally-emitted pollution, in contrast to the ESCAPE-LUR, which was specifically designed to assess mostly traffic-related differences in exposure concentration. For $PM_{2.5}$, however, we did not observe a clear distinction between background and traffic sites, whereas correlation coefficients for NO_2 were higher between EURAD-CTM and ESCAPE traffic sites ($r = 0.60$) than between EURAD-CTM and ESCAPE background sites ($r = 0.34$). One reason for the low to moderate correlation between $PM_{2.5}$ modeled by EURAD-CTM and $PM_{2.5}$ measured at ESCAPE sites could be the lack of the assimilation procedure within EURAD-CTM, since $PM_{2.5}$ has only been measured at routine monitoring sites since 2009. So, for the considered period of time, estimated $PM_{2.5}$ was only assimilated indirectly taking a (constant) proportion of PM_{10} and $PM_{2.5}$ into account.

Overall, correlations between EURAD-CTM estimates and measured concentrations at all ESCAPE measurement sites were moderate for $PM_{2.5}$ ($r = 0.45$) and NO_2 ($r = 0.55$), and high for PM_{10} ($r = 0.77$) and, therefore, slightly better than comparing residence-based modeled air pollution concentrations between EURAD-CTM and ESCAPE-LUR.

3.2.2. Comparison between Routinely-Monitored and Estimated Air Pollution Concentrations

Overall correlations between daily measurements at routine monitoring sites and EURAD-CTM estimations over one year (2008) were strong for PM_{10} and NO_2 (>0.8) and moderate for $PM_{2.5}$ (0.66–0.74) for both, background and traffic monitoring site (Table 4). This finding is a consequence of the assimilation procedure within EURAD-CTM for PM_{10} and NO_2.

Taking into account absolute annual values, we observed several findings: annual averages for January 2008 until December 2008 differ considerably from annual averages from 16 October 2008 to 15 October 2009 (ESCAPE measurement period), for PM (Table 4). Generally, PM concentrations throughout Germany were at a minimum in 2008, as reported by the Federal Environment Agency [40]. This finding points to the importance of a fine temporal resolution even in medium- and long-term exposure estimations.

Considering uncertainty, the EURAD-CTM estimations underestimated PM and overestimated NO_2 at background monitoring sites, while the ESCAPE-LUR estimations agreed well for $PM_{2.5}$ (all sites) and PM_{10} (background sites), but tended to underestimate NO_2 concentrations considerably (Table 4). The latter is supported by mean squared errors of the LOOCV, which were remarkably higher for NO_2 than for PM. Furthermore, we observed considerable disagreement between predicted ESCAPE-LUR PM_{10} and measured PM_{10} at the routine monitoring traffic-site. This finding might be a consequence of the disagreement between PM_{10} measured at the routine monitoring site and the measured PM_{10} at the closest ESCAPE site (26.64 *vs.* 32.70 $\mu g/m^3$), which were located only 2.2 m away from each other.

Table 4. Yearly mean air pollution concentrations measured at routine monitoring sites (background (BG) and traffic (TRAFFIC)), provided by LANUV, modeled by EURAD-CTM (for the respective grid cell), modeled by ESCAPE-LUR (at the location of the routine monitoring sites) and measured adjusted yearly mean at the closest ESCAPE site plus Pearson's correlation coefficient between LANUV daily measurements and EURAD-CTM daily estimations for $PM_{2.5}$, PM_{10}, and NO_2.

Air Pollutant (μg/m³)	LANUV Monitor (2008)	EURAD-CTM (2008)	Pearson Correlation Coefficient (LANUV*EURAD-CTM)	LANUV Monitor (October 2008–October 2009)	ESCAPE-LUR Prediction (October 2008-October 2009)	Closest ESCAPE-Measurement Site
			Mülheim-Styrum (BG) (grid cell: 679)			
$PM_{2.5}$	17.90	16.33	0.66	20.71	19.50	19.00 [a]
PM_{10}	25.24	23.21	0.88	28.20	28.86	29.00 [a]
NO_2	34.17	39.33	0.80	34.67	31.42	33.00 [a]
			Essen-Vogelheim (BG) (grid cell: 942)			
$PM_{2.5}$	22.08	16.21	0.74	20.18	19.31	18.50 [b]
PM_{10}	27.66	23.79	0.81	27.32	26.64	26.40 [b]
NO_2	35.17	41.56	0.76	35.70	28.75	53.30 [c]
			Essen-Ost city (TRAFFIC) (grid cell: 690)			
$PM_{2.5}$	20.08	14.72	0.69	20.51	21.05	20.90 [d]
PM_{10}	26.61	23.77	0.81	26.64	33.38	32.70 [d]
NO_2	46.36	44.97	0.87	47.65	42.01	43.50 [d]

[a] 6.7 m; [b] 2665.0 m; [c] 4060.1 m, [d] 2.2 m.

3.3. Temporal Resolution of Air Pollution Concentrations

Regarding different years (2006–2008) we saw a weak time-dependent decline in PM concentrations (Table 2), in line with the observed overall decline in PM concentrations from the year 2001 to 2008 within the HNR study area [29]. To examine the temporal resolution on a monthly basis, Figure 3 and Figure S3 present monthly distributions of EURAD-CTM estimated air pollution concentrations of PM_{10}, $PM_{2.5}$, and NO_2 respectively, in two grid cells, including one background grid cell (679) and one traffic routine monitoring site grid cell (690), presenting spatial variation. For the purpose of comparison, yearly mean air pollution concentrations estimated with EURAD-CTM for the two grid cells as well as the temporally-stable ESCAPE-LUR air pollution concentrations estimated at the locations of the monitoring sites, and monthly-based measured air pollution concentration at routine monitoring sites are presented as lines. Overall, we observed strong seasonal variation (high in winter and low in summer) for estimated EURAD-CTM air pollution concentrations and measured air pollution concentrations, which cannot be detected when using the temporally stable ESCAPE-LUR estimates. While ESCAPE-LUR estimates are primarily designed to yield long-term exposure estimates without temporal resolution, the integration of other measurements (*i.e.*, from routine monitoring sites), or other measurement periods (e.g., three month instead of one year), can be used to derive LUR-data for the analysis of medium-term health effects [41], although not covered in this manuscript.

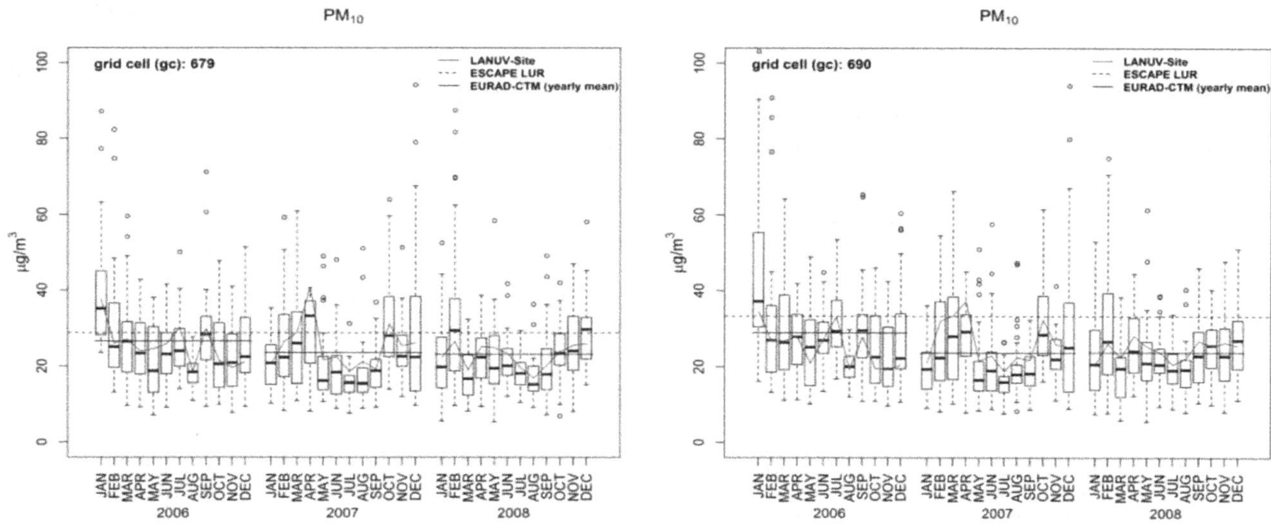

Figure 3. Box plots of air pollution concentrations of PM_{10} over time for two grid cells (gc), representing background (gc: 679) and traffic (gc: 690), estimated by EURAD-CTM on a monthly and yearly basis, long-term ESCAPE-LUR estimation and measured at monitoring sites on a monthly basis (median per month).

The seasonal patterns differed slightly across years and air pollutants (Figure 3 and Figure S3). Reasons for such differences might be specific meteorological conditions during the observation period as well as different chemical processes differentially influencing the concentration of the examined air pollutants, e.g., regarding transport, deposition or physical and chemical aging. These observed seasonal changes underscore the importance of time-dependent air pollution models for the analysis of short- and medium-term health effects. When using a LUR for short- and medium-term exposures, a finer temporal resolution can be achieved using back-extrapolation based on routine monitoring sites, as has been applied for birth outcomes in the framework of ESCAPE [41]. Furthermore, estimated $PM_{2.5}$ by EURAD-CTM, although following the seasonal pattern of measured $PM_{2.5}$, was considerably under-estimated, reflecting the lack of data assimilation within this modeling procedure. In contrast to the temporal variation over the considered time period, the spatial variation, presented by the two locations of a background and traffic site, is considerably smaller. This finding is in line with earlier findings, indicating a slightly higher temporal, than spatial, variation of particle number concentrations within the Ruhr area [42].

3.4. Source-Specific EURAD-CTM

Estimated local traffic-specific (TRA) and local industry-specific (IND) air pollution concentrations take up only a small amount of all sources: for $PM_{2.5}$ local traffic takes up 3.4% and local industry 9.6%; for PM_{10} it is 2.7% and 10.5%, respectively, and for NO_2 it is 21.4% and 2.4%, respectively. Correlation coefficients between PM concentrations, including all sources and including only local traffic, were weak (0.34–0.43), while all-sources PM and industry-specific PM correlated well (0.73–0.96) (Figure 4). Correlation coefficients for NO_2 were, in contrast to PM, higher between all sources and local-traffic (0.63) and lower for industry-specific (0.44). The rather small amount of local traffic-and industry-specific concentrations is not surprising considering that long-range transport and formation of secondary particles in the atmosphere can contribute considerably to the particle mass concentration in North-Rhine-Westphalia and the Ruhr area, sometimes more than 50% depending on the meteorological situation [28]. The spatial distribution within the study area, represented by quintiles of respective PM_{10} distributions (Figure 4), illustrates that the agreement between all sources and industry-specific sources is better than between all sources and traffic-specific PM. Due to substantial industrial emissions from the Duisburg inland harbor and the adjacent industrial area west of the study region, a strong west-east gradient can be observed for industry-specific PM and for all sources PM. The spatial distribution traffic-specific PM follows closely the population-density in the study area, with a strong north-to-south gradient.

The associations between residence-based exposure estimates derived from EURAD-CTM_{TRA} and ESCAPE-LUR are relatively high ($PM_{2.5}$: 0.69, PM_{10}: 0.58, and NO_2: 0.45), while they are expectedly considerably lower for EURAD-CTM_{IND} and ESCAPE-LUR ($PM_{2.5}$: 0.16, PM_{10}: 0.0, and NO_2: 0.25) (Table 5). Such patterns are displayed for PM_{10} in the spatial distribution of traffic-specific EURAD-CTM and ESCAPE-LUR and industry-specific EURAD-CTM and ESCAPE-LUR, respectively (Figure 4). A similar pattern is observed taking into account correlations for 14-day mean measurements at ESCAPE monitoring stations (background and traffic) and estimated 14-day mean EURAD-CTM_{TRA} within respective grid cells (Table 5).

Table 5. Spearman correlation coefficients between 14-day series of measurements at ESCAPE-LUR-monitoring stations and 14-day mean estimations of EURAD-CTM_{TRA} in respective grid cells.

EURAD-CTM_{TRA} (Traffic-Specific)	ESCAPE Background Sites	ESCAPE Traffic Sites	All ESCAPE Sites
$PM_{2.5}$	0.69 (n = 9)	0.88 (n = 6)	0.77 (n = 15)
PM_{10}	0.02 (n = 9)	0.83 (n = 6)	0.32 (n = 15)
NO_2	0.57 (n = 16)	0.79 (n = 13)	0.63 (n = 29)

These observations indicate that EURAD-CTM and ESCAPE-LUR do not represent identical aspects of air pollution: while EURAD-CTM represents an area average similar to urban background concentrations, the ESCAPE-LUR was designed to predominantly estimate variability in local traffic-related air pollution, leading to a comparatively high correlation with local traffic-specific air pollution concentrations modeled by EURAD-CTM. The very low correlation with local industry-specific air pollution concentration at the residences indicates, that ESCAPE-LUR represents industry rather poorly compared to EURAD-CTM, where the overall spatial distribution (Figure 3) is mainly driven by industrial sources as has been observed in a previous study [32].

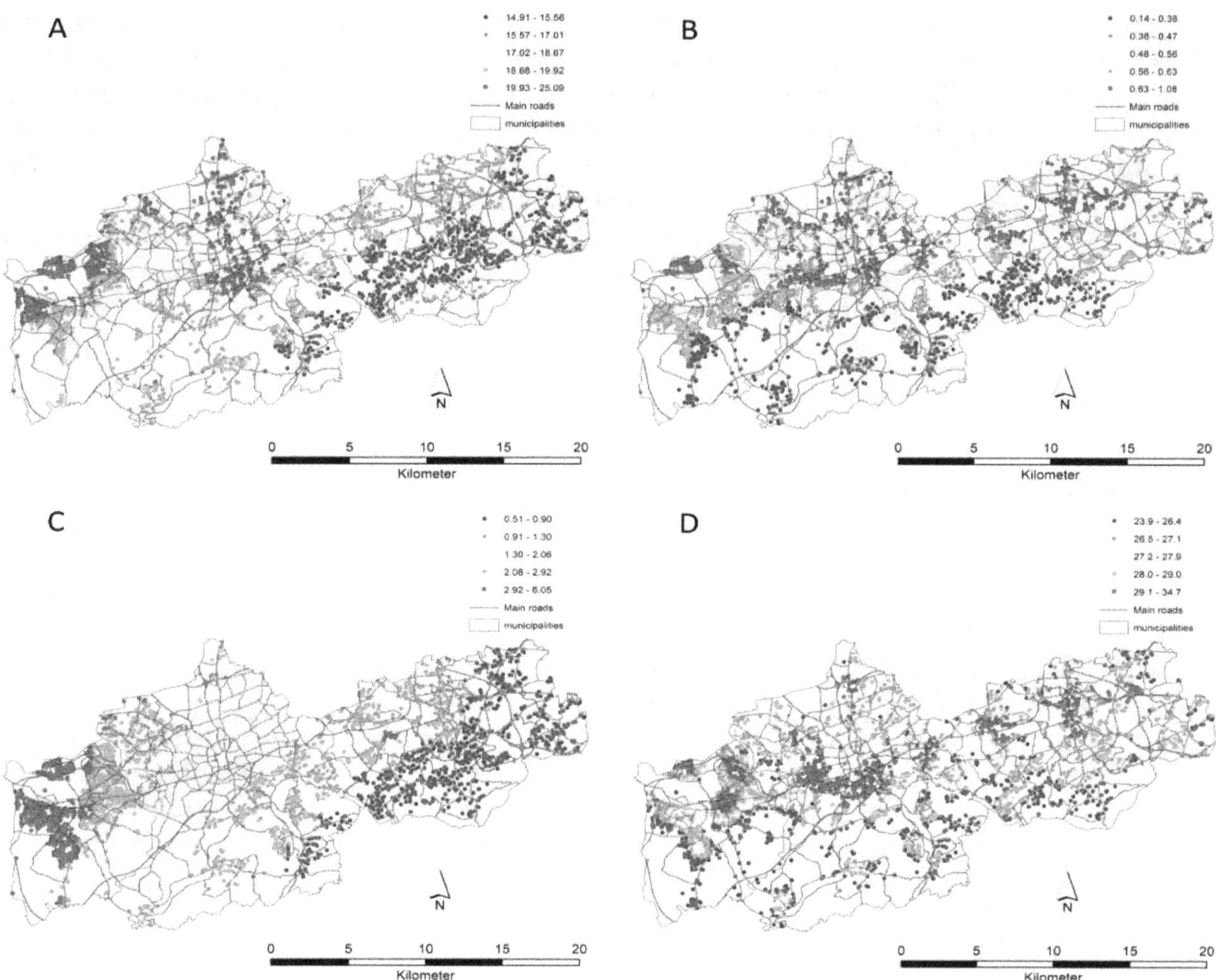

Figure 4. Residence-based spatial distribution of PM_{10} concentrations from EURAD-CTM: all-sources (**A**); local traffic (**B**) and local industry (**C**); and ESPCAPE-LUR (**D**).

4. Conclusions

Based on the comparison between air pollution concentrations modeled by ESCAPE-LUR and EURAD-CTM within the HNR study area, we showed that both model types have different input data as well as different temporal and spatial resolutions, driven by their different aims and application. While the point-specific ESCAPE-LUR primarily aims to estimate temporally stable and spatial variable long-term exposure to locally-emitted (mostly traffic-related) air pollution with a very high spatial resolution, the EURAD-CTM aims to estimate a spatio-temporal average air pollutant concentration in a small area (*i.e.*, 1 km^2), taking into account a range of major sources, e.g., traffic, industry, meteorological condition, and transport. While the observed weak to moderate overall agreement between the ESCAPE-LUR and the EURAD-CTM supports earlier findings [12], our analysis showed that the agreement between the two models improved considerably after restricting the EURAD-CTM to local traffic only. This finding was further supported by results comparing 14-day mean concentrations estimated by EURAD-CTM and measured at purpose-specific ESCAPE monitoring sites, yielding the highest correlations for traffic-specific EURAD-CTM estimates and measurements at traffic sites.

One of the principal strengths of the point-specific ESCAPE-LUR is to capture very small-scale variations in air pollution. Yet, this accuracy may be more error-prone than the coarser spatial resolution of 1 km^2 used by EURAD-CTM, regarding exposure assignment in cases of high personal mobility within small distances, like daily chores around the residence. The biggest strength of an LUR approach in general is the wide-ranging applicability, like the relatively small requirements on measurement sites (low cost), the individual location of measurement sites, the easy assessment of land use data, and the straight forward model building procedure, based on linear regression modeling. In contrast, the EURAD-CTM, or chemical transport and dispersion modeling approaches in general, are less accessible to changes by the user due to the highly complex underlying mathematical, physical, and chemical modelling procedures. These complex procedures are, however, accompanied with benefit of including chemical transport actions, which allow modeling air pollution components that have not been measured. The LUR, on the other hand, is limited to modelling measured air pollutants. Moreover, CTMs enable the investigation of the role of meteorology and the prediction of air pollutant concentrations under hypothetical emission situations.

The comparatively easy applicability of LUR modeling and statistical model building procedure may come along with potential costs of wrong decisions: the initial choice of locations of the measurement sites limits the specificity of the model to capture those emission sources, whose concentration gradients are well captured by the chosen sites and may fail to capture all important source-specific concentration gradients across a study area, especially if important sources change over time. Restricting predictors to land use data might neglect important predictors of air pollution concentrations from other sources and processes, like chemical interaction and transport. Similarly, CTMs are only valid if based on a comprehensive and detailed emission database. To overcome limitations of each of the models and optimally make use of the respective strengths, we propose to combine the two approaches into a hybrid model [43,44]. These hybrid models are usually based on the LUR model since LURs are by design much easier to modify.

To conclude, our results show that ESCAPE-LUR and the EURAD-CTM are constructed to estimate complementary aspects of air pollution and both approaches have respective strengths and limitations, which need to be considered especially when investigating health effects. The possibility of combining the strengths of both, e.g., using hybrid models will be the next step to enhance exposure assessment.

Supplementary Materials
Figure S1: Flowchart of the EURAD model system containing the meteorological driver MM5, the pre-processors ECP and PREP, the emission model EEM and the chemistry transport model EURAD (input parameters are shaded in blue, output parameters are shaded in yellow and procedural parts are shaded in green or magenta), Figure S2: Spatial distribution of EURAD-CTM (1 km^2, yearly mean 2008) and ESCAPE-LUR (point-specific yearly mean October 2008–October 2009) at 4809 residences within the HNR study area for PM_{10} (A+C) and NO_2 (B+D), Figure S3: Boxplots of air pollution concentrations of monthly-mean PM_{10} and NO_2 concentrations over three year for a traffic-specific (grid cell: 690), and a background-specific location (grid cell: 679) with annual mean ESCAPE-LUR estimates and annual measurements at LANUV monitoring sites, Table S1: Time and locations of the ESCAPE-measurement campaign, Table S2: ESCAPE-LUR for $PM_{2.5}$, PM_{10} and NO_2, Table S3: Time and Location of routine monitoring sites, provided by LANUV, within the HNR study area.

Acknowledgments: The research leading to these results has received funding from the European Community's Seventh Framework Program (FP7/2007–2011) under grant agreement number: 211250; The study was supported by the Heinz Nixdorf Foundation (chairman: M. Nixdorf; former chairman: G. Schmidt (deceased)) and the German Ministry of Education and Science, the German Research Foundation (DFG: HO 3314/2-1, HO 3314/2-3, HO 3314/4-3). Lilian Tzivian gratefully acknowledges support by a post-doctoral fellowship from the Environmental and Health Fund, Jerusalem, Israel.

Author Contributions: Frauke Hennig contributed to the design of study, the analysis and interpretation of the results and preparation of manuscript. Dorothea Sugiri contributed to the analysis and preparation of the manuscript. Kateryna Fuks, Lilian Tzivian, Danielle Vienneau, Kees de Hoogh, Susanne Moebus, Karl-Heinz Jöckel, Ulrich Quass and Thomas Kuhlbush contributed to the interpretation of results and preparation of manuscript. Hermann Jakobs and Michael Memmesheimer contributed to the preparation of the manuscript. Barbara Hoffmann contributed to the design of study, interpretation of the results and preparation of manuscript.

References

1. World Health Organization. *The World Health Report 2013: Research for Universal Health Coverage*; WHO: Geneva, Switzerland, 2013.

2. Dockery, D.W.; Pope, C.A.; Xu, X.; SPrengler, J.D.; Ware, J.H.; Fay, M.E.; Ferris, B.G.; Speizer, F.E. An association between air pollution and mortality in six U.S. cities. *New Engl. J. Med.* **1993**, *329*, 1753–1759. [CrossRef] [PubMed]
3. Jerrett, M.; Arain, A.; Kanaroglou, P.; Beckerman, B.; Potoglou, D.; Sahsuvaroglu, T.; Morrison, J.; Giovis, C. A review and evaluation of intraurban air pollution exposure models. *J. Expo. Anal. Environ. Epidemiol.* **2005**, *15*, 185–204. [CrossRef] [PubMed]
4. Health Effects Institute. *HEI Panel on the Health Effect of Traffic-Related Air Pollution: A Critical Review of the Literature on Emissions, Exposure, and Health Effects*; Health Effects Institute: Boston, MA, USA, 2010.
5. Özkaynak, H.; Baxter, L.K.; Dionisio, K.L.; Burke, J. Air pollution exposure prediction approaches used in air pollution epidemiology studies. *J. Expo. Sci. Environ. Epidemiol.* **2013**, *23*, 566–572. [CrossRef] [PubMed]
6. Briggs, D.J. The use of GIS to evaluate traffic-related pollution. *Occup. Environ. Med.* **2007**, *64*, 1–2. [CrossRef] [PubMed]
7. Hoek, G.; Beelen, R.; de Hoogh, K.; Vienneau, D.; Gulliver, J.; Fischer, P.; Briggs, D. A review of land-use regression models to assess spatial variation of outdoor air pollution. *Atmos. Environ.* **2008**, *42*, 7561–7578. [CrossRef]
8. Cyrys, J.; Hochadel, M.; Gehring, U.; Hoek, G.; Diegmann, V.; Brunekreef, B.; Heinrich, J. GIS-based estimation of exposure to particulate matter and NO_2 in an urban area: Stochastic *versus* dispersion modeling. *Environ. Health Perspect.* **2005**, *113*, 987–992. [CrossRef] [PubMed]
9. Marshall, J.D.; Nethery, E.; Brauer, M. Within-urban variability in ambient air pollution: Comparison of estimation methods. *Atmos. Environ.* **2008**, *42*, 1359–1369. [CrossRef]
10. Briggs, D.J.; Collins, S.; Elliott, P.; Fischer, P.; Kingham, S.; Lebret, E.; Pryl, K.; Van Reeuwijk, H.; Smallbone, K.; van Der Veen, A. Mapping urban air pollution using GIS: A regression-based approach. *Int. J. Geogr. Inf. Sci.* **1997**, *11*, 699–718. [CrossRef]
11. Beelen, R.; Voogt, M.; Duyzer, J.; Zandveld, P.; Hoek, G. Comparison of the performances of land use regression modelling and dispersion modelling in estimating small-scale variations in long-term air pollution concentrations in a Dutch urban area. *Atmos. Environ.* **2010**, *44*, 4614–4621. [CrossRef]
12. De Hoogh, K.; Korek, M.; Vienneau, D.; Keuken, M.; Kukkonen, J.; Nieuwenhuijsen, M.J.; Badaloni, C.; Beelen, R.; Bolignano, A.; Cesaroni, G.; *et al.* Comparing land use regression and dispersion modelling to assess residential exposure to ambient air pollution for epidemiological studies. *Environ. Int.* **2014**, *73*, 382–392. [PubMed]
13. Perez, L.; Wolf, K.; Hennig, F.; Penell, J.; Basagaña, X.; Aguilera, I.; Agis, D.; Beelen, R.; Brunekreef, B.; Cyrys, J.; *et al.* Air pollution and atherosclerosis: A cross-sectional analysis of four European cohort studies in the ESCAPE study. *Environ. Health Perspect.* **2015**, *123*, 597–605. [PubMed]
14. Duisburg-Essen University. Ruhr & Culture. Available online: https://www.uni-due.de/welcome-services/en/rk_index.php (accessed on 3 December2015).
15. Ebel, A. The Eurad Project. Available online: http://www.uni-koeln.de/math-nat-fak/geomet/eurad/index_e.html (accessed on 30 November 2015).
16. Hass, H.; Ebel, A.; Feldmann, H.; Jakobs, H.J.; Memmesheimer, M. Evaluation studies with a regional chemical transport model (EURAD) using air quality data from the EMEP monitoring network. *Atmos. Environ. A. Gen. Top.* **1993**, *27*, 867–887. [CrossRef]
17. Ebel, A.; Elbern, H.; Feldmann, H.; Jakobs, H.; Kessler, C.; Memmesheimer, M. *Air Pollution Studies with the EURAD Model System (3): EURAD-European Air Pollution Dispersion Model SYSTEM*; University of Cologne: Cologne, Germany, 1997.
18. Memmesheimer, M.; Friesse, E.; Ebel, A.; Jakobs, H.; Feldmann, H.; Kessler, C. Long-term simulations of particulate matter in Europe on different scales using sequential nesting of a regional model. *Int. J. Environ. Pollut.* **2004**, *22*, 108–132. [CrossRef]
19. Schell, B.; Ackermann, I.; Hass, H.; Binkowski, F.; Ebel, A. Modeling the formation of secondary organic aerosol within a comprehensive air quality modeling system. *J. Geophys. Res.* **2001**, *106*, 28275–28293. [CrossRef]
20. Büns, C.; Klemm, O.; Wurzler, S.; Hebbinghaus, H.; Steckelbach, I.; Friesel, J.; Ebel, A.; Friese, E.; Jakobs, H.; Memmesheimer, M. Comparison of four years of air pollution data with a mesoscale model. *Atmos. Res.* **2012**, *118*, 404–417. [CrossRef]

21. Grell, G.A.; Oceanic, N.; Administr, A.; Dudhia, J. *A Description of the Fifth-Generation Penn State/NCAR Mesoscale Model (MM5)*; University Corporation for Atmospheric Research (UCAR): Boulder, CO, USA, 2016.
22. Memmesheimer, M.; Tippke, J.; Ebel, A.; Hass, H.; Jakobs, H.; Laube, M. On the use of EMEP emission inventories for European scale air pollution modelling with the EURAD model. In Proceedings of the 1991 EMEP Workshop on Photooxidant Modelling for Long Range Transport in Relation to Abatement Strategies, Berlin, Germany, 16–19 April 1991; pp. 307–324.
23. Elbern, H.; Strunk, A.; Schmidt, H.; Talagrand, O. Emission rate and chemical state estimation by 4-dimensional variational inversion. *Atmos. Chem. Phys.* **2007**, *7*, 3749–3769. [CrossRef]
24. Petry, H.; Ebel, A.; Franzkowiak, V.; Hendricks, J.; Lippert, E.; Möllhoff, M. Impact of aircraft exhaust on atmosphere: Box model studies and 3-d mesoscale numerical case studies of seasonal differences. In Proceedings of 1996 Impact of Aircraft Emissions upon the Atmosphere, Paris, France, 15–18 October 1996; pp. 241–246.
25. Current concentrations of air pollutants in Germany. Available online: http://www.umweltbundesamt.de/daten/luftbelastung/aktuelle-luftdaten (accessed 30 November 30 2015).
26. Stationen und Messwerte. Available online: http://www.lanuv.nrw.de/umwelt/luft/immissionen/stationen-und-messwerte/ (accessed on 30 November 2015).
27. Elbern, H. The Objectives of Chemical Data Assimilation. Available online: http://db.eurad.uni-koeln.de/en/research/working_group_he/data_assimilation.php (accessed on 16 December 2015).
28. Hebbinghaus, H.; Wurzler, S.; Friese, E.; Jakobs, H.J.; Kessler, C.; Ebel, A. Determination of the contribution of different groups of emission sources on the concentration of PM10, PM2.5, and NO2 in North Rhine-Westphalia—A whodunnit. In Proceedings of the 2009 European Aerosol Conference, Karlsruhe, Germany, 6–11 September 2009.
29. Nonnemacher, M.; Jakobs, H.; Viehmann, A.; Vanberg, I.; Kessler, C.; Moebus, S.; Möhlenkamp, S.; Erbel, R.; Hoffmann, B.; Memmesheimer, M. Spatio-temporal modelling of residential exposure to particulate matter and gaseous pollutants for the Heinz Nixdorf Recall Cohort. *Atmos. Environ.* **2014**, *91*, 15–23. [CrossRef]
30. Fagerli, H.; Dutcheak, S.; Torseth, K.; QAmman, M.; Ritter, M. EMEP. Available online: http://www.emep.int/ (accessed on 30 November 2015).
31. European Environment Agency. Index to methodology chapters ordered by SNAP97 Activity. Available online: http://www.eea.europa.eu/publications/EMEPCORINAIR4/page009-a.html (accessed on 30 November 2015).
32. Hennig, F.; Fuks, K.; Moebus, S.; Weinmayr, G.; Memmesheimer, M.; Jakobs, H.; Bröcker-Preuss, M.; Führer-Sakel, D.; Möhlenkamp, S.; Erbel, R.; *et al.* Association between Source-Specific Particulate Matter Air Pollution and hs-CRP: Local Traffic and Industrial Emissions. *Environ. Health Perspect.* **2014**, *122*, 703–710. [PubMed]
33. Keil, M.; Bock, M.; Esch, T.; Metz, A.; Nieland, S.; Pfitzner, A. CORINE Land Cover Aktualisierung 2006 für Deutschland. Available online: http://www.uba.de/uba-info-medien/4086.html (accessed on 16 December 2015).
34. Eeftens, M.; Beelen, R.; de Hoogh, K.; Bellander, T.; Cesaroni, G.; Cirach, M.; Declercq, C.; Dedele, A.; Dons, E.; de Nazelle, A.; *et al.* Development of Land Use Regression Models for $PM_{2.5}$, $PM_{2.5}$ Absorbance, PM_{10} and PM_{coarse} in 20 European Study Areas; Results of the ESCAPE Project. *Environ. Sci. Technol.* **2012**, *46*, 11195–11205.
35. Cyrys, J.; Eeftens, M.; Heinrich, J.; Ampe, C.; Armengaud, A.; Beelen, R.; Bellander, T.; Beregszaszi, T.; Birk, M.; Cesaroni, G.; *et al.* Variation of NO_2 and NO_x concentrations between and within 36 European study areas: Results from the ESCAPE study. *Atmos. Environ.* **2012**, *62*, 374–390.
36. Beelen, R.; Hoek, G.; Vienneau, D.; Eeftens, M.; Dimakopoulou, K.; Pedeli, X.; Tsai, M.Y.; Künzli, N.; Schikowski, T.; Marcon, A.; *et al.* Development of NO_2 and NO_x land use regression models for estimating air pollution exposure in 36 study areas in Europe—The ESCAPE project. *Atmos. Environ.* **2013**, *72*, 10–23.
37. Wang, M.; Brunekreef, B.; Gehring, U.; Szpiro, A.; Hoek, G.; Beelen, R. A New Technique for Evaluating Land-use Regression Models and Their Impact on Health Effect Estimates. *Epidemiology* **2016**, *27*, 51–56. [CrossRef] [PubMed]
38. ESCAPE manuals. Available online: http://www.escapeproject.eu/manuals/ (accessed on 16 December 2015).
39. R Development Core Team. *R: A Language and Environment for Statistical Computing*; R Foundation for Statistical Computing: Vienna, Austria, 2015.

40. Auswertung der Luftbelastungssituation 2009. Available online: http://www.umweltbundesamt.de/sites/default/files/medien/515/dokumente/3895.pdf (accessed on 16 December 2015).
41. Pedersen, M.; Giorgis-Allemand, L.; Bernard, C.; Aguilera, I.; Andersen, A.M. N.; Ballester, F.; Beelen, R.M.J.; Chatzi, L.; Cirach, M.; Danileviciute, A.; *et al.* Ambient air pollution and low birthweight: A European cohort study (ESCAPE). *Lancet Respir. Med.* **2013**, *1*, 695–704. [PubMed]
42. Hertel, S.; Viehmann, A.; Moebus, S.; Mann, K.; Bröcker-Preuss, M.; Möhlenkamp, S.; Nonnemacher, M.; Erbel, R.; Jakobs, H.; Memmesheimer, M.; *et al.* Influence of short-term exposure to ultrafine and fine particles on systemic inflammation. *Eur. J. Epidemiol.* **2010**, *25*, 581–592. [PubMed]
43. Akita, Y.; Baldasano, J.M.; Beelen, R.; Cirach, M.; de Hoogh, K.; Hoek, G.; Nieuwenhuijsen, M.; Serre, M.L.; de Nazelle, A. Large Scale Air Pollution Estimation Method Combining Land Use Regression and Chemical Transport Modeling in a Geostatistical Framework. *Environ. Sci. Technol.* **2014**, *48*, 4452–4459. [CrossRef] [PubMed]
44. Vienneau, D.; de Hoogh, K.; Bechle, M.J.; Beelen, R.; van Donkelaar, A.; Martin, R.V.; Millet, D.B.; Hoek, G.; Marshall, J.D. Western European land use regression incorporating satellite- and ground-based measurements of NO_2 and PM_{10}. *Environ. Sci. Technol.* **2013**, *47*, 13555–13564.

11

Reconstructing Fire Records from Ground-Based Routine Aerosol Monitoring

Hongmei Zhao [1], Daniel Q. Tong [2,3,*], Pius Lee [2], Hyuncheol Kim [2,3] and Hang Lei [2]

1 Key Laboratory of Wetland Ecology and Environment, Northeast Institute of Geography and Agroecology, Chinese Academy of Sciences, Changchun 130102, China; zhaohongmei@iga.ac.cn

2 U.S. NOAA Air Resources Laboratory, College Park, MD 20740, USA; pius.lee@noaa.gov (P.L.); Hyuncheol.kim@noaa.gov (H.K.); hang.lei@noaa.gov (H.L.)

3 UMD/Cooperative Institute for Climate and Satellites, College Park, MD 20740, USA

* Correspondence: daniel.tong@noaa.gov

Abstract: Long-term fire records are important to understanding the trend of biomass burning and its interactions with air quality and climate at regional and global scales. Traditionally, such data have been compiled from ground surveys or satellite remote sensing. To obtain aerosol information during a fire event to use in analyzing air quality, we propose a new method of developing a long-term fire record for the contiguous United States using an unconventional data source: ground-based aerosol monitoring. Assisted by satellite fire detection, the mass concentration, size distribution, and chemical composition data of surface aerosols collected from the Interagency Monitoring of Protected Visual Environments (IMPROVE) network are examined to identify distinct aerosol characteristics during satellite-detected fire and non-fire periods. During a fire episode, elevated aerosol concentrations and heavy smoke are usually recorded by ground monitors and satellite sensors. Based on the unique physical and chemical characteristics of fire-dominated aerosols reported in the literature, we analyzed the surface aerosol observations from the IMPROVE network during satellite-detected fire events to establish a set of indicators to identify fire events from routine aerosol monitoring data. Five fire identification criteria were chosen: (1) high concentrations of $PM_{2.5}$ and PM_{10} (particles smaller than 2.5 and 10 in diameters, respectively); (2) a high $PM_{2.5}/PM_{10}$ ratio; (3) high organic carbon ($OC/PM_{2.5}$) and elemental carbon ($EC/PM_{2.5}$) ratios; (4) a high potassium ($K/PM_{2.5}$) ratio; and (5) a low soil/$PM_{2.5}$ ratio. Using these criteria, we are able to identify a number of fire episodes close to 15 IMPROVE monitors from 2001 to 2011. Most of these monitors are located in the Western and Central United States. In any given year within the study period fire events often occurred between April and September, especially in the two months of April and September. This ground-based fire climatology is also consistent with that derived from satellite retrievals. This study demonstrates that it is feasible to reconstruct historic records of fire events based on continuous ground aerosol monitoring. This dataset can provide not only fire activity information but also fire-induced aerosol surface concentrations and chemical composition that can be used to verify satellite-based products and evaluate air quality and climate modeling results. However, caution needs to be exercised because these indicators are based on a limited number of fire events, and the proposed methodology should be further tested and confirmed in future research.

Keywords: fire; aerosol; ground-based aerosol monitoring; fire identification

---

1. Introduction

Biomass burning, including both wildfires and prescribed burns, converts a sizeable amount of vegetation into burned ashes, fugitive gases, vapor, and particles [1]. The emitted gases, water vapor, and fine particles exert myriad effects on atmospheric chemistry, the Earth's radiative budget, and the hydrological cycle [2–4]. Due to the significant effects of wildfire on air quality and climate, wildfire biomass burning events have been extensively studied through ground observations [5,6], satellite sensor detection [5,7–9], and model simulations [10,11]. The continuous accumulation of fire-related data makes it possible for the scientific community to examine long-term trends in fire activity and the driving forces underlying these variations. Using satellite data and a biogeochemical model, van der Werf and colleagues [12] have examined the interannual variability in global biomass burning emissions, which exhibited large variations (with a range of more than 1 Pg C$\cdot$ year^{-1}) from 1997 to 2004. Westerling *et al.* [13] have compiled an extensive wildfire database and found that large wildfire activity in the Western United States (U.S.) has increased considerably since the mid-1980s, likely driven by increased spring/summer temperatures and an earlier snowmelt in the mid-elevation forests of the Northern Rockies. Development of long-term fire climatology, while holding great promise for climate analysis, presents substantial challenges due to the difficulty of obtaining accurate fire observations and the diverse requirements of fire indices. Polar-orbiting satellite sensors, such as the Moderate Resolution Imaging Spectroradiometer (MODIS) [5,8], the Along Track Scanning Radiometer (ATSR) [10], and the Visible and Infrared Scanner (VIIRS) [12], can provide global coverage of fire counts, burned areas, and fire's radiative power. High temporal resolution fire detection could be derived from National Oceanic and Atmospheric Administration (NOAA) Geostationary Operational Environmental Satellite (GOES)-based observations [9]. For both polar-orbiting and stationary satellites, wildfires can be difficult to observe due to cloud cover or fire induced convection. Furthermore, to quantify fire-related emissions, several important assumptions have to be made to convert satellite fire observations into fire emissions data, such as fuel loading, burn duration, emission factors, and plume vertical structure. However, the conversion process is not straightforward or intuitive. For instance, van der Werf *et al.* [12] reported that burned areas and total fire emissions are largely decoupled because forested areas dominate fire emissions, whereas savanna burning contribute disproportionally to burned area statistics globally. Uncertainty in these procedures hinders emission estimations and needs to be investigated with independent data sources, such as ground and aircraft observations. Ground surveys of fire counts and burned areas have been compiled (e.g., m [13]), but these datasets cannot be directly used to derive fire emissions without further processing. Finally, air quality research and regulatory communities are interested in surface concentrations of pollutants elevated by wildfires. However, there is no direct satellite observation of fire-induced $PM_{2.5}$ (particles smaller than 2.5 in diameters) surface concentrations and chemical composition. Lately, efforts have been made to convert column aerosol loading into surface concentrations [14]. A previous study focused on surface concentrations without chemical compositions, and sources of $PM_{2.5}$ were not considered. Clearly, there is need to develop ground-based fire observations.

This study proposes a new method of developing long-term wildfire records from traditional ground aerosol monitoring networks. This dataset, if successfully built, can provide not only fire activity information but also fire-induced aerosol surface concentrations and chemical composition data that can be used to verify satellite-based products, evaluate fire models (e.g., model fire occurrence and characteristics), evaluate biomass burning aerosol properties simulated by chemical transport, evaluate general circulation aerosol models using satellite-derived fire emissions, evaluate air quality and climate modeling results, and assess human exposure to fire pollution. The aerosol data are obtained from the Interagency Monitoring of Protected Visual Environments (IMPROVE) network. IMPROVE is a long-term, continuous aerosol monitoring network that measures the mass concentrations, size distribution, and chemical composition of ambient aerosols, and it provides 24-hour aerosol data every third day.

The main challenge of the new method lies in how to design effective criteria to identify fire events from mixed aerosol records. Routine aerosol networks such as IMPROVE were not designed exclusively for fire detection. Instead, IMPROVE observes aerosol levels originating from all sources, such as fires, dust storms, and anthropogenic sources. We use concurrent satellite fire detection and

IMPROVE data to examine the distinct physical and chemical characteristics of aerosols during fire episodes, so that a set of indicators can be established to separate fire-influenced samples from those dominated by other sources.

2. Data Sources

2.1. IMPROVE Aerosol Data

There are two reasons for choosing aerosol observation data from the IMPROVE network. First, the IMPROVE network is one of two national air quality monitoring networks in the U.S. It has been providing both mass concentrations and chemical compositions of aerosols every three days since 1988, which makes it ideal for long-term studies. Second, the IMPROVE monitors are mostly deployed in the national parks and remote areas in the U.S. [15], making it suitable for fire detection due to their distance from possible anthropogenic contamination. Locations of IMPROVE monitors over the contiguous United States (CONUS) are shown in Figure 1. There are other monitoring networks, such as the U.S. EPA Air Quality System (AQS) network, which has a national coverage but no aerosol composition data, and the Chemical Speciation Network (CSN), mostly deployed in urban areas and thus possibly affected by anthropogenic contamination. Hence, the IMPROVE network is selected for developing a ground-based fire detection method over the CONUS.

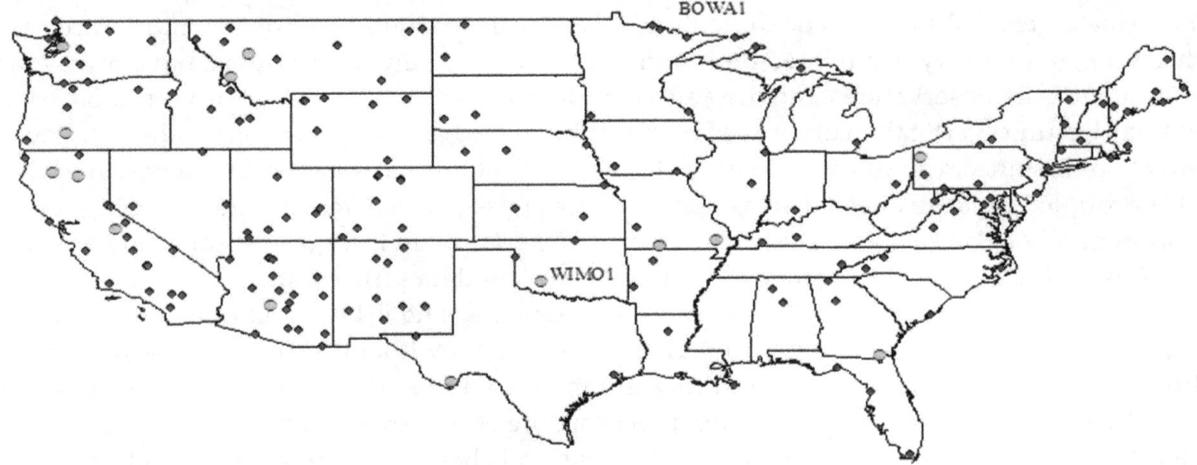

Figure 1. Locations of IMPROVE monitors (marked as ●) over the contiguous United States. The 15 sites (marked as ◦) indicate the locations where most fire events have been identified from 2001 to 2011 using the approach developed in this work.

2.2. Satellite Data

To train the fire detection algorithm, independent fire information is required. A typical source of fire data is satellite remote sensing. Fire events independently recorded by the U.S. National Aeronautics and Space Administration (NASA) Earth Observatory's Natural Hazards Products and other MODIS fire products are used here to assist with the ground-based analysis. These products provide fire information, including start time, end time, duration, burned scar area, and plume direction during fire events. These data are important for establishing fire identification criteria through analysis of the chemical and physical characteristics of fire events using concurrent ground observations. Therefore, we selected some satellite-detected fire events and examined the corresponding IMPROVE data collected around those fires to analyze the characteristics of the filter-based aerosol sample.

3. Identifying Fire Events

3.1. Selecting Fire Identification Criteria

Fire identification criteria were proposed based on the chemical and physical characteristics of biomass burning-dominated aerosol data reported in the literature. Prior studies have provided a collective view of particle mass emission factors [5,16–18], size distributions [19], and optical and physical properties of biomass burning emissions [2,3]. These studies formed the basis for particle mass and concentration characterization in accordance with: $PM_{2.5}$ and PM_{10} (particles smaller than 10 μm in diameter) mass concentrations, ratio of $PM_{2.5}$ to PM_{10}, percentages of Organic Carbon (OC), Elemental Carbon (EC), potassium (K), and soil in $PM_{2.5}$. During a fire episode, heavy smoke and low visibility are common, due to the large amount of fine and coarse particles emitted into the ambient air. Therefore, Particulate Matter (PM) concentration in the source region register elevated spikes when a fire event occurs. However, these spikes are not unique to fire events. Had there been a dust event or a volcanic plume, there would have been equal or higher PM concentration spikes. Therefore, to reliably attribute PM concentration spikes to a fire event rather than to a dust event, other indicators must be considered. A previous study documented that approximately 95% of the particles emitted from biomass burning are fine particles, and the dominant chemical components are carbonaceous [20]. Some reports also suggest that carbon accounts for 50% to 70% of the total mass of fire-emitted aerosols, with 55% and 8% of the fine particle mass attributed to OC and EC, respectively [3]. Therefore, a high $PM_{2.5}/PM_{10}$ ratio, dominated by high $OC/PM_{2.5}$ and $EC/PM_{2.5}$ ratios, are additional aerosol characteristics pertinent to a fire event. Furthermore, trace inorganic species account for approximately 10% of the fine mass of fresh smoke, mostly enriched in K [3,21]. Consequently, a high $K/PM_{2.5}$ ratio can also be considered as a fire event indicator.

3.2. Determining Threshold Values for Fire Identification Criteria

Next, we focus on a number of satellite-detected fire events to determine proper thresholds for each fire identification criterion. We checked the locations and times of some large fire events detected by satellites. We examined satellite imageries for fire events from the NASA Earth Observatory's Natural Hazards fire products [22], and MODIS fire maps [23]. We then merged these fire maps with the corresponding geographic coverage from IMPROVE network sites using the geo-spatial software in ArcGIS.

First, we focused on the BOWA1 site (marked in Figure 1) with a case study. A lightning strike in the Boundary Waters Canoe Area Wilderness Region of northeastern Minnesota started a forest fire. On 12 September 2011, fire event imageries were captured by MODIS aboard the Terra satellite (Figure 2). The fire was a plume-driven event, and it ultimately burned more than 60,000 acres. According to the IMPROVE aerosol data, during the fire event between September 9th and 18th (no data on 6th and 12th September 2011), the average concentrations of $PM_{2.5}$, PM_{10}, OC, and EC were 14.89, 18.31, 7.49, and 0.62 $\mu g \cdot m^{-3}$, respectively. By contrast, during average background conditions, these values were 4.39, 6.38, 1.77, and 0.13 $\mu g \cdot m^{-3}$, respectively. Although there were no data during part of this fire episode, these results suggest that there was a fire event detected by the IMPROVE monitor.

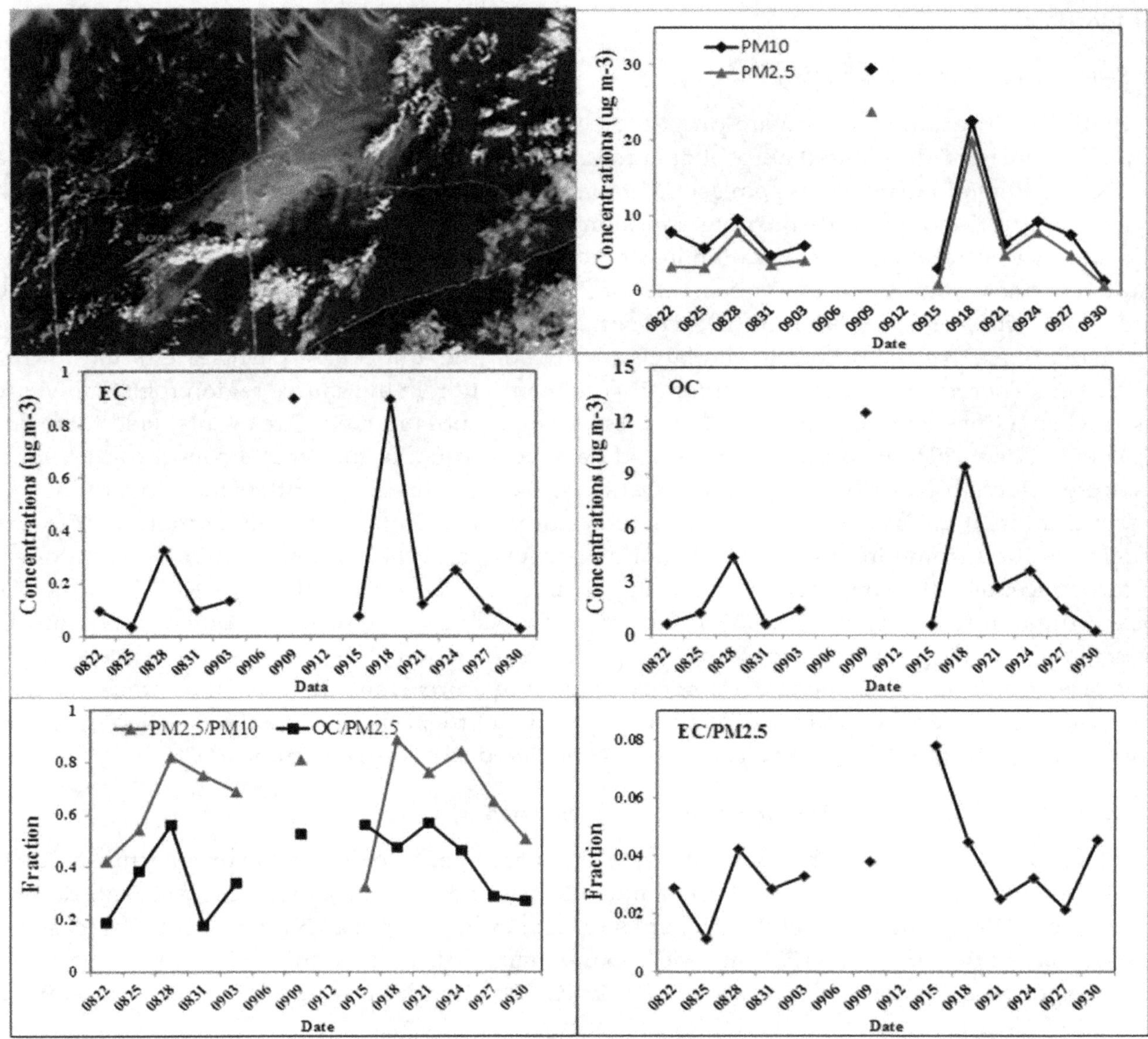

Figure 2. A case study of a fire event in September 2011 at the BOWA1 site. The MODIS on NASA's Terra satellite captured the top left image on 12th September 2011. Red outlines show areas of high surface temperatures associated with active burning, near the BOWA1 site. Characteristics of fire-dominated aerosol ($PM_{2.5}$, PM_{10}, OC, and EC concentrations, ratios of EC, OC in $PM_{2.5}$) during the fire event episode are shown in other figures.

We analyzed aerosol observation data from the IMPROVE sites during the satellite-detected fire episodes. We analyzed the temporal variability of these indicators (concentrations of PM_{10} and $PM_{2.5}$ and the ratios of $PM_{2.5}/PM_{10}$, $OC/PM_{2.5}$, $EC/PM_{2.5}$, $K/PM_{2.5}$, and $soil/PM_{2.5}$) before, during, and after the fire episode for a span of 15 days. Compared with the no-fire period conditions, the concentrations of PM_{10} and $PM_{2.5}$ and the ratios of $PM_{2.5}/PM_{10}$, $OC/PM_{2.5}$, $EC/PM_{2.5}$, and $K/PM_{2.5}$ were higher, and the ratio of $soil/PM_{2.5}$ was lower (Figure 3) during fire conditions. Finally, we determined the thresholds of fire identification criteria (Table 1) by summarizing aerosol characteristics from both prior studies and measurements from the IMPROVE sites.

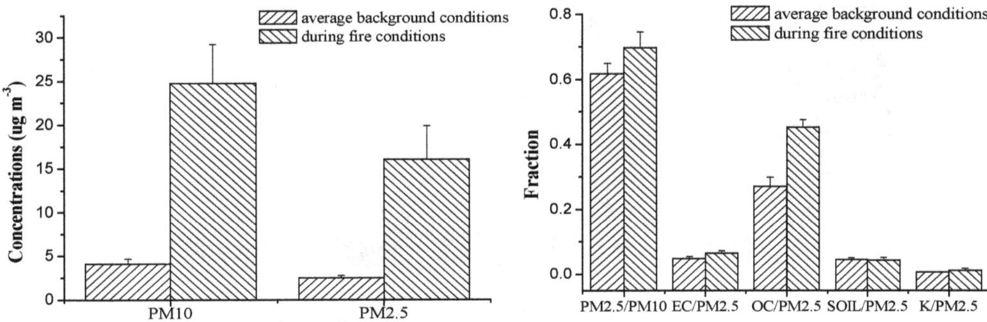

Figure 3. Comparison of fire identification criteria values during fire conditions and average background concentrations or conditions sampled by the monitoring site.

Table 1. Thresholds of fire identification criteria used in this study.

Indicator	$PM_{2.5}$ (μg/m^3)	PM_{10} (μg/m^3)	$PM_{2.5}/PM_{10}$	$OC/PM_{2.5}$	$EC/PM_{2.5}$	$K/PM_{2.5}$	$Soil/PM_{2.5}$
Static threshold	>15	>18	>0.6	>0.35	>0.05	>0.003	<0.03

3.3. Applying the Approach to Identify Fire Events

Finally, we applied the fire identification criteria and the corresponding thresholds to all IMPROVE sites to identify fire events between 2001 and 2011. A large number of fire events were identified at 15 IMPROVE sites (marked in Figure 1). Most of these events were located in the Western U.S., particularly in the states of California and Montana, and the Central U.S., especially in the states of Missouri, Oklahoma, and Texas. A previous study based on the GOES burned area product also documented that fires occurred most frequently in the Western and Southeastern U.S., and along the Central and Southern Mississippi Valley [9]. In the Western U.S., due to the dry climate and dense forests (or shrubs), the increased threat of larger, longer, and more intensive forest fires has become a concern [13,24]. A previous study documented that California, dominated by shrubland, was a high-intensity fire event area, with fires extending to sizes of 10,000 ha or more [25]. The surface-monitor-based methodology shows that fire has a great impact on local air quality. Unlike satellite-retrievals, our method can provide information about aerosol concentrations and chemical composition attributed to these fire events. The changes in $PM_{2.5}$ levels and composition caused by fire emissions vary over time and space [24]. For example, the concentrations of $PM_{2.5}$ were higher in the Eastern U.S. and lower in the central regions, with strong seasonal patterns [26]. In the Western U.S., the annual average percentage of OC in $PM_{2.5}$ is 40%, whereas in the eastern U.S. it is 25%, with an annual average of 28% for the whole U.S. For the entire year, the average in summer is higher than that in winter for the U.S., both regionally and CONUS-wide [26]. Therefore, the indicators should be specific for region and season.

Due to the differences in PM concentration and composition over time and space, regional and seasonal characteristics should be considered when applying fire indicators. Here we analyzed the temporal pattern of PM and the ratios of some species in $PM_{2.5}$ at the WIMO1 site (Figure 4). The results showed that $PM_{2.5}$, PM_{10} concentrations, and the $PM_{2.5}/PM_{10}$ ratio were seasonally distinctive: higher in summer and lower in winter, with annual means of 7.3 $\mu g \cdot m^{-3}$, 14.9 $\mu g \cdot m^{-3}$, and 0.5 $\mu g \cdot m^{-3}$, respectively. However, the $EC/PM_{2.5}$ ratio was higher in fall and lower in summer during the study period. Compared to the annual mean values, the data during the identified fire episodes were higher. This result suggested that the identification criteria worked well. Furthermore, we analyzed the monthly mean values of all indicators at the WIMO1 site (Figure 5). The results showed that $OC/PM_{2.5}$ and $EC/PM_{2.5}$ ratios are higher in March and October, but lower in July. Especially in July, the lower mean values of $OC/PM_{2.5}$ and $EC/PM_{2.5}$ ratios indicated that some fire events may have been missed. This result suggested that applying month-specific indicator values could enhance the method to generate more consistent reporting.

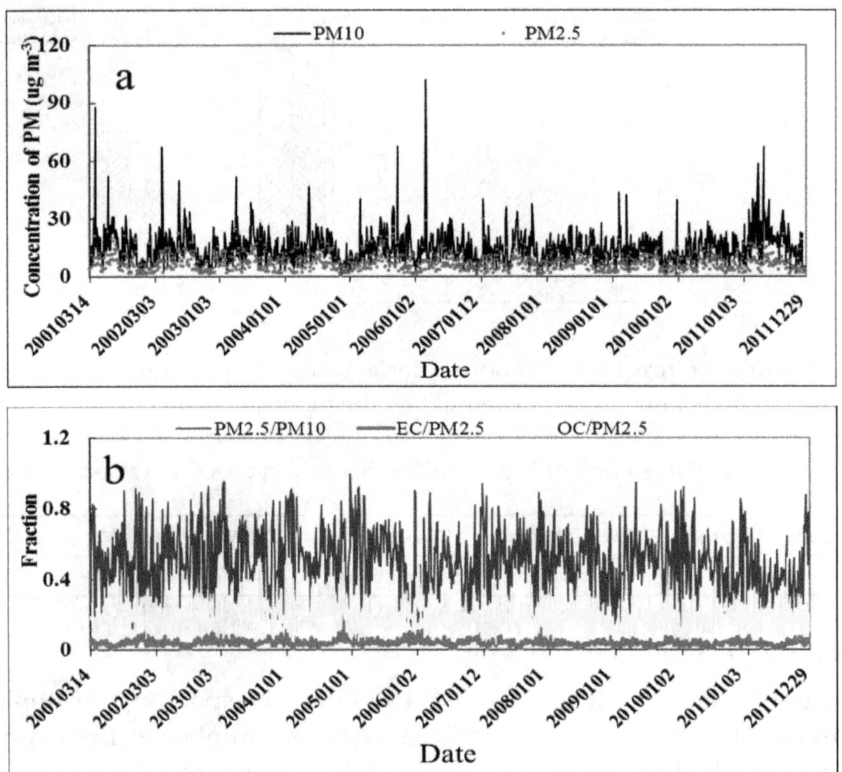

Figure 4. Characteristics of aerosols (**a**) Concentrations of $PM_{2.5}$ and PM_{10}; (**b**) Fractions of $PM_{2.5}/PM_{10}$, $EC/PM_{2.5}$, $OC/PM_{2.5}$) at the WIMO1 site from 2001 to 2011.

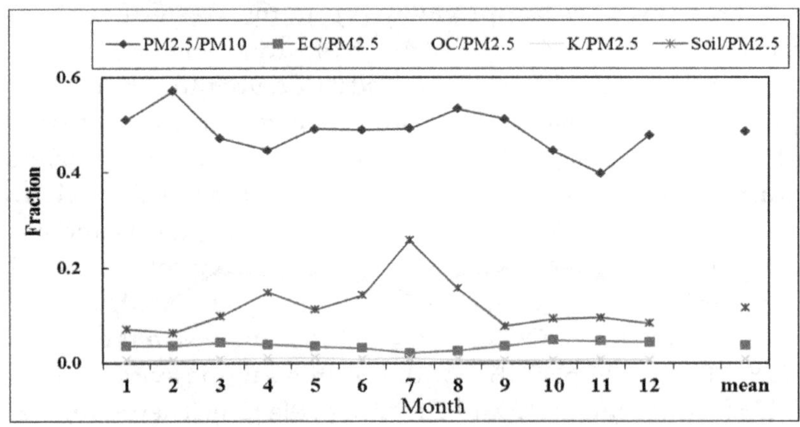

Figure 5. The monthly average of fire indicators at the WIMO1 site.

3.4. Testing Fire Identification Criteria

Finally, we analyzed the temporal and spatial characteristics of fire events to test this methodology. Following the suggested procedures, we were able to identify fire events in the proximity of 15 IMPROVE monitoring sites from 2001 to 2011. We compared those identified fire events with the HYbrid Single-Particle Lagrangian Integrated Trajectory (HYSPLIT) back-trajectory model [27,28] prediction, wildland fire summary and statistics annual report by the National Interagency Fire Center [29], and USGS's (United States Geological Survey) record of fire events [30] to confirm these events. We looked at trajectories from the locations of fires identified in satellite imageries and compared fire reports with these results. Most of fires identified by this method were consistent with the records. In fact, there were some fire events detected by this method but missed by satellite due to cloud cover. Furthermore, some fire events were missed by the annual report due to the small size of the burned areas, and only fires over 40,000 acres were marked in the wildland fire summary and statistics annual report. For example, we identified a fire event at the WIMO1 site on 30 May 2011, according to fire identification criteria. Then, we compared this result with the HYSPLIT simulation on the same day (provided by the NOAA ARL READY online platform). Fire points, light smoke, medium smoke, and heavy smoke can be found in the HYSPLIT simulation picture. From the picture, we found that there were fire events in Texas and the Gulf of Mexico USA, but no fire located near the WIMO1 site. Heavy smoke from Texas and the Gulf of Mexico may have caused this fire event to be missed by satellite. In addition, we analyzed the temporal and spatial characteristics of aerosols in identified fire events, and found that most of them were located in the Western U.S. (including California and Montana) and the Central U.S. (including Missouri, Oklahoma and Texas).

We chose the WIMO1 site (34.7315°N, 98.7155°W) in the case study for two reasons: (1) Frequent fires were detected at this site both by satellite and by this method; (2) this site is located in the Wichita Mountains in Oklahoma State, and Oklahoma and Texas are the two states with the most fire events recorded (Figure 1). Here, we calculated days of fire and numbers of fire events from 2001 to 2011. Because a fire can last for several days, if some fire events were identified by indicators for several consecutive days, we considered this one event. There were 83 fire events (161 days) identified between 2001 and 2011 at the WIMO1 site. We analyzed the characteristics of fire-dominated aerosol at the WIMO1 site between 2001 and 2011 (Figure 6). The results showed that concentrations of $PM_{2.5}$ and PM_{10} were 14.2 and 21.2 $\mu g \cdot m^{-3}$ during fire events, respectively. Compared with background conditions, these values were elevated by 42.7% and 94.6%, respectively. The mean fraction of $PM_{2.5}$ in PM_{10} was 66.4% during fire events, whereas the value was 50.0% during average background conditions, which also increased by 32.8%. Compared with the variations in PM concentrations, the fractions of some species in the aerosol were more stable. Because aerosol concentrations in smoke plumes were affected by fire intensity, severity, duration, fuel loading, wind direction, and site location, the ratios of species in aerosol were strongly dependent on the sources of aerosols (fire, dust or other natural and anthropogenic sources).

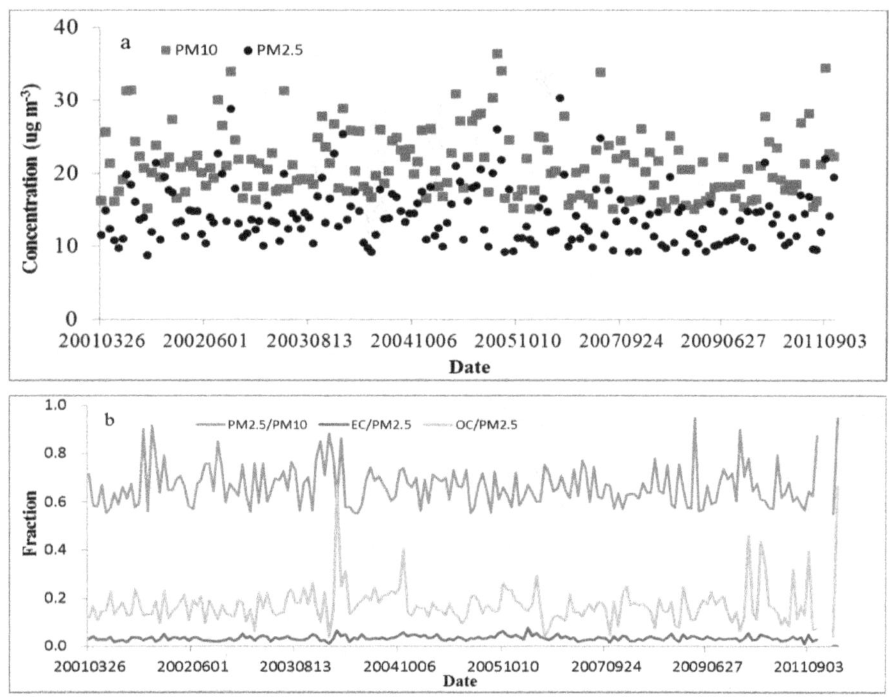

Figure 6. Characteristics of aerosols. (**a**) Concentrations of $PM_{2.5}$ and PM_{10}; (**b**) Fractions of $PM_{2.5}/PM_{10}$, $EC/PM_{2.5}$, $OC/PM_{2.5}$) during fire events at the WIMO1 site between 2001 and 2011.

Temporal patterns of fires at the WIMO1 site were distinctive. Between 2001 and 2011, the number of days with fires declined at the WIMO1 site (Figure 7). From 2001 to 2005, there were more than 16 days with fire records every year, including up to 26 days in 2005. However, from 2006 to 2011, the number of fire days declined, and there were fewer than 10 days in 2006. In any given year within the study period, fire events often occurred between April and September, especially in the two months of September and April (Figure 8). The spatial and temporal patterns of fires were often affected by topography, vegetation, climate, and human activity [31]. Previous studies of burned areas, which used data from the AVHRR (the Advanced Very High Resolution Radiometer) and GOES satellite, also documented that fire exhibits a distinctive seasonality, with a peak from June to August. In croplands, peak burning occurred from April to September because agricultural fires were set during pre-planting and post harvesting periods [9,32]. In this study, peak fire events in September may be related to agricultural burnings.

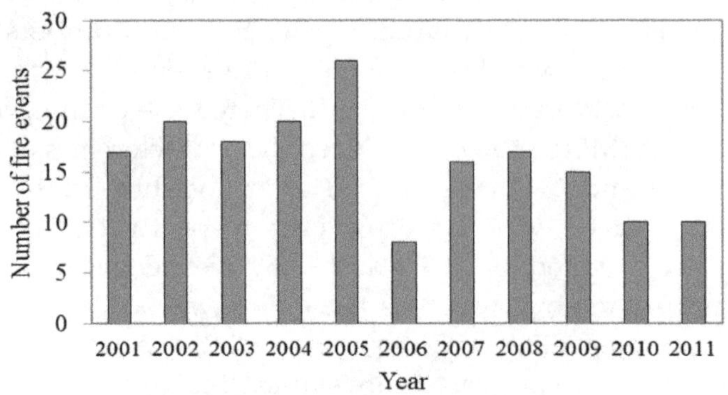

Figure 7. Number of fire events in every year from 2001 to 2011 at the WIMO1 site.

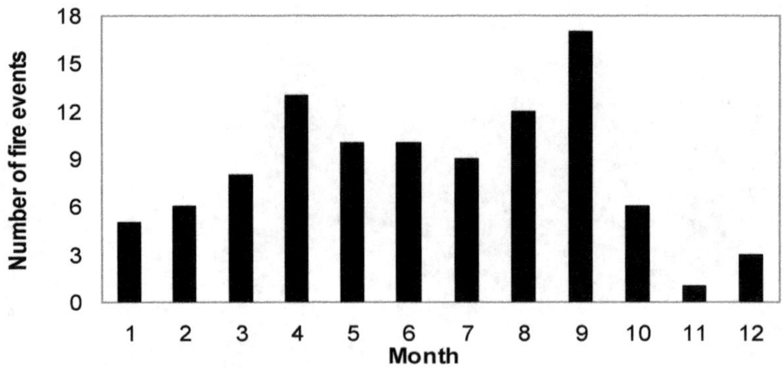

Figure 8. Number of fire events in every month between 2001 and 2011 at the WIMO1 site.

4. Discussion of Uncertainties and Limitations

The above results confirm that the proposed methodology has the potential to identify a fire event and quantify its impact on air quality. However, caution needs to be exercised because the indicators discussed here are based on a limited number of fire events. We should take note of the uncertainties and limitations of the methodology, which are caused by the following factors: (1) the sampling frequency (every third day) and the rather sparse distribution of the IMPROVE sites limit the capability of the IMPROVE network to detect all fires, especially for regions with high fire frequency but few IMPROVE sites; (2) neighboring monitoring sites may detect the same fires, resulting in double-counting; (3) excessive loading of smoke ash and other aerosols may disable the instruments; and finally (4) interference by background aerosols makes it difficult to clearly determine thresholds for the fire identification criteria. Aerosol concentrations and chemical composition are intrinsically highly variable over time and space. Consequently, one uniform threshold may not be applicable to all sites at all times. Low thresholds may cause false detection, whereas high thresholds may cause omission of some fire counts. For example, the results of this study showed that there were fewer fire events during winter compared with other seasons. This result can be explained by winter's vegetation and climate conditions, which are unfavorable for fires. Another reason may be that the concentrations of PM in winter are near the annual minimum. They cannot reach the threshold concentration level to trigger a fire count by the identification algorithm. If we use the average concentrations of the entire year as a cutoff value, the winter data will be omitted, and few fires will be identified. However, especially in the southern U.S., such as in Florida, fire events also occur in winter, but cannot be identified by this method due to low seasonal aerosol concentrations. Therefore, thresholds for these indicators should be adjusted according to region and season instead of adopting one uniform value.

5. Summary

Fire is a major source of aerosol. However, few observation sites are designed to record fire events and to track aerosol emissions from biomass burning. This work proposed a new approach to the reconstruction of historic fire records based on observations collected by a continuous ground-based aerosol monitoring network over the contiguous United States. Using five fire identification criteria, we were able to identify a number of fire episodes recorded by 15 IMPROVE monitors from 2001 to 2011. Most of these fire events were located in the Western and Central United States. In any given year within the study period, fire events often occurred between April and September, especially in the two months of April and September. There were 83 fire events (161 days) at WIMO1 sites between 2001 and 2011. This study demonstrates that it is feasible to reconstruct historic records of fire events based on continuous ground aerosol monitoring. This dataset would provide not only fire activity information but also fire-induced aerosol surface concentrations and chemical composition data that can be used to verify satellite-based products, evaluate air quality and climate modeling results, and assess human exposure to fire pollution. However, caution needs to be exercised because these indicators are based on a limited number of fire events, and the proposed methodology should be further tested and confirmed by future research.

Acknowledgments: We thank the three anonymous reviewers for their constructive comments and helpful suggestions. This work is financially supported by the National Key Technology R&D Program (No. 2014BAC16B03), the National Natural Science Foundation of China (No. 41201495), the Chinese Academy of Sciences/State Administration of Foreign Experts Affairs (CAS/SAFEA) International Partnership Program for Creative Research Teams (No. KZZD-EW-TZ-07), and the Natural Science Foundation of Jilin Province (No. 20150101010JC). The assistance of the U.S. NOAA Air Resources Laboratory is also gratefully acknowledged.

Author Contributions: The study was completed with cooperation among all authors: Daniel Q. Tong and Hongmei Zhao conceived and designed the research topic. Hongmei Zhao conducted the research and wrote the manuscript. Pius Lee, Hyuncheol Kim, and Hang Lei collaborated in discussing the results and providing editorial advice.

References

1. Crutzen, P.L.; Andteae, M.O. Biomass burning in the tropics: Impact on atmospheric chemistry and biogeochemical cycles. *Science* **1990**, *250*, 1669–1678. [CrossRef] [PubMed]
2. Reid, J.S.; Eck, T.F.; Christopher, S.A.; Koppmann, R.; Dubovik, O.; Eleuterio, D.P.; Holben, B.N.; Reid, E.A.; Zhang, J. A review of biomass burning emissions part III: Intensive optical properties of biomass burning particles. *Atmos. Chem. Phys.* **2005**, *5*, 827–849. [CrossRef]
3. Reid, J.S.; Koppmann, R.; Eck, T.F.; Eleuterio, D.P. A review of biomass burning emissions part II: Intensive physical properties of biomass burning particles. *Atmos. Chem. Phys.* **2005**, *5*, 799–825. [CrossRef]
4. Mielonen, T.; Aaltonen, V.; Lihavainen, H.; Hyvarunen, A.; Arola, A.; Komppula, M.; Kivi, R. Biomass burning aerosols abserved in Northern Finland during the 2010 wildfires in Russia. *Atmosphere* **2013**, *4*, 17–34. [CrossRef]

5. Wiedinmyer, C.; Quayle, B.; Geron, C.; Belote, A.; McKenzie, D.; Zhang, X.; O'Neill, S.; Wynne, K.K. Estimating emissions from fires in North America for air quality modeling. *Atmos. Environ.* **2006**, *40*, 3419–3432. [CrossRef]
6. Burling, I.R.; Yokelson, R.J.; Akagi, S.K.; Urbanski, S.P.; Wold, C.E.; Griffith, D.W.T.; Johnson, T.J.; Reardon, J.; Weise, D.R. Airborne and ground-based measurements of the trace gases and particles emitted by prescribed fires in the United States. *Atmos. Chem. Phys.* **2011**, *11*, 12197–12216. [CrossRef]
7. Hsu, N.C.; Herman, J.R.; Bhartia, P.K.; Seftor, C.J.; Torres, O.; Thompson, A.M.; Gleason, J.F.; Eck, T.F.; Holben, B.N. Detection of biomass burning from TOMS measurements. *Geophys. Res. Lett.* **1996**, *23*, 745–748. [CrossRef]
8. Ichoku, G.; Giglio, L.; Wooster, M.J.; Remer, L.A. Global characterization of biomass-burning patterns using satellite measurements of fire radiative energy. *Remote Sens. Environ.* **2008**, *112*, 2950–2962. [CrossRef]
9. Zhang, X.Y.; Kondragunta, S. Temporal and spatial variability in biomass burned area across the USA derived from the GOES fire product. *Remote Sens. Environ.* **2008**, *112*, 2886–2897. [CrossRef]
10. Duncan, B.N.; Martin, R.V.; Staudt, A.C.; Yevich, R.; Logan, J.A. Interannual and seasonal variability of biomass burning emissions constrained by satellite observations. *J. Geophys. Res.* **2003**, *108*. [CrossRef]
11. Wang, J.; Christopher, S.A.; Nair, U.S.; Reid, J.S.; Prins, E.M.; Szykman, J.; Hand, J. Mesoscale modeling of Central American smoke transport to the United States: 1. "Top-down" assessment of emission strength and diurnal variation impacts. *J. Geophys. Res. Atmos.* **2006**, *111*. [CrossRef]
12. Van der Werf, G.R.; Randerson, J.T.; Giglio, L.; Collatz, G.J.; Mu, M.; Kasibhstla, P.S.; Morton, D.C.; DeFries, R.S.; Jin, Y.; van Leeuwen, T.T. Global fire emission and the contribution of deforestation, savanna, forest, agricultural, and peat fires (1997–2009). *Atmos. Environ.* **2010**, *40*, 3419–3432.
13. Westerling, A.L.; Hidalgo, H.G.; Cayan, D.R.; Swetnam, T.W. Warming and earlier spring increase western U.S. forest wildfire activity. *Science* **2006**, *313*, 940–943. [CrossRef] [PubMed]
14. Boys, B.L.; Martin, R.V.; van Donkelaar, A.; MacDonell, R.J.; Hsu, N.C.; Cooper, M.J.; Yantosca, R.M.; Lu, Z.; Streets, D.G.; Zhang, Q.; *et al.* Fifteen-year global time series of satellite-derived fine particulate matter. *Environ. Sci. Technol.* **2014**, *48*, 11109–11118. [CrossRef] [PubMed]
15. Pitchford, M.L.; Malm, W.C. Development and applications of a standard visual index. *Atmos. Environ.* **1994**, *28*, 1049–1054. [CrossRef]
16. Akagi, S.K.; Yokelson, R.J.; Wiedinmyer, C.; Alvarado, M.J.; Reid, J.S.; Karl, T.; Crounse, J.D.; Wennberg, P.O. Emission factors for open and domestic biomass burning for use in atmospheric models. *Atmos. Chem. Phys.* **2011**, *11*, 4039–4072. [CrossRef]
17. Yokelson, R.J.; Burling, I.R.; Gilman, J.B.; Warneke, C.; Stockwell, C.E.; de Gouw, J.; Akagi, S.K.; Urbanski, S.P.; Veres, P.; *et al.* Coupling field and laboratory measurements to estimate the emission factors of identified and unidentified trace gases for prescribed fires. *Atmos. Chem. Phys.* **2013**, *13*, 89–116. [CrossRef]
18. Urbanski, S. Wildland fire emissions, carbon, and climate: Emission factors. *Forest Ecol. Manag.* **2014**, *317*, 51–60. [CrossRef]
19. Janhall, S.; Andreae, M.O.; Poschl, U. Biomass burning aerosol emissions from vegetation fires: Particle number and mass emission factors and size distributions. *Atmos. Chem. Phys.* **2010**, *10*, 1427–1439. [CrossRef]
20. Deng, C.R. Identification of Biomass Burning Source in Aerosols and the Formation Mechanism of Haze. Ph.D. Thesis, University of Fudan, Shanghai, China, 15 April 2011. (in Chinese).
21. Amodio, M.; Andriani, E.; Dambruoso, P.R.; Daresta, B.E.; de Gennaro, G.; di Gilio, A.; Intini, M.; Palmisani, J.; Tutino, M. Impact of biomass burning on PM_{10} concentrations. *Fresen. Environ. Bull.* **2012**, *21*, 3296–3300.
22. The NASA Earth Observatory's Natural Hazards fire products. Available online: http://earthobservatory.nasa.gov/NaturalHazards/category.php?cat_id=8&m=01&y=2013 (accessed on 23 July 2013).
23. MODIS fire maps. Available online: http://reverb.echo.nasa.gov/reverb/#utf8=%E2%9C%93&spatial_map=satellite&spatial_type=rectangle&keywords=MOD14 (accessed on 23 July 2013).
24. Schoennagel, T.; Veblen, T.T.; Romme, W.H. The interaction of fire, fuels, and climate across Rocky Mountain forests. *BioScience* **2004**, *54*, 661–676. [CrossRef]
25. Keeley, J.E.; Zedler, P.H. Large, high-intensity fire events in southern California shrublands: Debunking the fine-grain age patch model. *Ecol. Appl.* **2009**, *19*, 2254. [CrossRef]

26. Bell, M.L.; Dominici, F.; Ebisu, K.; Zeger, S.L.; Samet, J.M. Spatial and temporal variation in $PM_{2.5}$ chemical composition in the United States for health effects studies. *Environ. Health Perspect.* **2007**, *115*, 988–995. [CrossRef] [PubMed]
27. Draxler, R.R.; Rolph, G.D. HYbrid Single-Particle Lagrangian Integrated Trajectory (HYSPLIT) Model access via NOAA ARL READY. Available online: http://www.arl.noaa.gov/HYSPLIT_info.php (accessed on 23 July 2012).
28. Rolph, G.D. Real-time Environmental Applications and Display System (READY). Available online: http://ready.arl.noaa.gov/index.php (accessed on 23 July 2012).
29. The National Interagency Fire Center. Available online: http://www.nifc.gov/fireInfo/fireInfo_statistics.html (accessed on 23 July 2013).
30. The United States Geological Survey record of fire events. Available online: http://wildfire.cr.usgs.gov/firehistory/data.html (accessed on 23 July 2013).
31. Iniguez, J.M.; Swetnam, T.W.; Baisan, C.H. Spatially and temporally variable fire regime on Rincon Peak, Arizona, USA. *Fire Ecol.* **2009**, *5*, 3–21. [CrossRef]
32. Pu, R.; Li, Z.; Gong, P.; Csiszar, I.; Fraser, R.; Hao, W.; Kondragunta, S.; Weng, F. Development and analysis of a 12-year daily 1-km forest fire dataset across North America from NOAA/AVHRR data. *Remote Sens. Environ.* **2007**, *108*, 198–208. [CrossRef]

12

Detailed Source-Specific Molecular Composition of Ambient Aerosol Organic Matter using Ultrahigh Resolution Mass Spectrometry and ^{1}H NMR

Amanda S. Willoughby [†], Andrew S. Wozniak and Patrick G. Hatcher [*]

Department of Chemistry and Biochemistry, Old Dominion University, Norfolk, VA 23529, USA; awilloug@odu.edu (A.S.W.); awozniak@odu.edu (A.S.W.)

* Correspondence: phatcher@odu.edu

† Current address: Department of Physical and Environmental Sciences, Texas A & M University—Corpus Christi, Corpus Christi, TX 78412, USA.

Abstract: Organic aerosols (OA) are universally regarded as an important component of the atmosphere that have far-ranging impacts on climate forcing and human health. Many of these impacts are related to OA molecular characteristics. Despite the acknowledged importance, current uncertainties related to the source apportionment of molecular properties and environmental impacts make it difficult to confidently predict the net impacts of OA. Here we evaluate the specific molecular compounds as well as bulk structural properties of total suspended particulates in ambient OA collected from key emission sources (marine, biomass burning, and urban) using ultrahigh resolution mass spectrometry (UHR-MS) and proton nuclear magnetic resonance spectroscopy (^{1}H NMR). UHR-MS and ^{1}H NMR show that OA within each source is structurally diverse, and the molecular characteristics are described in detail. Principal component analysis (PCA) revealed that (1) aromatic nitrogen species are distinguishing components for these biomass burning aerosols; (2) these urban aerosols are distinguished by having formulas with high O/C ratios and lesser aromatic and condensed aromatic formulas; and (3) these marine aerosols are distinguished by lipid-like compounds of likely marine biological origin. This study provides a unique qualitative approach for enhancing the chemical characterization of OA necessary for molecular source apportionment.

Keywords: organic aerosols; molecular characterization; UHR-MS; NMR; principal component analysis; biomass burning aerosols; urban aerosols; marine aerosols; aerosol source apportionment

---

1. Introduction

Organic matter (OM) comprises a significant portion of total aerosol mass, as much as 90% in certain areas [1,2], and is generated by a number of anthropogenic and biogenic emission sources. Organic aerosol (OA) compounds, once emitted into the atmosphere as primary OA or formed *in situ* as secondary OA (SOA) from gas-phase precursors, can undergo a myriad of atmospheric reactions forming new compounds that have different chemical structures and associated physical properties. The immense complexity of OA contributes to the difficulty in understanding the net impacts OA has on human health, biogeochemical cycling, and net radiative forcing.

The composition and relative concentrations of OA are expected to vary spatially and temporally due to differences in emission inputs and in the extent to which OA are transformed in the atmosphere by secondary aging processes [1]. The molecular composition of OA resulting from these emissions and aging processes will, in part, determine its impacts on e.g., aerosol hygroscopicity [3,4], light absorption [5], and biogeochemical cycling upon deposition [6,7]. The ability to apportion molecular compositions and impacts to specific aerosol emission sources and aging processes is therefore an important goal of the atmospheric community. Many studies have addressed similarities and differences of OA molecular composition among various aerosol sources, as well as seasonal and diurnal variability [6,8–12], but considerable work remains for a full understanding of this complex problem.

Fortunately, the adoption of new powerful analytical techniques such as nuclear magnetic resonance spectroscopy (NMR) and ultrahigh resolution mass spectrometry (UHR-MS) have enabled the characterization of important OA molecular and structural details. Solution-state proton NMR (^{1}H NMR) enables the characterization of soluble extracts of OA without extensive sample preparation and has been thoroughly reviewed [13,14]. A ^{1}H NMR spectrum can be analyzed to determine the relative contributions of major proton groups (e.g., alkyl protons) within OA, and also to identify specific compounds such as acetate, methanesulfonic acid, and levoglucosan that feature as sharp peaks in a spectrum [6,15–18]. Where NMR succeeds in providing these general structural details, it is unable to provide specific molecular details due to the overlapping signals from hydrogens associated with atmospherically relevant functional groups, and spectral interpretations must be simplified.

Pairing NMR with UHR-MS, which provides detailed molecular composition but only limited structural information, is therefore an attractive approach. UHR-MS, particularly Fourier transform ion cyclotron resonance mass spectrometry (FTICR-MS), allows for the determination of molecular formulas for thousands of high-molecular weight (>200 Da) compounds present within a single sample providing important chemical properties of OA. Its ultrahigh resolution (~500,000 over m/z 200–800) and mass accuracy (<1 ppm) can be used to obtain vital fingerprints, in the form of specific related molecular compound classes, that may be diagnostic for OA from specific emission sources and/or that have undergone molecular transformations. Numerous studies have used UHR-MS to reveal the molecular details of atmospheric OM [17,19–33]. UHR-MS techniques are generally regarded as non-quantitative, and structure can only be inferred from the molecular formula information, but when used in tandem with NMR extensive molecular and structural information can be achieved with minimal sample preparation. Such a pairing can provide fingerprints for the qualitative apportionment of OA molecular features that determine OA impact in the atmosphere and depositional environments.

The work presented here pairs these two powerful techniques (UHR-MS and ^{1}H NMR) to provide source-specific molecular characteristics for ambient aerosol samples from key anthropogenic and biogenic emission sources that can be used to aid more traditional source apportionment studies. The emission sources chosen here (marine, mixed source, biomass burning, urban) have significant regional and global quantitative importance on atmospheric aerosol loadings. Water-soluble and pyridine-soluble extracts for multiple samples from each source were evaluated for molecular characteristics using FTICR-MS and ^{1}H NMR. Compositional differences were elucidated using the chemometric approach, principal component analysis (PCA), to identify characteristic molecular features for each source type. Though qualitative in nature, this study sheds light on the sources of OA present in the ambient atmosphere and will be valuable for assessing the source-specific environmental impacts of OA as relationships between molecular characteristics and environmental impacts are strengthened.

2. Experiments

2.1. Aerosol Sample Collection

Ambient aerosol total suspended particulates (TSP, n = 14) were collected from four different locations to represent different emission source types. Air was drawn through pre-combusted (4 h, 475 °C) quartz microfiber filters (Whatman QM/A, 20.3 × 25.4 cm, 419 cm^2 exposed area, 0.6 μm effective pore size) using a TSP high-volume air sampler (model GS2310, Thermo Andersen, Smyrna, GA, USA) at flow rates ranging between 0.7 and 0.9 m^3· min^{-1}. Air particles were collected for 8–29 h with total air volumes ranging between 410 and 1170 m^3. The filters were transferred to combusted foil pouches immediately after collection and stored at 8 °C until analysis. Exact sampling dates and locations can be found in the supplemental methods. Briefly, marine TSP (n = 4) were collected aboard the R/V Knorr (Woods Hole, MA, USA) as part of the 2011 US GEOTRACES program cruise (GA03) [34] and as part of the 2014 second Western Atlantic Climate Study [35]. The US GEOTRACES samples (n = 3) are a subset of samples described in Wozniak *et al.* [7,31] where their WSOM FTICR-MS (negative ionization) and ^{1}H NMR characteristics were described. Mixed source TSP samples (n = 3) were collected at sea level at the Virginia Institute of Marine Science in Gloucester Point, Virginia, USA. Data for these mixed source samples are also presented in Willoughby *et al.* [30], a study demonstrating the utility of pyridine as a method for extracting and analyzing water-insoluble aerosol OM. Biomass burning TSP samples (n = 2) were collected at sea level in Suffolk, VA downwind of heavy smoke pollution from a fire burning at the Great Dismal Swamp. Urban TSP samples (n = 5) were collected ~60 m above sea level on the roof of an academic building at Drexel University in downtown Philadelphia, PA, USA. A storage or field blank filter was analyzed for each respective aerosol sample.

2.2. *Aerosol Mass and Carbon Measurements*

The QM/A filters were weighed before and after sampling to determine the TSP mass loadings (Table 1). A portion of each aerosol filter was analyzed in triplicate for total carbon (TC) using a FlashEA 1112 elemental analyzer (Thermo Scientific, Waltham, Massachusetts, USA). Black carbon (BC) amounts were determined using the chemothermal oxidation at 375 °C method (CTO-375) [36] and measured on the same elemental analyzer. Solvent extracts of the aerosols and respective filter blanks were obtained by combining aerosol filter plugs of known OC masses with ultrapure water (Millipore Synergy Ultrapure Water System, Darmstadt, Germany) or pyridine (Sigma-Aldrich, St. Louis, Missouri, USA, ⩾99.9%), and insoluble particles were removed using a syringe with a 0.45 μm PTFE filter cartridge. Percent water-soluble organic carbon (WSOC) for each water filtrate was determined by evaluating the non-purgeable organic carbon using a Shimadzu (Kyoto, Japan) TOC-VCPH analyzer. The pyridine-soluble organic carbon percentage (%PSOC) was determined by dissolving each of the aerosol samples into pyridine-D_5 and comparing spectral signals determined by ^{1}H NMR to that of a glucose standard [30]. These methods are described in greater detail in the supplementary methods. Due to limited sample availability, the marine aerosols were only measured for TC and WSOC.

Table 1. Average total suspended particulates (TSP) and total carbon (TC) concentrations and carbon percentages (relative to TC) for each aerosol source type.

Aerosol Source	n	TSP (μg· m^{-3})	TC (μg· m^{-3})	%BC	%WSOC	%PSOC
* Marine	4	-	0.5 ± 0.7	-	39.6 ± 25.1	-
Biomass burning	2	73.2 ± 5.3	24.8 ± 4.4	6.5 ± 0.7	33.6 ± 2.6	66 ± 20
Urban	5	47.1 ± 11.0	6.3 ± 1.3	3.4 ± 3.2	40.8 ± 5.5	44 ± 6
Mixed source	3	24.1 ± 2.9	5.7 ± 0.7	1.9 ± 2.5	50.5 ± 10.6	45 ± 13

* Some parameters for the marine aerosols could not be evaluated due to limited sample availability.

2.3. *FTICR-MS Analysis*

For FTICR-MS analyses, the water extracts were desalted using an established procedure for Agilent PPL solid-phase extraction cartridges [37]. The desalted sample was eluted in methanol (Acros Organics, Geel, Belgium, 99.9%), and will be referred to as $WSOM_{PPL}$ to differentiate it from WSOM. All $WSOM_{PPL}$ samples were analyzed in both positive and negative electrospray ionization (ESI) mode (WSOM+ and WSOM−, respectively), and PSOM samples were analyzed in negative mode (PSOM−) only due to poor signal observed in the positive mode. A respective field blank extract was prepared identically and analyzed immediately prior to each of the sample extracts to obtain a representative experimental blank spectrum. Each of the samples was analyzed on a Bruker Daltonics (Bremen,

Germany) 12 Tesla Apex Qe FTICR-MS with an Apollo II ESI source housed at the College of Sciences Major Instrumentation Cluster at Old Dominion University. Samples were infused to the ESI source at 120 nL$\cdot$ min^{-1}, and spray voltages were optimized for each sample. Ions were accumulated in the hexapole for 0.5–2.0 s before transfer into the ICR cell, where exactly 300 transients were co-added. The instrument was externally calibrated daily using a polyethylene glycol standard. Each spectrum was internally calibrated using the naturally occurring molecules including fatty acids and other homologous series of compounds containing only carbon, hydrogen and oxygen [38]. Peaks consistent with salts (mass defect 0.4–0.98 for $m/z < 400$, and mass defect 0.6–0.97 for $m/z > 400$), blank peaks (those found in the respective filter blank), and ^{13}C isotopologue peaks were subtracted from the mass list and not considered for formula assignments.

2.4. Molecular Formula Assignments

A unique molecular formula was assigned to a majority (82% $\pm$ 9%) of the measured peaks having a *S/N* ratio of at least 3 using an in-house generated MatLab (The MathWorks Inc., Natick, MA, USA) code according to the criteria $^{12}C_{5\text{-}80}{}^{1}H_{5\text{-}200}{}^{16}O_{1\text{-}30}{}^{14}N_{0\text{-}5}{}^{32}S_{0\text{-}2}{}^{34}P_{0\text{-}2}$ for negative ESI and $^{12}C_{5\text{-}80}{}^{1}H_{5\text{-}200}{}^{16}O_{1\text{-}30}{}^{14}N_{0\text{-}5}{}^{32}S_{0\text{-}2}{}^{23}Na_{0\text{-}1}$ for positive ESI, where the subscripts indicate the range of atoms allowed in a single formula. The assigned formulas were screened to remove any chemically unreasonable formulas for natural OM molecules according to previously published criteria (e.g., studies by Stubbins *et al.* [39] and Wozniak *et al.* [20]), and Kendrick mass defect homologous series (CH_2, H_2, and CO_2) were used to verify ambiguous assignments. Each of the assigned formulas has a calculated mass within 1 ppm agreement with the measured *m/z*, where a large majority of the formulas (88% $\pm$ 7%) have less than 0.5 ppm error.

2.5. ^{1}H NMR Spectroscopy

Each WSOM extract was diluted immediately before ^{1}H NMR analysis using D_2O (100% atom D, Acros Organics) at a ratio of 90:10 WSOM:D_2O. The deuterated WSOM solutions were analyzed via ^{1}H NMR spectroscopy using a Bruker Daltonics 400 MHz NMR with a BBI probe. Each sample was scanned 4000 (mixed source, biomass burning, and urban WSOM) or 8000 times (marine WSOM) using a standard Bruker water-suppression pulse program, where the 90° pulse and the transmitter offset were optimized individually for each sample. The signals obtained from ^{1}H NMR spectra were integrated over the entire spectral range to obtain the total signal response, and were also integrated over four specific chemical shift ranges to determine contributions from major proton types [16,18]. The signal response was normalized to the total signal in these regions (*i.e.*, total signal = $Area_{0.6\text{–}4.4\ ppm}$ + $Area_{6.0\text{–}9.0\ ppm}$) to determine the average relative contributions for each region. The regions are defined based on the chemical environment of protons exhibiting signal at those chemical shifts: (1) aliphatic hydrogen (H-C, 0.6–1.8 ppm); (2) unsaturated alkyl hydrogen (H-C-C=, 1.8–3.2 ppm); (3) oxygenated aliphatic hydrogen (H-C-O, 3.2–4.4 ppm), and aromatic hydrogen (Ar-H, 6.0–9.0 ppm). Aldehyde and carboxylic acid hydrogen (H-C=O, and HO-C=O) would appear downfield of the aromatic protons (*i.e.*, >9 ppm), but were not detected because these protons readily exchange with the deuterium in the D_2O required for analysis.

2.6. Principal Component Analysis

Principal component analysis was applied separately to the molecular formulas assigned to FTICR mass spectra and peaks present in the ^{1}H NMR spectra in order to reveal the components that describe the greatest amount of variance between the source types. The PCA was performed using an in-house MatLab script. The first PC (PC1) explains the most amount of variance, and the second PC (PC2) is orthogonal to PC1 and explains the second largest portion of the variance. Each successive PC explains less variance until a point of diminishing returns is reached (*i.e.*, <1% variance explained).

2.6.1. FTICR-MS PCA

Similar to positive matrix factorization, PCA is a factor analysis technique for explaining the observed composition in complex mixtures. It uses an eigenvector analysis of a correlation matrix to calculate principal components and variable loadings. The molecular formulas from each FTICR mass spectrum for the WSOM−, WSOM+, and PSOM− were compiled into a master formula list containing the formulas present in 2–13 of the 14 samples (14,808 formulas). Formulas present in only one sample, and formulas present in all 14 samples were removed to avoid biasing the PCA toward rare formulas and to eliminate formulas that do not contribute to the sample variance, respectively [31,40,41]. A 14,808 × 14 matrix was created by using an input value of 1 if a formula is present and an input value of 0 for a formula not present within a given sample.

2.6.2. ^{1}H NMR PCA

The ^{1}H NMR spectra were analyzed in a PCA following previous studies [6,7]. Briefly, all of the peaks in the ^{1}H NMR spectrum for each of the aerosol WSOM extracts between 0.0 and 11.0 ppm were binned to a resolution of 1 data point per 0.005 ppm from an initial resolution of 0.0008 ppm between data points. The discrete signal (peak area) at each chemical shift was normalized to the total area in the given spectrum, and the normalized area was used as the data input variables ($n = 2769$) for the PCA. The ^{1}H NMR spectra were evaluated initially using all of the aerosol samples ($n = 14$), and a second time after omitting the marine aerosols ($n = 10$).

3. Results and Discussion

3.1. Aerosol Loadings

The aerosol samples from each of the emission sources show OM characteristics that distinguish the sources from one another. TSP and TC concentrations and BC, WSOC and PSOC percentages (relative to TC) were determined for each of the aerosol samples (Table 1). TSP, %BC, and %PSOC were not determined for the marine samples as discussed in the methods section. TSP loadings were highest for the biomass burning samples (73.2 $\mu g \cdot m^{-3}$) followed by the urban (47.1 $\mu g \cdot m^{-3}$) and mixed source (24.1 $\mu g \cdot m^{-3}$) samples. The marine samples show TC loadings (0.5 $\mu g \cdot m^{-3}$) one order of magnitude lower than the mixed source (5.7 $\mu g \cdot m^{-3}$) and urban (6.3 $\mu g \cdot m^{-3}$) samples and two orders of magnitude lower than the biomass burning samples (24.8 $\mu g \cdot m^{-3}$) as one would expect for samples collected over the middle of the ocean far from major terrestrial and anthropogenic sources. In spite of the urban samples having TSP loadings that are approximately twice that of the mixed source samples, they show similar TC concentrations indicating that the urban samples contain high amounts of inorganic materials. The biomass burning samples showed the highest %BC (6.5%) and lowest %WSOC (33.6%) values of the samples which is expected for samples collected in proximity (<30 km) to biomass combustion processes that produce BC and have not been exposed to aging processes known to increase water solubility. The marine samples show low but variable %WSOC values (39.6 ± 25.1%). The urban aerosol samples contained a higher %BC (3.4%) and lower %WSOC (40.8%) than the mixed source aerosol (%BC = 1.9, %WSOC = 50.5) samples. Like all elemental and BC measurements, the CTO-375 method used here is subject to artifacts; it is thought to be selective for highly condensed soot BC and is susceptible to potential positive bias from the charring of melanoidin-like species [42], and the BC results should be viewed with this in mind. In a multi-laboratory comparison study, CTO-375 of aerosol particulate materials yielded lower %BC (3.7%–14.3%) relative to elemental carbon measured using thermal optical reflectance and thermal optical transmittance methods (16%–50%) frequently used in atmospheric studies likely due to the latter two methods inclusion of a portion of OC and the CTO-375 method's selectivity for highly condensed soot BC [42].

The %PSOC amounts were calculated from ^{1}H NMR data (Supplemental Table S1) after Willoughby *et al.* [30] and are useful for understanding the amount of material analyzed in PSOM extracts for FTICR-MS analyses of water-insoluble OM. Because the pyridine extractions were conducted in parallel to the water extractions and some carbon compounds are soluble in both solvents and others are soluble in neither, the sum of these two percentages (%WSOC + %PSOC) may be more or less than 100%. The biomass burning samples can be expected to contain significant amounts of water-insoluble primary OA, and this is reflected in the high %PSOC and low %WSOC values. The calculation for PSOC percentage by the ^{1}H NMR technique omits aromatic peaks due to interference by the exchanged pyridine protons. Because the biomass burning samples are expected to have high aromatic contributions, as indicated by the high %BC and increased signal in aromatic region of the ^{1}H NMR spectrum for the WSOM (discussed in Section 3.4), this %PSOC value may be considered a low estimation. The urban and mixed source aerosols have considerably lower %PSOC (urban = 44%; mixed = 45%) reflecting their higher water solubility and suggesting that these two sample types have more influence from secondary and aging reactions that produce OA insoluble in pyridine.

3.2. Mass Spectra and Average Source Molecular Characteristics

Averaged values calculated for properties determined using FTICR-MS molecular formulas for each of the sample types demonstrate some source-specific characteristics that differentiate them from one another. Each of the ESI-FTICR mass spectra for the ambient aerosol extracts average thousands of peaks across a broad range of 200–800 *m/z* (Figure 1). The average number of formulas found in each sample's master formula list (*i.e.*, WSOM−, WSOM+, and PSOM− combined) followed the same trend as was observed for the TSP and TC concentrations (Table 2). The biomass burning aerosols showed the highest TC loads and averaged 6579 (±173) formulas assigned to each sample and a total of 7891 formulas. The urban aerosols were the next most molecularly abundant sample type averaging 6527 (±173) formulas per sample and a total of 10,701 formulas in all samples. An average of 4104 (±467) formulas were assigned to each mixed source sample, and a total of 6134 formulas were identified. To go along with their lowest TC load (Table 1), the marine aerosol samples totaled just 4570 formulas, averaging 2569 (±736) formulas per aerosol sample. Higher quantities of organic carbon increase the overall number of compounds and the probability of a more diverse suite of compounds, and the similar trends in molecular formula abundances and TC loads are thus logical and likely related.

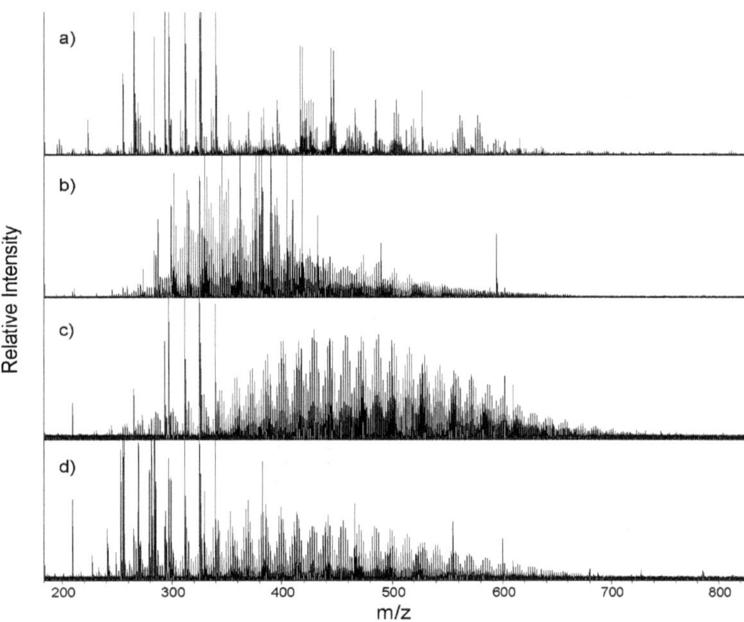

Figure 1. Representative full ESI(−) FTICR mass spectra for $WSOM_{PPL}$ extracts of (**a**) marine; (**b**) biomass burning; (**c**) urban; and (**d**) mixed source aerosols between 200 and 800 *m/z*. Some intense peaks are shown off scale.

Table 2. Total formulas and average elemental properties for aerosol $WSOM_{PPL}$ and PSOM from each emission source determined using FTICR mass spectra.

	Marine Aerosols $n = 4$	Biomass Burning Aerosols $n = 2$	Urban Aerosols $n = 5$	Mixed Source Aerosols $n = 3$
Average formulas	2569 ± 736	6579 ± 173	6527 ± 173	4104 ± 467
Total formulas	4570	7891	10,701	6134
Average O/C	0.32 ± 0.19	0.32 ± 0.19	0.45 ± 0.23	0.44 ± 0.23
Average H/C	1.56 ± 0.39	1.35 ± 0.39	1.44 ± 0.37	1.46 ± 0.35
Average AI_{mod}	0.18 ± 0.26	0.29 ± 0.27	0.18 ± 0.24	0.18 ± 0.23

The marine aerosols share the lowest average O/C ratio (0.32 ± 0.19; Table 2) with the biomass burning aerosols, differentiating them from the urban and mixed source aerosols which had considerably higher average O/C ratios (0.45 ± 0.23 and 0.44 ± 0.23, respectively). The lower average O/C values for the biomass burning and marine aerosol samples suggest they are less oxidized than the mixed source and urban aerosols due to less post-emission atmospheric processing and/or lower characteristic O/C OM at emission. Atmospheric aging processes are known to increase the average O/C ratio of aerosol OM [43,44]. The average O/C ratio (0.45 ± 0.23) for the urban aerosols is the highest of all the samples, supporting previous work showing highly oxidized OM near urban regions due to active photochemistry and abundant inorganic oxidants [45,46] in spite of proximity to primary OA sources including vehicle exhausts that are expected to have a hydrocarbon-like (low O/C) profile. The similarly high average O/C ratio measured for the mixed source samples, however, demonstrates that the high O/C ratios and presumed high extent of atmospheric oxidation are not unique to urban environments.

The marine aerosols have the highest average H/C ratio (1.56 ± 0.39). The modified aromaticity index (AI_{mod}) was calculated for each molecular formula according to the formula proposed by Koch and Dittmar [47]. The average AI_{mod} for the marine aerosols (0.18) is equal to the mixed source and urban aerosols and indicative of the prevalence of olefinic/alicyclic ($0 < AI_{mod} < 0.5$) compounds. These ratios differentiate the marine aerosols from the biomass burning aerosols where the O/C ratios did not. The average H/C ratios for the biomass burning samples are the lowest (1.35 ± 0.39) of any of the aerosol sources indicating a large number of unsaturated molecules, and this is verified by its average AI_{mod} (0.29 ± 0.27), which is much higher than those of the other three types of samples. As was the case for the O/C ratios, the average H/C ratio (1.46 ± 0.35), and AI_{mod} (0.18 ± 0.23) of the mixed source aerosols are very similar to those calculated for the urban samples (H/C ratio = 1.44 ± 0.37, AI_{mod} = 0.18 ± 0.24) making them almost indistinguishable based on these properties alone (Table 2). The biomass burning aerosols have substantially more aromatic ($0.5 \leqslant AI_{mod} < 0.67$, 16.7%, 1318 formulas) and condensed aromatic ($AI_{mod} \geqslant 0.67$, 6.6%, 523 formulas), and fewer aliphatic ($AI_{mod} = 0$, 23.6%, 1864 formulas) formulas than any of the other emission sources but a similar amount of olefinic/alicyclic formulas (53.0%; Supplemental Table S2). The urban, marine, and mixed source aerosols all showed similar distributions with regard to AI_{mod} classifications.

The averaged contributions of elemental formula combinations to each master formula list also present a broad brush method for distinguishing among the four sample types. Details regarding the properties for each ionization source can be found in the supplementary information (Supplemental Table S3), but will not be discussed. CHO formulas were always the most abundant (marine, mixed source) or second most abundant (biomass burning, urban) formula type for each group (Supplemental Table S2). The relative distribution of CHOS and CHONS formulas are similar among all the sources (CHOS = 17.8%–20.9%, CHONS = 12.2%–15.4%) with the mixed source aerosols having the highest contributions from both formula types. The marine aerosols show the largest relative contributions from P-containing molecular formulas. The van Krevelen diagram enables a representation of all of the molecular formulas assigned to the samples including the variation of H/C and O/C ratios as well as a visualization of some of the differences and commonalities described by the averaged ratio values (Supplemental Figure S1), and can be used to characterize differences among the four sample types. However, any of several molecular formulas can plot at the same H/C and O/C ratios, and it is laborious and inefficient to evaluate similarities and differences among samples by looking at tens of van Krevelen diagrams. It is more efficient and statistically valid to identify the defining features among several samples with large molecular formula datasets using a factor analysis such as PCA.

3.3. FTICR-MS PCA

The majority of the variance (66.0%) in the FTICR-MS formula identifications among the four aerosol sample types is explained by the first three principal components (PC1, 32.5%; PC2, 22.4%; PC3, 11.1%; Figure 2). The aerosol samples show distinctive PC1–PC3 values based on their source characterization. Each of the marine samples have a negative PC1 score, a positive PC2 score, and a positive PC3 score. The mixed source samples have a positive PC1, a positive PC2, and a negative PC3 score. The biomass burning samples have a negative PC1, a negative PC2, and a positive PC3. The urban samples have a positive PC1, a negative PC2, and a positive PC3. These PC score classifications were used to identify the loadings (the molecular formulas used as PCA input variables; Supplemental Figure S2) diagnostic for each source. For example, if a molecular formula has negative PC1 loadings, and positive PC2 and PC3 loadings, it is classified as a formula characteristic of marine sources. This resulted in the identification of 1078 formulas characteristic for marine aerosols, 693 formulas for mixed source aerosols, 4174 formulas for biomass burning aerosols, and 3484 formulas for urban aerosols. The remaining 5379 formulas contain characteristics that are represented by multiple sources indicating that they are not diagnostic of a particular source and may be ubiquitous in aerosol OM or inconsistently present in a given source.

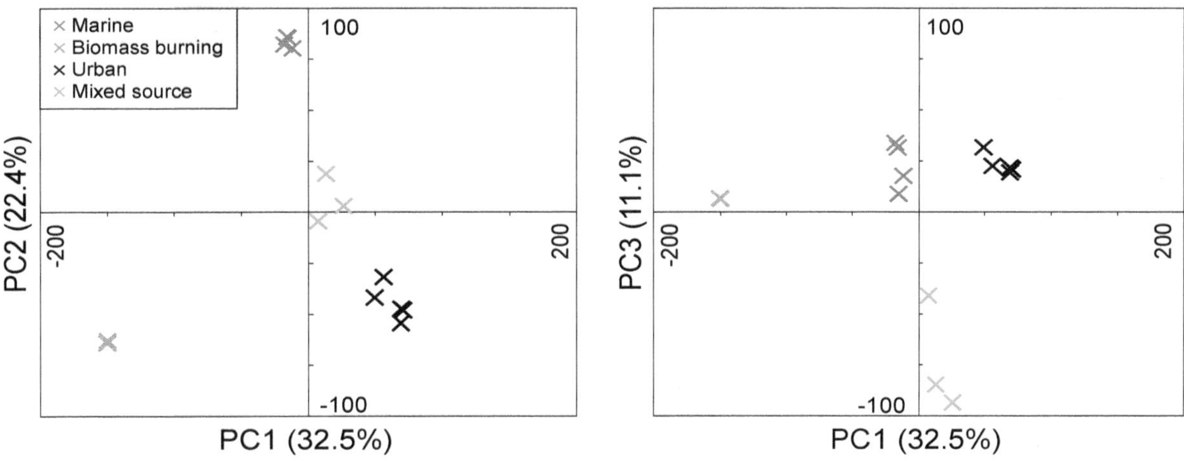

Figure 2. PCA score plots for PC2 *versus* PC1 (**left**) and PC3 *versus* PC1 (**right**). PCA was performed using molecular formulas identified in FTICR mass spectra for aerosol OM extracts from marine, biomass burning, urban, and mixed source areas.

3.3.1. Marine Aerosols

In contrast with the average CHO contributions for the master formula list (41.1%), the PCA-identified formulas for the marine aerosols showed the lowest contributions from CHO formulas (31.6%, Table 3) of all the sample types. CHO formulas are still the most abundant formula type of the PC loadings associated with the marine aerosols, and the majority of these formulas are present at high H/C and low O/C ratios (Figure 3a). In fact, most of the formulas identified as specific to marine sources by PCA are localized to the upper-left region of the van Krevelen diagram (O/C $\leqslant$ 0.6, H/C $\geqslant$ 1.5), a region where many biologically-relevant compounds (lipids, fatty acids, proteins), suggesting that biological activity is an important source for marine aerosols (Figure 3a).

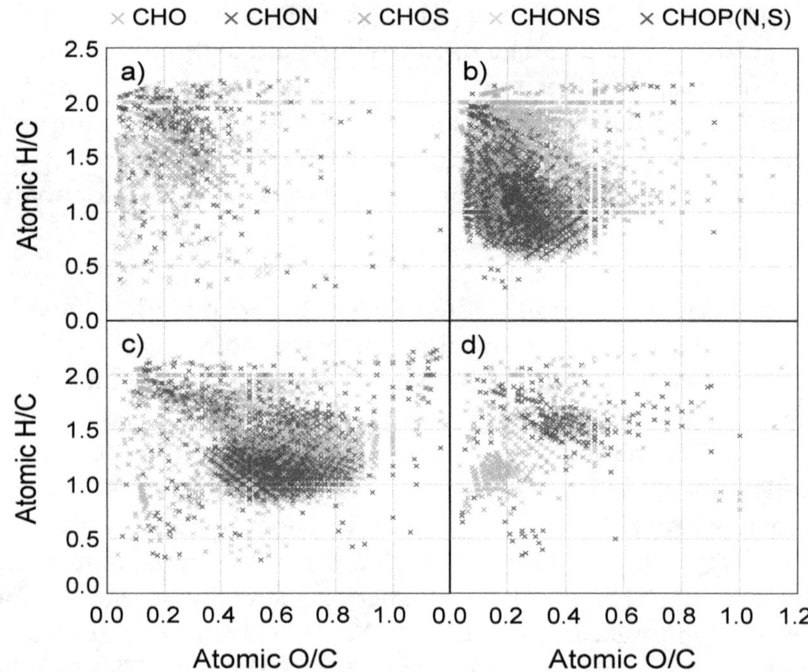

Figure 3. Van Krevelen diagrams for molecular formulas identified by PCA for (**a**) marine; (**b**) biomass burning; (**c**) urban; and (**d**) mixed source aerosols. Each data point is colored according to the atomic content of the molecular formula.

Table 3. Total formulas and average elemental properties for molecular formulas identified by PCA. Distributions of formulas based on atomic content and AI_{mod} structure type are listed as percentage of total formulas.

	Marine Aerosols	Biomass Burning Aerosols	Urban Aerosols	Mixed Source Aerosols
Total formulas	1078	4174	3484	693
Average O/C	0.27 ± 0.17	0.26 ± 0.13	0.55 ± 0.21	0.33 ± 0.18
Average H/C	1.55 ± 0.41	1.24 ± 0.37	1.36 ± 0.34	1.48 ± 0.34
Average AI_{mod}	0.20 ± 0.25	0.37 ± 0.23	0.18 ± 0.22	0.20 ± 0.21
Atomic Content (%)				
CHO	31.6	34.7	33.4	45.0
CHON	13.4	45.4	38.0	24.4
CHOS	27.4	12.0	13.3	18.3
CHONS	19.0	7.8	12.5	6.6
CHOP(N,S)	8.6	0.1	2.8	5.6
Structure Type (%)				
Aliphatic	31.4	12.0	34.1	24.0
Olefinic/alicyclic	56.6	55.6	60.6	67.8
Aromatic	5.5	26.8	3.1	4.9
Condensed aromatic	6.6	5.6	2.3	3.3

The PC loadings for the marine aerosols have the largest fraction of CHOS (27.4%), CHONS (19.0%), and CHOP (N,S) (8.6%) formulas of all the samples (Table 3). It appears thus, that the heteroatom containing formulas that are present in marine aerosols are very distinctive. The abundance of P-containing formulas in this region of the van Krevelen diagram is consistent with inputs of biologically-derived phospholipids as has been observed previously for marine aerosols [31]. Membrane phospholipids have characteristic fatty acid tails with hydrophobic alkyl chains and hydrophilic phosphate heads that can impart amphipathic characteristics, and these P-containing formulas are found in both the PSOM− and WSOM− (Supplemental Table S4; Figure S3) in agreement with this partial solubility. The large number of CHOS and CHONS formulas plotting in these regions have O/S ratios >4 and are suggestive of organosulfate compounds which have also been observed in marine aerosols [31,48,49]. Organosulfates are formed via photochemical aging reactions with the acid-catalyzed ring opening reactions of precursor molecules being the most kinetically favorable reaction mechanism [50–52]. In this instance, the sulfate available for reaction is likely to be marine-derived from biological emissions of dimethyl sulfide that is oxidized in the atmosphere or from sea salt sulfates emitted with sea spray. The low O/C ratios of the marine CHOS and CHONS molecular formulas suggest that any precursor organic compounds were lipid-like in nature and also had low O/C ratios.

Many studies have demonstrated the influence of biological activity on marine aerosols citing the importance of carbohydrate-like and amino-acid-like compounds, for example [53–55]. Carbohydrates are not strongly ionizable compounds under electrospray and are not observed in these samples, and it is important to note that these characteristics represent only the polar, ionizable fraction of marine aerosols. The high H/C and low O/C CHO, CHOP(N,S), CHOS, and CHONS molecular formulas are characteristic of that polar, ionizable fraction, and studies apportioning the sources of OA to a coastal site, for example, can use them as evidence for a marine input to coastal OA. Likewise, studies apportioning sources to OA collected in the marine environment can take large contributions from aromatic or highly oxygenated compounds (O/C > 0.6) as evidence for terrestrial sources.

3.3.2. Biomass Burning Aerosols

The biomass burning aerosol PC loadings are characterized by having the lowest average O/C and H/C values, and the highest AI_{mod} values (Table 3). These average ratios are consistent with the bulk properties discussed in Section 3.2 (Supplemental Table S2) but exaggerate the extreme ratios in the biomass burning aerosols to even lower H/C (1.24 *vs.* 1.35) and O/C (0.26 *vs.* 0.32) ratios and a higher AI_{mod} (0.37 *vs.* 0.29). The PC loadings that define the biomass burning aerosols thus have low O/C and H/C and high AI_{mod} and show the largest fraction of aromatic and condensed aromatic formulas (32.4%) of all the sample types (Table 3). Additionally, more than 80% of the biomass burning PC loading formulas are CHON (45.4%) and CHO (34.7%), suggesting that these two elemental formula groups are most responsible for distinguishing biomass burning aerosols from the other sample groups. Interestingly, the CHOS and CHONS formulas for the biomass burning aerosols master list of formulas made up nearly 20% and 12% of all assigned formulas and spanned a very wide O/C (0.10–1.20) and H/C (1–2) range (Figure 2). However, CHOS and CHONS formulas accounted for just 12% and 7.8% of the PC loadings associated with the biomass burning aerosols and plot in two clusters of the vK diagram (1) at H/C between 1.5 and 2.0 and O/C between 0.1 and 0.5; and (2) at H/C between 1.0 and 1.5 and O/C between 0.3 and 0.6. The much lower contributions of S-containing formulas attributable to biomass burning in the PC loadings suggests that the majority of the CHOS and CHONS formulas in the biomass burning samples (at high H/C and high O/C) are found in multiple sample types, and those particular CHOS and CHONS formulas may not be particularly useful as diagnostic of any one sample type. Studies of S-containing OA demonstrate the ability of existing OA and VOCs to form organosulfates via atmospheric oxidation reactions [17,56–58]. Emissions of inorganic sulfur are commonly associated with anthropogenic (e.g., fossil fuel emissions) [59,60] and marine sources (e.g., gas and aerosol-phase marine emissions) [61], but the abundance of CHOS and CHONS formulas in the mixed source and biomass burning aerosols demonstrates the ubiquity of organosulfur compounds in and beyond those environments.

The van Krevelen diagram shows the CHON and CHO formulas to make up a large portion of the aromatic and condensed aromatic PC loading formulas for the biomass burning aerosols, as indicated by their presence in the low O/C and low H/C regions (Figure 3b). Aromatic and condensed aromatic compounds have higher potential for absorbing light than more saturated molecules [5] and make up portions of the BC and brown carbon pools. The definitions of these two carbon pools are, by necessity, operational and remain a topic of debate [5,42,62]. Traditional definitions of BC assume it to be insoluble in water, but this has been challenged, and BC is recognized to exist on a continuum of solubility. Recent studies have shown that BC can become soluble upon oxidation [63,64] and have detected dissolved forms of BC in aerosols [20] and aquatic dissolved OM [65]. The definitive study relating dissolved BC determined by chemical techniques to thermal optically-defined BC has not to our knowledge been performed. However, the polyaromatic structures that are required for formulas defined as condensed aromatic using the AI_{mod} classification system [47] are consistent with structures proposed for BC [66]. The biomass burning PC loadings are distributed relatively evenly among the three extract-ionization pairings (Supplemental Figure S3). Interestingly, however, the majority of the aromatic and condensed aromatic formulas assigned as important to the biomass burning samples by the PCA were detected in the WSOM− and WSOM+ analyses (Supplemental Table S4). This may result because pyridine does not efficiently extract BC or because less oxygenated BC that is extracted into PSOM does not ionize efficiently in ESI, and it is unclear what fraction of the biomass burning BC is soluble.

A great many CHON formulas have very low O/C (0.05–0.40) and H/C (<1.0) ratios, indicating that they may contain reduced nitrogen functional groups (e.g., amines) or have heterocyclic rings. A subset of these compounds contributes to the black nitrogen (heterocyclic aromatic nitrogen produced during biomass combustion or derivatives of BC that have undergone reactions with inorganic nitrogen) compounds that have recently suggested to be important components of the nitrogen and carbon cycles in aquatic systems [65]. Indeed, the WSOM from these biomass burning aerosols do contain fluorophoric compounds, and the fluorescence intensity is substantially higher than the mixed source and urban aerosols (Willoughby *et al.*, unpublished data). The brown color of the sample filters suggests that these samples do contain significant amounts of brown carbon which, like BC, has no unequivocal chemical definition [5]. Brown carbon is formed alongside BC in combustion processes [5], and has also been formed through model reactions of aqueous SOA with NH_3 [67,68]. Though the global importance of brown carbon is still a topic of debate, it has recently been suggested that biomass burning is the predominant source of brown carbon to the world's atmosphere [69], and the results presented here and elsewhere [70,71] suggest that aromatic N-containing compounds are major components of that brown carbon and diagnostic of biomass burning.

3.3.3. Urban Aerosols

The molecular formulas with PC loadings that identify them as important to the urban aerosols are also characterized by a large number of CHON (38.0%) and CHO (33.4%) formulas. The majority of these PC loadings are found in WSOM− measurements (Supplemental Figure S3) and have significantly higher O/C ratios than those found in the biomass burning aerosols (Figure 3c). The average O/C (0.55 ± 0.21) is double and the average AI_{mod} (0.18 ± 0.22) is half that of the biomass burning aerosols. The average O/C value for the urban aerosol PC loadings is much higher than the average value for the entire urban aerosol master list (0.45, Table 2) highlighting the distinguishing nature of those high O/C components. Many of the CHO and CHON formulas that the PCA associates as important to the urban aerosols have O/C ratios between 0.35 and 0.85 and H/C ratios between 0.8 and 1.8, an area of the van Krevelen diagram that overlaps with regions previously described for lignin-like or carboxyl-rich alicyclic molecules (CRAM) in the soil and aquatic literature (Figure 3c) [72,73]. In the atmosphere, formulas in this region are likely to be the result of atmospheric aging reactions which tend to increase the oxidation state of carbon [44] and transform OA compounds into new compounds that plot downwards and to the right on a van Krevelen diagram [43].

Aliphatic CHOS and CHONS molecular formulas plotting at O/C between 0.3 and 1.2 and H/C between 1.3 and 2.0 were also abundant among the PC loadings for the urban aerosols. Fossil fuel combustion and biological emissions are major sources for aliphatic compounds in the atmosphere [11,74], and the presence of heavy traffic and heavy industrial activity in short and long range proximity of the urban sampling site suggests that fossil fuel combustion as strong candidates for these compounds. The high O content in these formulas is suggestive of organosulfate (O/S > 4) and nitrooxyorganosulfate (O/S > 7, contains N) compounds as potential identities. The abundance of oxidized inorganic nitrogen and sulfur emitted in urban areas combined with emissions of aliphatic biological and anthropogenic compounds make for an excellent environment for organosulfate and nitrooxyorganosulfate compound formation [75,76].

Those two major distinguishing features (CHO and CHON formulas at $0.35 < O/C < 0.85$ and CHOS and CHONS formulas at $0.8 < O/C < 1.8$) can be used to apportion OA as anthropogenically-influenced urban aerosols. In fact, they have been used in a study of OA collected on a transect in the North Atlantic Ocean to apportion OA to North American terrestrial air mass influences [31]. The CHON, CHONS, and CHOS formulas identified as important to the urban OA have distinct characteristics from those identified as important to the marine and biomass burning aerosols. The carbon backbones to which N and S functional groups are bound (or incorporated) are very different, and the PCA has made this clear.

3.3.4. Mixed Source Aerosols

The mixed source aerosols contain the fewest formulas identified by PCA, indicating those samples have few compounds specific to a unique source and share many molecular features with the other aerosols. Given the potential influences to these aerosols, the lack of an abundance of truly defining features is perhaps expected. Nonetheless, the PCA did identify a few OA components specific to the mixed source aerosols. Like the urban aerosols, the majority of the PC loadings important to the mixed source samples were water-soluble. Unlike the urban aerosols, many of these PC loadings were in the WSOM+ measurements suggesting relatively higher amounts of compounds with basic functional groups in the mixed source environment (Supplementary Table S4, Figure S3).

A cluster of CHOS formulas at low O/C and H/C ratios ($O/C < 0.4$, $H/C < 1.2$) is evident in the PC loadings (Figure 3d). Most of these formulas do not contain sufficient oxygen to be organosulfates, and therefore must represent more reduced forms of organic sulfur (e.g., thiols). Examination of the master list van Krevelen diagram shows that these types of compounds are also present in the urban and biomass burning aerosols but not nearly to the extent of the mixed source OA (Figure S1). Sulfonates are a common anthropogenic pollutant and are ubiquitous in personal care products, and have been previously identified in aerosol OM [75]. Non-sulfate aromatic S-containing compounds have been identified in fossil fuels [77,78]. The presence of these reduced sulfur compounds in North American continental relative to marine aerosols lends support to the idea that this group of compounds may be anthropogenically-derived. A tight cluster of P-containing formulas ($0.35 < O/C < 0.45$, $1.45 < H/C < 1.55$; Figure 3d) is also evident in the mixed source aerosols. OA compounds containing P have been attributed to biological sources [31,79], and a distinct local biological source is speculated for these formulas in the mixed source OA. The remaining CHO, CHON, and CHOS formulas with PC loadings assigned to the mixed source OA are scattered at $O/C < 0.6$ and $H/C > 1.0$, less oxygenated than the urban aerosol PC loadings and less aromatic than the biomass burning aerosols. The defining features for the mixed source aerosols are, as expected, indicative of the multiple potential sources to the sampling site.

As was found for each of the other sources, a portion of the mixed source PC loading formulas can be defined as condensed aromatic. The presence of condensed aromatic species, which may also be characterized as BC and functionalized derivatives of BC, in all of these samples reflects their ubiquity in the atmosphere even over the North Atlantic Ocean. BC is traditionally regarded as a product of combustion, as in the combustion of fossil fuel or biomass. Numerous pyrogenic sources exist on the east coast of the US (vehicular exhaust, shipping, biomass burning, and long-range transport from various anthropogenic activities), and are probably contributors to BC even in the rural environment where the mixed source aerosols were sampled and over the marine environment. It has been recently shown that BC-like compounds can also be produced from non-pyrogenic sources [80,81], but the prevalence of this mechanism has yet to be established for atmospheric systems. Regardless of origin, these aromatic and condensed aromatic species are capable of absorbing ultraviolet radiation resulting in a positive radiative forcing (*i.e.*, climate warming) [5]. While ESI-FTICR-MS provides an immense amount of molecular information regarding complex OM mixtures, the limitations are well-established. The ability of a compound to be analyzed is highly dependent on its ability to ionize, and non-polar and non-ionizable compounds (e.g., hydrocarbons and carbohydrates) are largely omitted from this analysis. Additionally, this analysis is necessarily qualitative because a combination of charge competition, concentration, and ionization efficiency drive the peak intensities. However, ^{1}H NMR does not have the same bias for the detection of OM, and provides a complementary set of information regarding the chemical makeup of these complex samples.

3.4. ^{1}H NMR Analysis

The WSOM for each of the aerosols display an array of proton types spanning the spectral region between 0 and 11 ppm (Figure 4). The observed chemical shifts are related to the chemical environment of each proton, and provides clues to the structural connectivity of the atoms within each sample. The region between 0 and 4.4 ppm contains at least 90% of the signal in each of the spectra. Most of the spectra contain broad peaks indicating similar proton types attached to varying carbon chain lengths or located in varying proximity to polar functional groups. These broad peaks exemplify the complexity of aerosol WSOM, and make it difficult to identify individual compounds. However, the region in which a proton signal is detected can indicate the general class of compounds to which that proton belongs. For example, the most intense peak in each spectrum is located at 1.2 ppm and is indicative of protons that are part of a CH_2 group, which represent those that are part of an alkyl chain. This peak is broadest in the urban aerosols, suggesting the alkyl chains are longer, and the CH_2 protons in these aerosols are attached to the widest variety of structural entities. Conversely, the narrower CH_2 peak in the marine, mixed source, and biomass burning aerosols suggests shorter chain lengths and less diversity regarding chemical environments among CH_2 groups. Smaller chain lengths could indicate decomposition of larger molecules by photochemical degradation, or suggest that the molecules have not undergone substantial oligomerization reactions, such as those that add small volatile species like isoprene.

Dividing the spectrum into key proton regions and evaluating the relative contributions of the total spectral intensity can reveal important differences among the different sources (Table 4). The region where signal from aromatic protons are found (6–9 ppm) is the most variable among the sources. A broad signal is observed in the aromatic region of the biomass burning aerosols spectra (Figure 4b inset) and it makes up 9.2% of the total intensity, a percentage more than 4 times greater than the other sources (Table 4). Protons in this region are attached to aromatic or condensed aromatic rings, and the broadness of the peak indicates a high degree of structural diversity among the aromatic protons in these aerosols. There is little signal in the aromatic region in the urban and mixed source aerosols, and essentially no signal in the marine aerosols, indicating there are either very few aromatic compounds in these aerosols or that only water-insoluble aromatic species are present. The observation of a much larger intensity in the aromatic region of the biomass burning aerosols supports the detection of more aromatic and condensed aromatic formulas in the FTICR mass spectra as well as the higher concentration of BC, and suggests that the biomass burning aerosols contain a larger quantity of chromophoric OM than the other aerosol sources investigated here.

As previously mentioned, the majority of the signal in each NMR spectrum falls in the region between 0.6 and 4.4 ppm, and the relative signal in each of the major proton regions does not vary greatly between the sources. At least 50% of the signal falls in the H-C region (0.6–1.8 ppm) in all cases, indicating that a majority of the protons are part of alkyl groups. The larger H-C fractions observed in the urban and mixed source aerosols suggest that OM contains larger carbon backbones (linear or branched). Carbon can be added to existing OM via oligomerization reactions, and the larger carbon chains present in the mixed and urban aerosols suggests that they have undergone oligomerization reactions more extensively than the marine and biomass burning aerosols. Protons that are more downfield in this region (1.4–1.8 ppm) are often attributed to an H-C group that is β to a carbon attached to heteroatoms (H-C-C-C-X, where X = N, S, or O) [14]. The presence of a large number of heteroatomic compounds identified in each of the mass spectra (CHON, CHOS, and CHONS) indicates some portion of the signal in this region represents these species as can be expected for OA and was noted previously (Section 3.2).

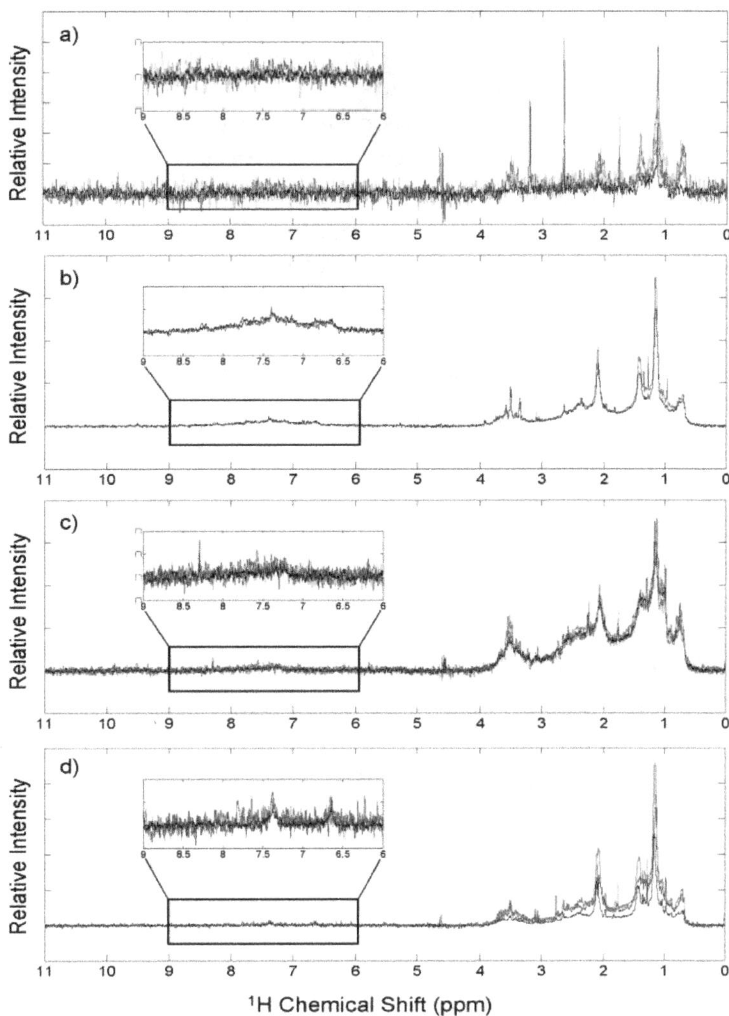

Figure 4. ^{1}H NMR spectra for WSOM of (**a**) marine; (**b**) biomass burning; (**c**) urban; and (**d**) mixed source aerosols, where each colored line represents a different sample spectrum. The region between 6 and 9 ppm represents aromatic protons, and is expanded in the inset of each spectrum.

Table 4. Average relative contributions of total spectral intensity for integrations of major proton regions in ^{1}H NMR spectra for each of the aerosol sources. Standard deviations of the relative signal in each region among aerosols from the same source are provided.

Aerosol Source	H-C (0.6–1.8 ppm)	H-C-C= (1.8–3.2 ppm)	H-C-O-R (3.2–4.4 ppm)	Ar-H (6.0–9.0 ppm)
Marine	50.3 ± 5.5	34.2 ± 3.9	14.5 ± 3.6	1.0 ± 0.3
Biomass burning	50.9 ± 3.4	31.1 ± 0.6	9.4 ± 0.9	9.2 ± 1.9
Urban	53.2 ± 1.5	35.0 ± 2.0	9.4 ± 1.1	2.4 ± 0.6
Mixed source	57.0 ± 1.3	32.9 ± 0.4	8.4 ± 0.8	1.7 ± 0.3

Approximately one-third of the signal intensity falls in the unsaturated alkyl region (H-C-C=; 1.8–3.2 ppm), which includes carbonyl, carboxyl, alkenes, and also hydrogen attached to carbons adjacent to a nitrogen or sulfur (*i.e.*, amines, thiols, *etc.*). The marine and urban aerosols have the highest relative percentage of proton signal in this region, but are also the most variable. On average, approximately 9% of the proton signals are found in the oxygenated aliphatic region (3.2–4.4 ppm) for the aerosols collected over terrestrial environments and a much higher percentage is observed in the marine aerosols (14.5%). The greater signal intensity in this region of the marine aerosols is surprising given the low O/C ratios of the molecules identified in the FTICR mass spectra, but this could be due to the presence of oxygenated species that are outside of the ESI-FTICR-MS analytical window (*i.e.*, nonionizable compounds, compounds < 200 *m/z*) as suggested by the presence of a sharp, intense peak in this region. This region includes protons attached to carbons that are singly bound to an oxygen atom such as ethers, esters, alcohols, and carbohydrates. Carbohydrates are thought to be important components of aerosols produced via bubble bursting [54] but do not ionize efficiently via electrospray ionization. The complexity of aerosol OM limits the ability to identify each of the individual components within the mixture, but ^{1}H NMR provides valuable information regarding the connectivity of the compounds present and because of its complementarity to FTICR-MS we can gather complementary chemical features contained within these complex mixtures. The important and distinguishing features can be observed more clearly with the help of PCA.

3.5. ^{1}H NMR PCA

PCA was performed on the whole ^{1}H NMR spectra for each of the aerosol WSOM extracts, similar to previous studies [6,7,82]. The initial PCA results indicate that the marine aerosols are significantly different than the mixed source, biomass burning, and urban aerosols and inclusion of the marine aerosols does not adequately explain the variance between the three terrestrial sources (Figure 5a). The main differences include the fact that the marine aerosols contain a few sharp peaks (e.g., methanesulfonic acid at 2.7 ppm, and acetate at 1.8 ppm), and the other aerosols contain multiple broad signals throughout each spectrum. The PCA was evaluated a second time using only the mixed source, biomass burning, and urban aerosols, and key differences among those sources are discussed further. The key features of marine aerosol WSOM that distinguish them from continentally-influenced air masses have been discussed at length in previous studies [6,7]. Briefly, the marine aerosols differ from those influenced by the North American continent in that the marine aerosols have more saturated aliphatic chains and are less structurally diverse.

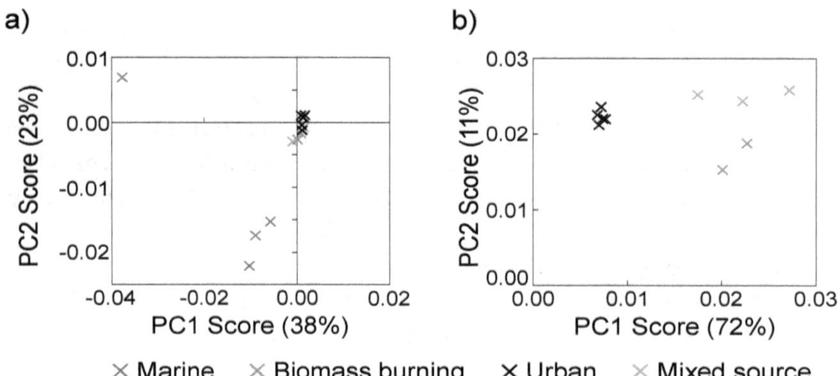

Figure 5. Aerosol WSOM (**a**) PC1 and PC2 scores for PCA of full ^{1}H NMR spectra of marine, mixed source, biomass burning, and urban aerosols and (**b**) PC1 and PC2 scores for PCA omitting marine aerosols. The amount of variation explained by each PC is indicated in parentheses on each axis.

The first two principal components (PC1 and PC2) explain more than 80% of the variance between the mixed source, biomass burning, and urban aerosol WSOM (Figure 5b). The urban aerosols have lower PC1 scores than the mixed source and biomass burning aerosols, the biomass burning aerosols have lower PC2 scores than the urban and mixed source aerosols, and the mixed source aerosols do not have unique PC values (Figure 5b). Thus, PC1 shows the spectral characteristics that differentiate the urban aerosols from the biomass burning and mixed source aerosols, and PC2 shows the spectral characteristics that differentiate the biomass burning aerosols from the other aerosols.

The most intense peak in the variable loadings plot for PC1 (Figure 6a) is positive and represents a CH_2 group (1.2 ppm), suggesting that alkyl chain length is important in distinguishing aerosol emission sources. The CH_2 peak in the mixed source and biomass burning aerosols is more narrow than in the urban aerosols, so the positive peak in the PC1 loadings may be indicative of shorter alkyl chains (*i.e.*, less variability). A splitting of the CH_3 peak (0.9 ppm) is observed, and the negative peak is more downfield than the positive peak. This splitting suggests that the urban aerosols (represented by negative PC1 variable loadings) contain terminal methyl groups in closer proximity to polar functional groups, which is supported by the presence of broad peaks in the region where polar functional groups appear (1.8–3.2 ppm). Another intense positive peak is found around 2.1 ppm, which can represent hydrogen bound to a carbon α to a carbonyl group (Figure 6a). There is an intense peak in this area in each of the whole sample spectra, but the width of the peaks vary (Figure 4) indicating varying structural diversity of similar functional groups. This peak is sharper in the mixed source and biomass burning aerosols, and broad in the urban aerosols. In fact, the edges of the peak at 2.1 ppm can be found as negative peaks in PC1 indicating that the broadness of that peak is characteristic of the urban aerosols. While protons bound to carbon adjacent to a carbonyl group are widespread among the aerosols as indicated by a strong signal around 2.1 ppm in each ^{1}H NMR spectrum, the urban aerosols contain protons that are in more diverse chemical environments.

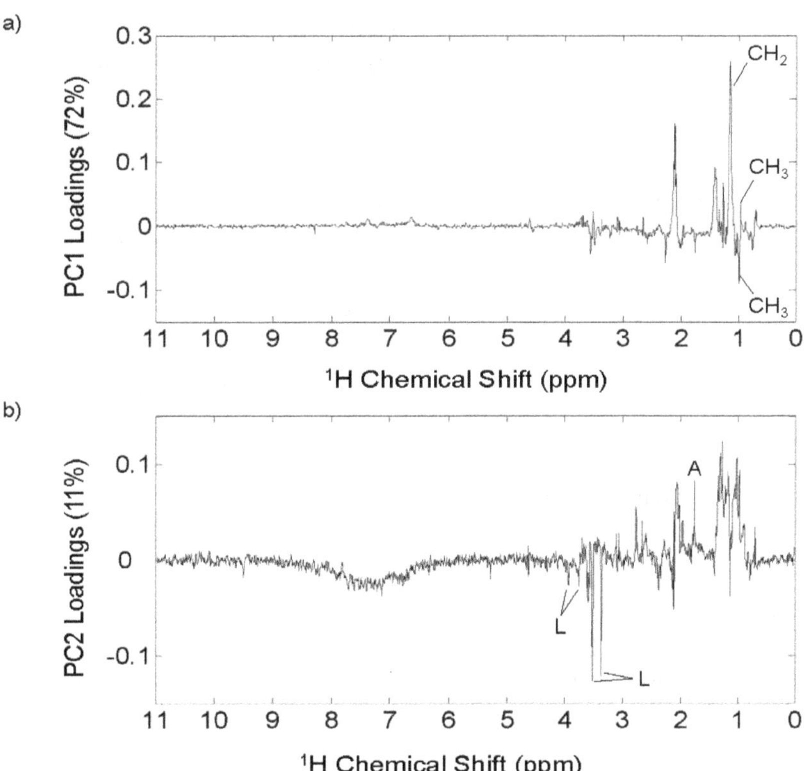

Figure 6. Variable loadings plots for (**a**) PC1 and (**b**) PC2 resulting from PCA using full ^{1}H NMR spectra of mixed source, biomass burning, and urban aerosol WSOM. Some peaks are labeled with the functional group region and some peaks labeled with a single letter to indicate a specific compound, where A = acetic acid or acetate and L = levoglucosan.

The variable loadings plot for PC2 (Figure 6b) shows the spectral characteristics important to the biomass burning aerosols. A broad negative peak in the aromatic region (6–9 ppm) is apparent, and is equivalent to the aromatic signal observed in the individual spectra of the biomass burning aerosols. Peaks consistent with levoglucosan (3.36, 3.52, 3.75, and 3.93 ppm) [15] are identified in the negative PC2 loadings, indicating that they are important in distinguishing the biomass burning aerosols from the other source. Levoglucosan is a common product of the combustion of cellulose material and is widely used as a tracer compound for biomass burning aerosols [83]. Levoglucosan does produce additional ^{1}H NMR peaks, but they are obscured by the water-suppression pulse program due to their proximity to the peak generated by water (~4.7 ppm). A slight negative peak at 2.3 ppm is consistent with hydrogen attached to nitrogen. The peak is broad indicating that there are many compounds with the same functional group attached to varying carbon structures. Without standards, it cannot be confirmed that this peak represents amino groups, but the large number of CHON formulas identified in the FTICR mass spectra supports the presence of a diverse suite of N-containing OM in the biomass burning aerosols. Positive peaks in the aliphatic region (0.8–1.5 ppm) and at 1.8 ppm representing acetate or acetic acid demonstrate that the higher PC2 scores for the urban and mixed source aerosols have a more oxygenated and aliphatic WSOM composition.

Overall, there is good consistency and complementarity between the ^{1}H NMR and FTICR-MS techniques. Both techniques identify aromatic compounds and nitrogen-containing compounds that are not extensively oxidized as important for distinguishing the OM from biomass burning aerosols from the other sources. The lower degree of oxidation in the OM found in the biomass burning and also the marine aerosols suggests that these aerosols are collected close to their source and/or have not been exposed to conditions that promote extensive oxidative or oligomeric transformations. The urban aerosols are characteristically more polar as indicated by ^{1}H NMR peaks that are shifted more slightly downfield than in the other aerosols and also the higher O/C ratios of the formulas, with more structurally diverse carbon chains. This structural diversity may be represented by longer and/or more branched carbon chains indicative of more oligomeric or other chain elongation reactions. While some of these features were readily apparent, the PCA was able to highlight and confirm some of these variations.

The combination of ^{1}H NMR and FTICR-MS provides an incredible amount of molecular-level information regarding the chemical composition of aerosol OM. While ^{1}H NMR overcomes some of the bias introduced by ESI-FTICR-MS, it comes with its own limitations. The ^{1}H NMR analysis used here requires a liquid sample, so the water-insoluble OM is not analyzed by this method. Solid-state NMR techniques that analyze a whole solid sample do exist but require very large samples that require extremely large sample volumes which typically can only be obtained over weeks of sampling. Additionally, the water-suppression removes all signal in the region of 4.7 ppm and may reduce the signal in the region within ±0.5 ppm including protons directly attached to carbon-carbon double bonds (H-C=C) which are typically found around 5 ppm.

4. Conclusions

The ^{1}H NMR and FTICR-MS results described here, while not fully quantitative in nature, provide detailed, source-specific distinguishing characteristics of the water- and pyridine- extractable and ionizable components of marine, urban, biomass burning, and mixed source OA. PCA efficiently highlights the molecular formulas and ^{1}H NMR structural components that distinguish the OM contained within aerosols from each of these sources. Marine aerosols contain molecules consistent with biological inputs including lipid- and phospholipid-like compounds as well as a large fraction of low O/C ratio organosulfur compounds, likely derived from reactions with sulfate derived from biological dimethyl sulfide emissions. The carbon backbones of these molecules are more aliphatic and less diverse than terrestrial OA, and the OM is overall less oxidized. The OM in biomass burning aerosols has more aromatic character and less oxygen content than the other OA groups. Aromatic, nitrogen containing compounds are a defining feature of biomass burning aerosols, and nitrogen-incorporation reactions play a major role in the formation or transformation of these aerosols. The urban aerosols are also characteristically more oxygenated and more structurally complex, suggesting they have been subject to more extensive atmospheric aging. Nitrogen-containing formulas of lesser aromaticity than described for the biomass burning OA are an important feature of urban aerosols. These compounds likely reflect the prevalence of OA transformations that incorporate inorganic NOx, either through NOx additions to aliphatic/olefinic compounds or ring-opening reactions of biomass burning OA. The combined results of the FTICR-MS and ^{1}H NMR PCAs show the mixed source OA to be very similar to but less chemically diverse than the urban aerosols.

These defining OA molecular features can be used in future work to provide clues to the primary inputs and degree of atmospheric transformation. In environments impacted by multiple emission sources, these molecular features can help identify key sources. Further, as the relationships between molecular composition and atmospheric impacts (e.g., hygroscopicity, ice and cloud condensation formation, light absorption) and environmental fates (e.g., photochemical and microbial lability, metal complexing ability) are strengthened, these source-specific molecular characteristics can be used to partition the relative importance of each of these emission sources to these impacts and fates. The highly aliphatic nature of and biological source for the marine aerosols suggests them to be highly susceptible to microbial components, while the aromatic nature of biomass burning OA are likely to be less microbially labile and more susceptible to photodegradation. Though the exact relationship is yet to be fully understood, the extent of oxidation is associated with hygroscopicity and the ability of aerosols to act as cloud condensation nuclei, and the high O/C content of urban and mixed source aerosols present them as high hygroscopic candidates. The additional molecular formula and structural details provided here may help understand deviations from the expected O/C ratio—hygroscopic behavior relationship. The highly aromatic, nitrogen compounds are likely an important component of atmospheric brown carbon, which makes a significant contribution to light-absorbing carbon in the atmosphere. Any extensive light absorption found for other source components may be due to higher amounts of inorganic components or olefinic or non-ionizable OA.

Aerosol FTICR-MS and ^{1}H NMR analyses provide a complementary set of information regarding OA chemical composition, and PCA provides a useful tool for deconstructing the important components that define each of the aerosol sources. While this study presents several aerosols from some key sources, application of this method to a larger number of samples from more emission sources are needed for a more comprehensive inventory of atmospherically-relevant OA. Additionally, pairing these analysis with highly time-resolved methods, such as aerosol mass spectrometry, would provide an excellent accounting of the inorganic ions and volatile OM that influence aerosol atmospheric and environmental impacts and fates.

Supplementary Materials: The following are available online at http://www.mdpi.com/2073-4433/7/6/79/s1,

Supplementary methods.

Table S1: Percent area contributions from the major proton regions and calculated H/C ratios in ^{1}H NMR spectra for aerosol PSOM.

Table S2: Total formulas and average elemental properties for aerosol $WSOM_{PPL}$ and PSOM from each emission source determined using FTICR mass spectra. The distribution of molecular formulas based on atomic content and AI_{mod} structure type are listed as number of formulas with the percentage of total formulas in parentheses directly below.

Table S3: Total formulas and average elemental properties for aerosol WSOM−, WSOM+, and PSOM− from each emission source determined using FTICR-MS. Atomic content and structure type values are expressed as the number of formulas. The values in parentheses are the percentage of total molecular formulas in each sample, an average for each source.

Table S4: Total formulas and average elemental properties for aerosol WSOM−, WSOM+, and PSOM from each emission source identified by PCA. Distributions of formulas based on atomic content and AI_{mod} structure type are listed as percentage of total formulas.

Figure S1: Van Krevelen diagrams for molecular formulas identified in the FTICR mass spectra for the marine, biomass burning, urban, and mixed source aerosols. Each row represents a different source, and each column represents only those formulas with a specific elemental makeup (CHO, CHON, or CHOS). Each "×" represents one or more molecular formulas.

Figure S2: The loadings for (a) PC1 and PC2 and (b) PC1 and PC3 from the PCA analysis of the FTICR-MS molecular formulas.

Figure S3: Venn diagrams showing the relative distribution of PCA molecular formulas present in any of the three solvent/ionization methods (WSOM−, WSOM+, and PSOM) for each aerosol source. Areas of overlap represent percentages of molecular formulas that appear in two or more of those samples. Areas with no overlap represent the percentage of molecular formulas unique to that individual solvent/ionization method.

Acknowledgments: This work was funded by the Frank Batten Endowment to Patrick Hatcher. Amanda Willoughby was supported by a US EPA STAR graduate fellowship program (Grant: FP 91736001). The EPA has not officially endorsed this publication and the views expressed herein may not reflect the views of the EPA. The authors thank Peter DeCarlo and Anita Johnson at Drexel University for assistance in the collection of the urban aerosol samples, Siddhartha Mitra for BC analyses, and Susan Hatcher and Jared Callan of the COSMIC facility at Old Dominion University for assistance with the FTICR-MS. We thank the crew of the R/V Knorr for assistance in collecting the marine samples and William Landing for sharing the GEOTRACES marine samples collected as part of NSF grants OCE-0752351, 0929919, and 1132766. Any opinions, findings, and conclusions or recommendations expressed in this material are those of the authors and do not necessarily reflect the views of the National Science Foundation. Finally, we thank two anonymous reviewers for their valuable suggestions that have improved this manuscript.

Author Contributions: All authors conceived and designed the experiments; Amanda Willoughby performed the experiments; Amanda Willoughby and Andrew Wozniak analyzed the data; Patrick Hatcher contributed reagents/materials/analysis tools; Amanda Willoughby and Andrew Wozniak wrote the paper.

Abbreviations

The following abbreviations are used in this manuscript:

AI_{mod}	modified aromaticity index
BC	black carbon
CTO-375	chemothermal oxidation at 375 °C
ESI	electrospray ionization
FTICR-MS	Fourier transform ion cyclotron resonance mass spectrometry
H/C	hydrogen-to-carbon atomic ratio
^{1}H NMR	proton nuclear magnetic resonance spectroscopy
NMR	nuclear magnetic resonance spectroscopy
O/C	oxygen-to-carbon atomic ratio
OA	organic aerosols
OM	organic matter
PC	principal component
PCA	principal component analysis
PPL	solid phase extraction medium
PSOC	pyridine-soluble organic carbon
PSOM	pyridine-soluble organic matter
SOA	secondary organic aerosol
TC	total carbon
TSP	total suspended particulates
UHR-MS	ultrahigh resolution mass spectrometry
WSOC	water-soluble organic carbon
WSOM	water-soluble organic matter
$WSOM_{PPL}$	PPL-extracted water-soluble organic matter

References

1. Jimenez, J.L.; Canagaratna, M.R.; Donahue, N.M.; Prevot, A.S.H.; Zhang, Q.; Kroll, J.H.; DeCarlo, P.F.; Allan, J.D.; Coe, H.; Ng, N.L.; *et al.* Evolution of organic aerosols in the atmosphere. *Science* **2009**, *326*, 1525–1529. [CrossRef] [PubMed]
2. Kanakidou, M.; Seinfeld, J.; Pandis, S.; Barnes, I.; Dentener, F.; Facchini, M.; Van Dingenen, R.; Ervens, B.; Nenes, A.; Nielsen, C.; *et al.* Organic aerosol and global climate modelling: A review. *Atmos. Chem. Phys.* **2005**, *5*, 1053–1123. [CrossRef]
3. Chan, M.N.; Choi, M.Y.; Ng, N.L.; Chan, C.K. Hygroscopicity of water-soluble organic compounds in atmospheric aerosols: Amino acids and biomass burning derived organic species. *Environ. Sci. Technol.* **2005**, *39*, 1555–1562. [CrossRef] [PubMed]
4. Cerully, K.; Bougiatioti, A.; Hite, J., Jr.; Guo, H.; Xu, L.; Ng, N.; Weber, R.; Nenes, A. On the link between hygroscopicity, volatility, and oxidation state of ambient and water-soluble aerosols in the southeastern united states. *Atmos. Chem. Phys.* **2015**, *15*, 8679–8694. [CrossRef]
5. Andreae, M.O.; Gelencsér, A. Black carbon or brown carbon? The nature of light-absorbing carbonaceous aerosols. *Atmos. Chem. Phys.* **2006**, *6*, 3131–3148. [CrossRef]
6. Wozniak, A.S.; Shelley, R.U.; Sleighter, R.L.; Abdulla, H.A.; Morton, P.L.; Landing, W.M.; Hatcher, P.G. Relationships among aerosol water soluble organic matter, iron and aluminum in European, North African, and marine air masses from the 2010 US GEOTRACES cruise. *Mar. Chem.* **2013**, *54*, 557–898. [CrossRef]
7. Wozniak, A.S.; Shelley, R.U.; McElhenie, S.D.; Landing, W.M.; Hatcher, P.G. Aerosol water soluble organic matter characteristics over the North Atlantic Ocean: Implications for iron-binding ligands and iron solubility. *Mar. Chem.* **2015**, *173*, 162–172. [CrossRef]
8. Samburova, V.; Szidat, S.; Hueglin, C.; Fisseha, R.; Baltensperger, U.; Zenobi, R.; Kalberer, M. Seasonal variation of high-molecular-weight compounds in the water-soluble fraction of organic urban aerosols. *J. Geophys. Res. Atmos.* **2005**, *110*. [CrossRef]
9. Takegawa, N.; Miyakawa, T.; Kondo, Y.; Jimenez, J.L.; Zhang, Q.; Worsnop, D.R.; Fukuda, M. Seasonal and diurnal variations of submicron organic aerosol in Tokyo observed using the Aerodyne aerosol mass spectrometer. *J. Geophys. Res. Atmos.* **2006**, *111*. [CrossRef]
10. Baduel, C.; Voisin, D.; Jaffrezo, J. Seasonal variations of concentrations and optical properties of water soluble HULIS collected in urban environments. *Atmos. Chem. Phys.* **2010**, *10*, 4085–4095. [CrossRef]
11. Alves, C.; Vicente, A.; Pio, C.; Kiss, G.; Hoffer, A.; Decesari, S.; Prevôt, A.S.H.; Cruz Minguillón, M.; Querol, X.; Hillamo, R. Organic compounds in aerosols from selected European sites—Biogenic *versus* anthropogenic sources. *Atmos. Environ.* **2012**, *59*, 243–255. [CrossRef]
12. Kieber, R.J.; Adams, M.B.; Willey, J.D.; Whitehead, R.F.; Avery, G.B.; Mullaugh, K.M.; Mead, R.N. Short term temporal variability in the photochemically mediated alteration of chromophoric dissolved organic matter (CDOM) in rainwater. *Atmos. Environ.* **2012**, *50*, 112–119. [CrossRef]
13. Chalbot, M.G.; Kavouras, I.G. Nuclear magnetic resonance spectroscopy for determining the functional content of organic aerosols: A review. *Environ. Pollut.* **2014**, *191*, 232–249. [CrossRef] [PubMed]
14. Duarte, R.M.; Freire, S.M.; Duarte, A.C. Investigating the water-soluble organic functionality of urban aerosols using two-dimensional correlation of solid-state ^{13}C NMR and FTIR spectral data. *Atmos. Environ.* **2015**, *116*, 245–252. [CrossRef]
15. Decesari, S.; Facchini, M.; Matta, E.; Lettini, F.; Mircea, M.; Fuzzi, S.; Tagliavini, E.; Putaud, J. Chemical features and seasonal variation of fine aerosol water-soluble organic compounds in the Po Valley, Italy. *Atmos. Environ.* **2001**, *35*, 3691–3699. [CrossRef]
16. Moretti, F.; Tagliavini, E.; Decesari, S.; Facchini, M.; Rinaldi, M.; Fuzzi, S. NMR determination of total carbonyls and carboxyls: A tool for tracing the evolution of atmospheric oxidized organic aerosols. *Environ. Sci. Technol.* **2008**, *42*, 4844–4849. [CrossRef] [PubMed]
17. Schmitt-Kopplin, P.; Gelencser, A.; Dabek-Zlotorzynska, E.; Kiss, G.; Hertkorn, N.; Harir, M.; Hong, Y.; Gebefugi, I. Analysis of the unresolved organic fraction in atmospheric aerosols with ultrahigh-resolution mass spectrometry and nuclear magnetic resonance spectroscopy: Organosulfates as photochemical smog constituents. *Anal. Chem.* **2010**, *82*, 8017–8026. [CrossRef] [PubMed]

18. Shakya, K.M.; Place, P.F.; Griffin, R.J.; Talbot, R.W. Carbonaceous content and water-soluble organic functionality of atmospheric aerosols at a semi-rural New England location. *J. Geophys. Res. Atmos.* **2012**, *117*, D03301. [CrossRef]
19. Altieri, K.E.; Seitzinger, S.P.; Carlton, A.G.; Turpin, B.J.; Klein, G.C.; Marshall, A.G. Oligomers formed through in-cloud methylglyoxal reactions: Chemical composition, properties, and mechanisms investigated by ultra-high resolution FT-ICR mass spectrometry. *Atmos. Environ.* **2008**, *42*, 1476–1490. [CrossRef]
20. Wozniak, A.; Bauer, J.; Sleighter, R.; Dickhut, R.; Hatcher, P. Technical note: Molecular characterization of aerosol-derived water soluble organic carbon using ultrahigh resolution electrospray ionization fourier transform ion cyclotron resonance mass spectrometry. *Atmos. Chem. Phys.* **2008**, *8*, 5099–5111. [CrossRef]
21. Heaton, K.J.; Sleighter, R.L.; Hatcher, P.G.; Hall, W.A., IV; Johnston, M.V. Composition domains in monoterpene secondary organic aerosol. *Environ. Sci. Technol.* **2009**, *43*, 6950–6955.
22. Laskin, J.; Laskin, A.; Roach, P.; Slysz, G.; Anderson, G.; Nizkorodov, S.; Bones, D.; Nguyen, L. High-resolution desorption electrospray ionization mass spectrometry for chemical characterization of organic aerosols. *Anal. Chem.* **2010**, *82*, 2048–2058. [CrossRef] [PubMed]
23. Mazzoleni, L.; Ehrmann, B.; Shen, X.; Marshall, A.; Collett, J., Jr. Water-soluble atmospheric organic matter in fog: Exact masses and chemical formula identification by ultrahigh-resolution Fourier transform ion cyclotron resonance mass spectrometry. *Environ. Sci. Technol.* **2010**, *44*, 3690–3697. [CrossRef] [PubMed]
24. Nguyen, T.; Bateman, A.; Bones, D.; Nizkorodov, S.; Laskin, J.; Laskin, A. High-resolution mass spectrometry analysis of secondary organic aerosol generated by ozonolysis of isoprene. *Atmos. Environ.* **2010**, *44*, 1032–1042. [CrossRef]
25. LeClair, J.P.; Collett, J.L.; Mazzoleni, L.R. Fragmentation analysis of water-soluble atmospheric organic matter using ultrahigh-resolution FT-ICR mass spectrometry. *Environ. Sci. Technol.* **2012**, *46*, 4312–4322. [CrossRef] [PubMed]
26. Mazzoleni, L.R.; Saranjampour, P.; Dalbec, M.M.; Samburova, V.; Hallar, A.G.; Zielinska, B.; Lowenthal, D.H.; Kohl, S. Identification of water-soluble organic carbon in non-urban aerosols using ultrahigh-resolution FT-ICR mass spectrometry: Organic anions. *Environ. Chem.* **2012**, *9*, 285–297. [CrossRef]
27. Mead, R.; Mullaugh, K.; Brooks Avery, G.; Kieber, R.; Willey, J.; Podgorski, D. Insights into dissolved organic matter complexity in rainwater from continental and coastal storms by ultrahigh resolution Fourier transform ion cyclotron resonance mass spectrometry. *Atmos. Chem. Phys.* **2013**, *13*, 4829–4838. [CrossRef]
28. Mitra, S.; Wozniak, A.S.; Miller, R.; Hatcher, P.G.; Buonassissi, C.; Brown, M. Multiproxy probing of rainwater dissolved organic matter (DOM) composition in coastal storms as a function of trajectory. *Mar. Chem.* **2013**, *154*, 67–76. [CrossRef]
29. Kourtchev, I.; Fuller, S.J.; Giorio, C.; Healy, R.M.; Wilson, E.; O'Connor, I.; Wenger, J.C.; McLeod, M.; Aalto, J.; Ruuskanen, T.M. Molecular composition of biogenic secondary organic aerosols using ultrahigh-resolution mass spectrometry: Comparing laboratory and field studies. *Atmos. Chem. Phys.* **2014**, *14*, 2155–2167. [CrossRef]
30. Willoughby, A.S.; Wozniak, A.S.; Hatcher, P.G. A molecular-level approach for characterizing water-insoluble components of ambient organic aerosol particulates using ultrahigh-resolution mass spectrometry. *Atmos. Chem. Phys.* **2014**, *14*, 10299–10314. [CrossRef]
31. Wozniak, A.S.; Willoughby, A.S.; Gurganus, S.C.; Hatcher, P.G. Distinguishing molecular characteristics of aerosol water soluble organic matter from the 2011 Trans-North Atlantic US GEOTRACES cruise. *Atmos. Chem. Phys.* **2014**, *14*, 8419–8434. [CrossRef]
32. Kourtchev, I.; Doussin, J.-F.; Giorio, C.; Mahon, B.; Wilson, E.; Maurin, N.; Pangui, E.; Venables, D.; Wenger, J.; Kalberer, M. Molecular composition of fresh and aged secondary organic aerosol from a mixture of biogenic volatile compounds: A high-resolution mass spectrometry study. *Atmos. Chem. Phys.* **2015**, *15*, 5683–5695. [CrossRef]
33. Reinhardt, A.; Emmenegger, C.; Gerrits, B.; Panse, C.; Dommen, J.; Baltensperger, U.; Zenobi, R.; Kalberer, M. Ultrahigh mass resolution and accurate mass measurements as a tool to characterize oligomers in secondary organic aerosols. *Anal. Chem.* **2007**, *79*, 4074–4082. [CrossRef] [PubMed]
34. Geotraces: An international study of the marine biogeochemical cycles of trace elements and their isotopes. Available online: http://www.geotraces.org/ (accessed on 1 June 2016).

35. National Oceanic and Atmospheric Administration: Pacific Marine Environmental Laboratory. WACS2 (2014) Western Atlantic Climate Study. Available online: http://saga.pmel.noaa.gov/field_WACS2 (accessed on 1 June 2016).
36. Gustafsson, Ö.; Haghseta, F.; Chan, C.; MacFarlane, J.; Gschwend, P. Quantification of the dilute sedimentary soot phase: Implications for pah speciation and bioavailability. *Environ. Sci. Technol.* **1997**, *31*, 203–209. [CrossRef]
37. Dittmar, T.; Koch, B.; Hertkorn, N.; Kattner, G. A simple and efficient method for the solid-phase extraction of dissolved organic matter (SPE-DOM) from seawater. *Limnol. Oceanogr. Methods* **2008**, *6*, 230–235. [CrossRef]
38. Sleighter, R.; McKee, G.; Liu, Z.; Hatcher, P. Naturally present fatty acids as internal calibrants for Fourier transform mass spectra of dissolved organic matter. *Limnol. Oceanogr. Methods* **2008**, *6*, 246–253. [CrossRef]
39. Stubbins, A.; Spencer, R.G.M.; Chen, H.; Hatcher, P.G.; Mopper, K.; Hernes, P.J.; Mwamba, V.L.; Mangangu, A.M.; Wabakanghanzi, J.N.; Six, J. Illuminated darkness: Molecular signatures of Congo River dissolved organic matter and its photochemical alteration as revealed by ultrahigh precision mass spectrometry. *Limnol. Oceanogr.* **2010**, *55*, 1467–1477. [CrossRef]
40. Sleighter, R.; Liu, Z.; Xue, J.; Hatcher, P. Multivariate statistical approaches for the characterization of dissolved organic matter analyzed by ultrahigh resolution mass spectrometry. *Environ. Sci. Technol.* **2010**, *44*, 7576–7582. [CrossRef] [PubMed]
41. Gurganus, S.C.; Wozniak, A.S.; Hatcher, P.G. Molecular characteristics of the water soluble organic matter in size-fractionated aerosols collected over the North Atlantic Ocean. *Mar. Chem.* **2015**, *170*, 37–48. [CrossRef]
42. Hammes, K.; Schmidt, M.W.; Smernik, R.J.; Currie, L.A.; Ball, W.P.; Nguyen, T.H.; Louchouarn, P.; Houel, S.; Gustafsson, Ö.; Elmquist, M. Comparison of quantification methods to measure fire-derived (black/elemental) carbon in soils and sediments using reference materials from soil, water, sediment and the atmosphere. *Glob. Biogeochem. Cycles* **2007**, *21*. [CrossRef]
43. Heald, C.; Kroll, J.; Jimenez, J.; Docherty, K.; DeCarlo, P.; Aiken, A.; Chen, Q.; Martin, S.; Farmer, D.; Artaxo, P. A simplified description of the evolution of organic aerosol composition in the atmosphere. *Geophys. Res. Lett.* **2010**, *37*. [CrossRef]
44. Kroll, J.H.; Donahue, N.M.; Jimenez, J.L.; Kessler, S.H.; Canagaratna, M.R.; Wilson, K.R.; Altieri, K.E.; Mazzoleni, L.R.; Wozniak, A.S.; Bluhm, H. Carbon oxidation state as a metric for describing the chemistry of atmospheric organic aerosol. *Nat. Chem.* **2011**, *3*, 133–139. [CrossRef] [PubMed]
45. Zhang, Q.; Alfarra, M.R.; Worsnop, D.R.; Allan, J.D.; Coe, H.; Canagaratna, M.R.; Jimenez, J.L. Deconvolution and quantification of hydrocarbon-like and oxygenated organic aerosols based on aerosol mass spectrometry. *Environ. Sci. Technol.* **2005**, *39*, 4938–4952. [CrossRef] [PubMed]
46. Aiken, A.C.; Decarlo, P.F.; Kroll, J.H.; Worsnop, D.R.; Huffman, J.A.; Docherty, K.S.; Ulbrich, I.M.; Mohr, C.; Kimmel, J.R.; Sueper, D. O/C and OM/OC ratios of primary, secondary, and ambient organic aerosols with high-resolution time-of-flight aerosol mass spectrometry. *Environ. Sci. Technol.* **2008**, *42*, 4478–4485. [CrossRef] [PubMed]
47. Koch, B.; Dittmar, T. From mass to structure: An aromaticity index for high-resolution mass data of natural organic matter. *Rapid Commun. Mass Spectrom.* **2006**, *20*, 926–932. [CrossRef]
48. Claeys, M.; Wang, W.; Vermeylen, R.; Kourtchev, I.; Chi, X.; Farhat, Y.; Surratt, J.D.; Gómez-González, Y.; Sciare, J.; Maenhaut, W. Chemical characterisation of marine aerosol at Amsterdam Island during the austral summer of 2006–2007. *J. Aerosol Sci.* **2010**, *41*, 13–22. [CrossRef]
49. Hawkins, L.; Russell, L.; Covert, D.; Quinn, P.; Bates, T. Carboxylic acids, sulfates, and organosulfates in processed continental organic aerosol over the southeast Pacific Ocean during VOCALS-REX 2008. *J. Geophys. Res.* **2010**, *115*. [CrossRef]
50. Iinuma, Y.; Müller, C.; Berndt, T.; Böge, O.; Claeys, M.; Herrmann, H. Evidence for the existence of organosulfates from β-pinene ozonolysis in ambient secondary organic aerosol. *Environ. Sci. Technol.* **2007**, *41*, 6678–6683. [CrossRef] [PubMed]
51. Minerath, E.C.; Casale, M.T.; Elrod, M.J. Kinetics feasibility study of alcohol sulfate esterification reactions in tropospheric aerosols. *Environ. Sci. Technol.* **2008**, *42*, 4410–4415. [CrossRef] [PubMed]
52. Minerath, E.C.; Elrod, M.J. Assessing the potential for diol and hydroxy sulfate ester formation from the reaction of epoxides in tropospheric aerosols. *Environ. Sci. Technol.* **2009**, *43*, 1386–1392. [CrossRef] [PubMed]
53. Kuznetsova, M.; Lee, C.; Aller, J. Characterization of the proteinaceous matter in marine aerosols. *Mar. Chem.* **2005**, *96*, 359–377. [CrossRef]

54. Hawkins, L.N.; Russell, L.M. Polysaccharides, proteins, and phytoplankton fragments: Four chemically distinct types of marine primary organic aerosol classified by single particle spectromicroscopy. *Adv. Meteorol.* **2010**, *2010*, 1–14. [CrossRef]
55. Russell, P.; Bergstrom, R.; Shinozuka, Y.; Clarke, A.; DeCarlo, P.; Jimenez, J.; Livingston, J.; Redemann, J.; Dubovik, O.; Strawa, A. Absorption angstrom exponent in AERONET and related data as an indicator of aerosol composition. *Atmos. Chem. Phys.* **2010**, *10*, 1155–1169. [CrossRef]
56. Surratt, J.D.; Kroll, J.H.; Kleindienst, T.E.; Edney, E.O.; Claeys, M.; Sorooshian, A.; Ng, N.L.; Offenberg, J.H.; Lewandowski, M.; Jaoui, M. Evidence for organosulfates in secondary organic aerosol. *Environ. Sci. Technol.* **2007**, *41*, 517–527. [CrossRef] [PubMed]
57. Noziere, B.; Ekström, S.; Alsberg, T.; Holmström, S. Radical-initiated formation of organosulfates and surfactants in atmospheric aerosols. *Geophys. Res. Lett.* **2010**, *37*. [CrossRef]
58. McNeill, V.F.; Woo, J.L.; Kim, D.D.; Schwier, A.N.; Wannell, N.J.; Sumner, A.J.; Barakat, J.M. Aqueous-phase secondary organic aerosol and organosulfate formation in atmospheric aerosols: A modeling study. *Environ. Sci. Technol.* **2012**, *46*, 8075–8081. [CrossRef] [PubMed]
59. Cullis, C.; Hirschler, M. Atmospheric sulphur: Natural and man-made sources. *Atmos. Environ.* **1980**, *14*, 1263–1278. [CrossRef]
60. Möller, D. Estimation of the global man-made sulphur emission. *Atmos. Environ.* **1984**, *18*, 19–27. [CrossRef]
61. Andreae, M.O. Dimethylsulfoxide in marine and freshwaters. *Limnol. Oceanogr.* **1980**, *25*, 1054–1063. [CrossRef]
62. Masiello, C. New directions in black carbon organic geochemistry. *Mar. Chem.* **2004**, *92*, 201–213. [CrossRef]
63. Decesari, S.; Facchini, M.; Matta, E.; Mircea, M.; Fuzzi, S.; Chughtai, A.; Smith, D. Water soluble organic compounds formed by oxidation of soot. *Atmos. Environ.* **2002**, *36*, 1827–1832. [CrossRef]
64. Kamegawa, K.; Nishikubo, K.; Kodama, M.; Adachi, Y.; Yoshida, H. Oxidative degradation of carbon blacks with nitric acid II. Formation of water-soluble polynuclear aromatic compounds. *Carbon* **2002**, *40*, 1447–1455. [CrossRef]
65. Wagner, S.; Dittmar, T.; Jaffé, R. Molecular characterization of dissolved black nitrogen via electrospray ionization Fourier transform ion cyclotron resonance mass spectrometry. *Org. Geochem.* **2015**, *79*, 21–30. [CrossRef]
66. Goldberg, E. *Black Carbon in the Environment: Properties and Distribution*; John Wiley and Sons: New York, NY, USA, 1985.
67. Bateman, A.P.; Nizkorodov, S.A.; Laskin, J.; Laskin, A. High-resolution electrospray ionization mass spectrometry analysis of water-soluble organic aerosols collected with a particle into liquid sampler. *Anal. Chem.* **2010**, *82*, 8010–8016. [CrossRef] [PubMed]
68. Lee, H.J.; Aiona, P.K.; Laskin, A.; Laskin, J.; Nizkorodov, S.A. Effect of solar radiation on the optical properties and molecular composition of laboratory proxies of atmospheric brown carbon. *Environ. Sci. Technol.* **2014**, *48*, 10217–10226. [CrossRef] [PubMed]
69. Washenfelder, R.A.; Attwood, A.R.; Brock, C.A.; Guo, H.; Xu, L.; Weber, R.J.; Ng, N.L.; Allen, H.M.; Ayres, B.R.; Baumann, K. Biomass burning dominates brown carbon absorption in the rural southeastern United States. *Geophys. Res. Lett.* **2015**, *42*, 1–12. [CrossRef]
70. Knicker, H.; Almendros, G.; González-Vila, F.J.; Martín, F.; Lüdemann, H.-D. ^{13}C- and ^{15}N-NMR spectroscopic examination of the transformation of organic nitrogen in plant biomass during thermal treatment. *Soil Biol. Biochem.* **1996**, *28*, 1053–1060. [CrossRef]
71. Laskin, A.; Smith, J.; Laskin, J. Molecular characterization of nitrogen-containing organic compounds in biomass burning aerosols using high-resolution mass spectrometry. *Environ. Sci. Technol.* **2009**, *43*, 3764–3771. [CrossRef] [PubMed]
72. Hertkorn, N.; Benner, R.; Frommberger, M.; Schmitt-Kopplin, P.; Witt, M.; Kaiser, K.; Kettrup, A.; Hedges, J.I. Characterization of a major refractory component of marine dissolved organic matter. *Geochim. Cosmochim. Acta* **2006**, *70*, 2990–3010. [CrossRef]
73. Ohno, T.; He, Z.; Sleighter, R.L.; Honeycutt, C.W.; Hatcher, P.G. Ultrahigh resolution mass spectrometry and indicator species analysis to identify marker components of soil-and plant biomass-derived organic matter fractions. *Environ. Sci. Technol.* **2010**, *44*, 8594–8600. [CrossRef] [PubMed]
74. Wozniak, A.S.; Bauer, J.E.; Dickhut, R.M. Characteristics of water-soluble organic carbon associated with aerosol particles in the eastern United States. *Atmos. Environ.* **2012**, *46*, 181–188. [CrossRef]

75. Altieri, K.E.; Turpin, B.J.; Seitzinger, S.P. Oligomers, organosulfates, and nitrooxy organosulfates in rainwater identified by ultra-high resolution electrospray ionization FT-ICR mass spectrometry. *Atmos. Chem. Phys.* **2009**, *9*, 2533–2542. [CrossRef]
76. Surratt, J.D.; Gómez-González, Y.; Chan, A.W.H.; Vermeylen, R.; Shahgholi, M.; Kleindienst, T.E.; Edney, E.O.; Offenberg, J.H.; Lewandowski, M.; Jaoui, M. Organosulfate formation in biogenic secondary organic aerosol. *J. Phys. Chem. A* **2008**, *112*, 8345–8378. [CrossRef] [PubMed]
77. Nishioka, M. Aromatic sulfur compounds other than condensed thiophenes in fossil fuels: Enrichment and identification. *Energy Fuels* **1988**, *2*, 214–219. [CrossRef]
78. Panda, S.K.; Andersson, J.T.; Schrader, W. Mass-spectrometric analysis of complex volatile and nonvolatile crude oil components: A challenge. *Anal. Bioanal. Chem.* **2007**, *389*, 1329–1339. [CrossRef] [PubMed]
79. Anderson, L.; Faul, K.; Paytan, A. Phosphorus associations in aerosols: What can they tell us about P bioavailability? *Mar. Chem.* **2010**, *120*, 44–56. [CrossRef]
80. Chen, H.; Abdulla, H.A.; Sanders, R.L.; Myneni, S.C.; Mopper, K.; Hatcher, P.G. Production of black carbon-like and aliphatic molecules from terrestrial dissolved organic matter in the presence of sunlight and iron. *Environ. Sci. Technol. Lett.* **2014**, *1*, 399–404. [CrossRef]
81. Waggoner, D.C.; Chen, H.; Willoughby, A.S.; Hatcher, P.G. Formation of black carbon-like and alicyclic aliphatic compounds by hydroxyl radical initiated degradation of lignin. *Org. Geochem.* **2015**, *82*, 69–76. [CrossRef]
82. Decesari, S.; Finessi, E.; Rinaldi, M.; Paglione, M.; Fuzzi, S.; Stephanou, E.; Tziaras, T.; Spyros, A.; Ceburnis, D.; O'Dowd, C. Primary and secondary marine organic aerosols over the North Atlantic Ocean during the map experiment. *J. Geophys. Res. Atmos.* **2011**, *116*. [CrossRef]
83. Simoneit, B.R.T.; Schauer, J.J.; Nolte, C.; Oros, D.R.; Elias, V.O.; Fraser, M.; Rogge, W.; Cass, G.R. Levoglucosan, a tracer for cellulose in biomass burning and atmospheric particles. *Atmos. Environ.* **1999**, *33*, 173–182. [CrossRef]

13

Effect of Pollution Controls on Atmospheric $PM_{2.5}$ Composition during Universiade in Shenzhen, China

Nitika Dewan [1], Yu-Qin Wang [2], Yuan-Xun Zhang [2,3], Yang Zhang [2], Ling-Yan He [4], Xiao-Feng Huang [4] and Brian J. Majestic [1,*]

1. Department of Chemistry and Biochemistry, University of Denver, Denver, CO 80208, USA; dewannitika@yahoo.com
2. College of Resources and Environment, University of Chinese Academy of Sciences, Beijing 100049, China; wangyuqin11@mails.ucas.ac.cn (Q.W.); yxzhang@ucas.ac.cn (X.Z.); zhangyang@ucas.ac.cn (Y.Z.)
3. Huairou Eco-Environmental Observatory, Chinese Academy of Sciences, Beijing 101408, China
4. Key Laboratory for Urban Habitat Environmental Science and Technology, School of Environment and Energy, Peking University Shenzhen Graduate School, Shenzhen 518055, China; hely@pkusz.edu.cn (Y.H.); huangxf@pkusz.edu.cn (F.H.)

* Correspondence: brian.majestic@du.edu

Abstract: The 16th Universiade, an international multi-sport event, was hosted in Shenzhen, China from 12 to 23 August 2011. During this time, officials instituted the Pearl River Delta action plan in order to enhance the air quality of Shenzhen. To determine the effect of these controls, the current study examined the trace elements, water-soluble ions, and stable lead isotopic ratios in atmospheric particulate matter (PM) collected during the controlled (when the restrictions were in place) and uncontrolled periods. Fine particles ($PM_{2.5}$) were collected at two sampling sites in Shenzhen: "LG"—a residential building in the Longgang District, with significant point sources around it and "PU"—Peking University Shenzhen Graduate School in the Nanshan District, with no significant point sources. Results from this study showed a significant increase in the concentrations of elements during the uncontrolled periods. For instance, samples at the LG site showed (controlled to uncontrolled periods) concentrations (in $ng \cdot m^{-3}$) of: Fe (152 to 290), As (3.65 to 8.38), Pb (9.52 to 70.8), and Zn (98.6 to 286). Similarly, samples at the PU site showed elemental concentrations (in $ng \cdot m^{-3}$) of: Fe (114 to 301), As (0.634 to 8.36), Pb (4.86 to 58.1), and Zn (29.5 to 259). Soluble Fe ranged from 7%–15% for the total measured Fe, indicating an urban source of Fe. Ambient $PM_{2.5}$ collected at the PU site has an average $^{206}Pb/^{204}Pb$ ratio of 18.257 and 18.260 during controlled and uncontrolled periods, respectively. The LG site has an average $^{206}Pb/^{204}Pb$ ratio of 18.183 and 18.030 during controlled and uncontrolled periods, respectively. The $^{206}Pb/^{204}Pb$ ratios at the PU and the LG sites during the controlled and uncontrolled periods were similar, indicating a common Pb source. To characterize the sources of trace elements, principal component analysis was applied to the elements and ions. Although the relative importance of each component varied, the major sources for both sites were identified as residual oil combustion, secondary inorganic aerosols, sea spray, and combustion. The $PM_{2.5}$ levels were severely decreased during the controlled period, but it is unclear if this was a result of the controls or change in meteorology.

Keywords: water-soluble ions; iron (Fe) speciation; enrichment factor (EF); lead (Pb) isotopes; ICP-MS

---

1. Introduction

Due to an increase in urbanization and economic growth in China, air pollution has become a severe problem. $PM_{2.5}$ is a key pollutant strongly impacted by the rapid development in

China [1–3]. High $PM_{2.5}$ levels are associated with human mortality [4–6], climate change [7], visibility degradation [8,9], and agricultural yield reduction [10]. Increased morbidity and mortality rates and the adverse health effects of particle exposure are predominantly linked to chemical composition of PM [11,12]. From a toxicological viewpoint, the trace metals play an important role in increasing the redox activity of ambient PM [13–15]. Metals are hard to eliminate and they therefore accumulate in organisms and plants and can cause severe human health related problems and environmental pollution [16,17]. The inhalation of metals is associated with disruption of the nervous system and the functioning of internal organs [18–20].

Shenzhen (22°33′N, 114°06′E), home to a population of 10.62 million residents, is one of the most important industrial centers in China. It is a coastal city in the Guangdong Province located at the mouth of the Pearl River Delta (PRD), bordering Hong Kong. Previous studies measured $PM_{2.5}$ mass concentrations during winter and summer months in Shenzhen and Hong Kong. The $PM_{2.5}$ levels (in $\mu g \cdot m^{-3}$) were higher at Shenzhen, 47 ± 17 and 61 ± 18, relative to Hong Kong, 31 ± 17 and 55 ± 23, during summer and winter months, respectively [21,22]. Overall, Shenzhen displayed maximum $PM_{2.5}$ levels in winter months relative to summer months [23], both of which exceed the 24-h mean ambient air quality standards of the World Health Organization (WHO) of 25 $\mu g \cdot m^{-3}$ [24] and the annual ambient air quality standards of People's Republic of China (GB 3095-2012) of 35 $\mu g \cdot m^{-3}$ [25].

The 16th Universiade, an international multi-sport and cultural event organized for university athletes by the International University Sports Federation, was hosted in Shenzhen, China from 12 to 23 August 2011. During this time, officials instituted several restrictions: (a) the PRD action plan [26], which includes the prevention and control of industrial pollution, flow source pollution, and dust and point source pollution; (b) an ozone controlling plan which includes control of emissions of volatile organic compounds (VOCs) and promotion of an oil to gas project for thermal power plants; and (c) traffic-control actions such as restricting access within the region in order to enhance the air quality of Shenzhen. In this study, airborne $PM_{2.5}$ (aerodynamic diameter <2.5 μm) was collected at two sampling sites in Shenzhen during the controlled period (when the restrictions were implied) and during the uncontrolled period (when the restrictions were released). A previous study had evaluated the impact of emission controls and traffic intervention measures during the 29th Olympic and Paralympics games in Beijing [27], where significant reductions in vehicle emissions and ambient traffic-related air pollutants were observed.

In this study, we employed several chemical and statistical methods to determine the impact of the emission restrictions on $PM_{2.5}$ and trace elements. We report trace elements as well as water-soluble major ions. For the first time in the region, Pb isotopic ratios, as well as soluble iron oxidation state speciation, are reported. In addition to quantification, we employ principal component analysis (PCA) to determine the source of trace elements, allowing a unique interpretation of the quantitative data.

2. Materials and Methods

2.1. Sample Collection

Airborne $PM_{2.5}$ was collected at two sampling sites (LG and PU) in Shenzhen in 2011 both during the controlled period (12 August–23 August) and uncontrolled period (24 August–4 September) of Universiade. The map of the region showing the two sampling sites is shown in Figure 1 [28]. The "LG" site (22.70°N, 114.21°E), about 500 m away from the main venue, is located on top of a 31 floor Lotus residential building in the Longgang District, with significant point sources (e.g., plastic processing plants, glass factories, papermaking and painting industries) nearby. During the controlled periods, these point sources were supposed to be closed. However, we note that there was no accountability and no way of verification. The "PU" site (22.60°N, 113.97°E) is located at the top of Building E of the Peking University Shenzhen Graduate School in the Nanshan District, with no significant point sources around it and about 33 km away from the main venue. The LG site was located at 161 m and the PU site was located at 50 m above ground level, with no major geological features between the sites. The distance between the two sampling sites is about 45 km. A previous study during the same time period at these sites showed that both PM mass and PM composition (EC/OC) were significantly (and similarly) altered when comparing the controlled and uncontrolled periods [26].

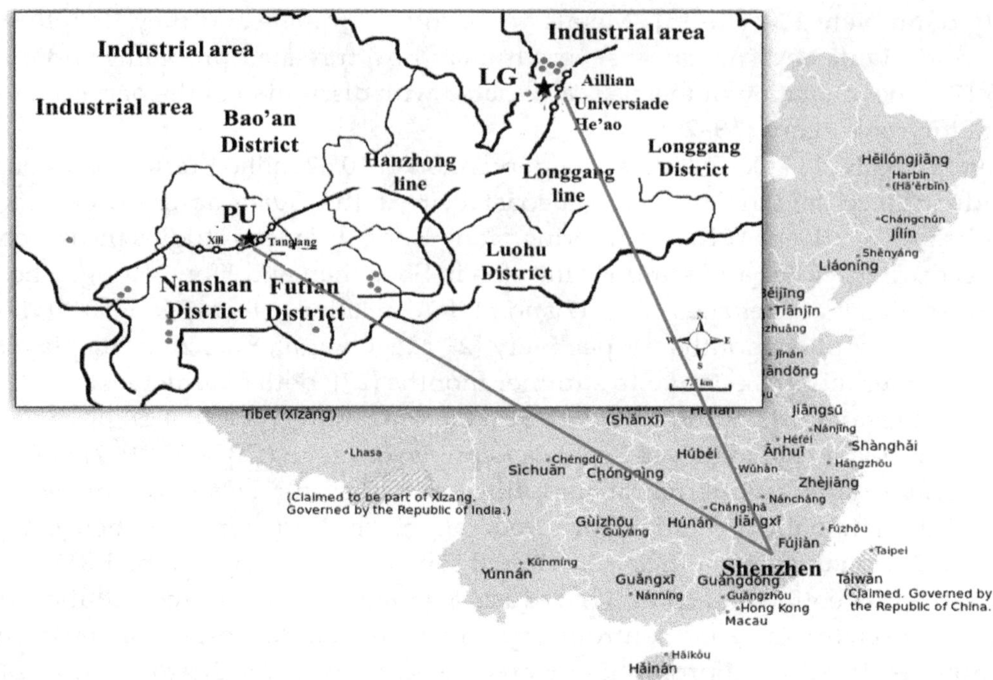

Figure 1. Map showing the geographical location of the PU and LG sampling sites (shown as stars) relative to the Universiade center in Shenzhen. The red dots in the map represent the stadiums where the events were held [28].

Co-located samples were collected on both 47 mm quartz and Teflon filters (Whatman, Pittsburgh, PA, USA) for 24-h from 12 August to 4 September with a flow rate of 21 L$\cdot$ min^{-1} using dual channel samplers (GUCAS 1.0) [29]. The quartz filters were used for EC/OC analysis following the protocol mentioned in EPA/NIOSH [30] and these results are reported elsewhere [26]. All sample preparation was performed under positive pressure HEPA filtered air. A microbalance (Mettler Toledo AX105DR, Columbus, OH, USA) was used for determination of mass (estimated total uncertainty of ± 6 µg). Prior to weighing, the filters were equilibrated in a constant humidity (40% $\pm$ 3%) and temperature (20 $\pm$ 1 °C) environment for 48 h.

The temperature, pressure, wind directions, and relative humidity were constant during the controlled periods. Based on the 72-h backward HYSPLIT (Hybrid Single Particle Lagrangian Integrated Trajectory) model [31], the air mass was transported from the South Sea at both sampling sites during the controlled periods. During the uncontrolled periods, the wind directions were more variable, but northern winds were more prominent and, based on trajectory analysis, air mass was transported from an industrial zone to both sites. Additional details of the meteorology during this time can be found in a previous manuscript [26]. There were two minor rain events during the controlled and three during the uncontrolled periods, all less than 12 mm. There were no reductions observed in the overall PM mass during those days.

2.2. Total Elemental Analysis

Prior to digestion, the polypropylene ring was removed from the Teflon filter using a ceramic blade. The Teflon filters were digested in sealed, pre-cleaned Teflon digestion bombs in a 30-position Microwave Rotor (Milestone Ethos, Shelton, CT, USA) with trace metal grade acid matrix (Fisher,

Waltham, MA, USA) consisting of 1.5 mL of nitric acid (16 M), 750 μL of hydrochloric acid (12 M), 200 μL of hydrofluoric acid (28 M), and 200 μL of 30% hydrogen peroxide. Digestates were diluted to 30 mL with high purity water (>18 MΩcm, MQ) and elemental concentrations (Al, As, Ba, Ca, Cd, Cr, Cu, Fe, K, Mg, Mn, Mo, Na, Ni, Pb, Rb, Sb, Se, Sr, Ti, V, and Zn) were quantified by quadrupole inductively coupled plasma-mass spectrometry (ICP-MS, Agilent 7700, Santa Clara, CA, USA) with indium (In) as an internal standard. The accuracy of the results from the elemental analysis was verified by National Institute of Standards and Technology (NIST) Standard Reference Materials (SRM). The SRMs, San Joaquin Soil (NIST 2709) and Urban Dust (NIST 1649a), were digested and analyzed with every batch of 25 samples. The percent recovery of the reported elements from these SRMs was 85%–120%. Data were also blank-corrected using the average of multiple field filter blanks. Blank concentrations (in $\mu g \cdot L^{-1}$) ranged from 0.0041 (2.4% of the total) to 14.65 (5.3% of the total) for the elements during the controlled and uncontrolled periods at the two sampling sites. The uncertainty associated with each element in every sample was calculated from an error propagation analysis, which included uncertainty in the field blanks and in the air flow rates.

2.3. *Soluble Ions Analysis*

For soluble ion analysis, the Teflon filters were leached in 10 mL high purity water for 2 h. Water-soluble ions were analyzed from the unacidified portion using ion exchange chromatography (Dionex-ICS5000, Bannockburn, IL, USA) followed by a self-regeneration suppressor (model CSR 300 for cations and ASR 300 for anions) and coupled with conductivity detector (Thermo). Cations (NH_4^+, K^+, Na^+, and Mg^{2+}) were separated by Dionex IonPac CS12A column and using 20 mM of methanesulfonic acid as a mobile phase at a flowrate of 0.5 $mL \cdot min^{-1}$. For anions (Cl^-, NO_3^-, and SO_4^{2-}), a Dionex IonPac AS22 column was used for separation along with a mixture of 4.5 mM sodium carbonate and 1.4 mM sodium bicarbonate as a mobile phase with a flowrate of 0.5 $mL \cdot min^{-1}$. A calibration curve of authentic standard (Dionex) for target ions was used to identify and quantify cations and anions in the samples. Details about calibration, method and instrument detection limits, and other measurement parameters have been previously reported [32].

2.4. *Iron Oxidation State Analysis*

Iron (Fe) speciation analysis was also performed with the water-soluble extracts. 1.8 mL of the soluble extract was mixed with 0.2 mL of 5.88 μM Ferrozine reagent ((3-(2-pyridyl)-5,6-diphenyl-1,2,4-triazine-4′,4″-disulfonic acid sodium salt), Sigma, St. Louis, MO, USA). The absorbance of the Fe(II)–Ferrozine complex was measured at 560 nm using a 1 m liquid waveguide capillary cell spectrophotometer [33,34]. The pH for all water extracts (each site, controlled *vs.* uncontrolled) ranged between 4.28 and 4.41. This suggests that, despite an unbuffered extract solution, pH effects were not important. The calibration curve generated using known Fe(II)-Ferrozine solutions provided Fe(II) concentration. Soluble Fe(III) was then determined by subtracting total soluble Fe concentration (from ICP-MS) from the soluble Fe(II) concentration.

2.5. *Stable Pb Isotope Analysis*

Stable Pb isotopic ratios were measured in the digested extracts with no further purification using high-resolution magnetic sector inductively coupled plasma mass spectrometer (MC)-ICP-MS (Thermo-Finnigan Neptune Plus). For the Pb isotope analysis, the digests were evaporated in Teflon vials and diluted to 2 mL using 2% optima grade HNO_3 acid (Fisher, Waltham, MA, USA). The Pb content of the digests ranged from 15 to 30 ng. The uncertainties for Pb isotope ratios depended on the isotope system and were in the range of 0.0025 and 0.0034 for $^{206}Pb/^{204}Pb$ and $^{207}Pb/^{206}Pb$, respectively. Analysis of the common Pb isotopic standard (NIST 981) yielded $^{206}Pb/^{204}Pb = 16.937 \pm 0.018$ ($n = 18$) and $^{207}Pb/^{206}Pb = 0.9145 \pm 0.0001$ ($n = 17$) *versus* the certified values of 16.944 ± 0.015 and 0.9146 ± 0.0003, respectively.

2.6. Principal Component Analysis (PCA)

To identify the potential origin of the elements in $PM_{2.5}$, PCA was conducted using SPSS statistical software (SPSS, version 22). A varimax rotation was employed for interpretation of the principal components and factors with eigenvalues greater than unity were retained in the analysis [35]. Given that the bulk properties, Pb isotope ratios, and elements trends were similar at each site, it is clear that similar sources affect each site. Therefore, the sites were grouped together for the PCA analysis. Prior to conducting factor analysis, we performed a Pearson correlation matrix of 52 samples collected at two sampling sites during controlled and uncontrolled periods. Based on correlations matrix, we had a total of 26 samples with OC [26], 8 elements, and 6 ions as variables, resulting in a sample/variable ratio consistent with recommended criteria for a robust PCA analysis with KMO test of sampling adequacy >0.5 [36]. In addition, the significance value (0.000) for Bartlett's test of sphericity indicates that the correlations are appropriate for this data set.

3. Results and Discussion

3.1. Trace Element Concentrations

One previous study [26] has shown that during the controlled periods, the monitored $PM_{2.5}$ mass concentrations were 12.9 ± 3.7 $\mu g \cdot m^{-3}$ at the PU site and 25.2 ± 5.2 $\mu g \cdot m^{-3}$ at the LG site. During uncontrolled periods, significant increases in the $PM_{2.5}$ mass concentrations were observed (PU = 48.0 ± 8.7 $\mu g \cdot m^{-3}$ and LG = 54.0 ± 6.5 $\mu g \cdot m^{-3}$). The fact that the wind directions were drastically different during controlled and uncontrolled periods complicate the effects of the controls and this may be another reason why such drastic differences in PM were observed between the two periods. Consistent with the increased PM mass, results from our study showed an increase in the concentrations of most abundant and trace elements in $PM_{2.5}$ collected at the PU and LG sites during uncontrolled periods (Figure 2a–d). For instance, average Ca and Fe concentrations were ~360 (range, 264–467) and ~290 (range, 216–412) $ng \cdot m^{-3}$, respectively, during uncontrolled periods and ~270 (range, 169–340) and ~152 (range, 90–352) $ng \cdot m^{-3}$ during controlled periods at the LG site. Similarly, at the PU site, average Ca and Fe concentrations were ~250 (range, 114–469) and ~300 (range, 201–588) $ng \cdot m^{-3}$, respectively, during uncontrolled periods and ~100 (range, 29–160) and ~115 $ng \cdot m^{-3}$ (range, 55–169) during controlled periods. The elemental analysis showed that both Al and K were the dominant elements at both sampling sites. Also, all trace elements, except Ni and V, had higher concentrations during uncontrolled periods at both sites. Both Ni and V are markers for the residual oil combustion, which suggests that oil combustion sources were essentially unaffected by the controls. The significant increase in V during the controlled periods could be attributed to emissions from ships around the Shenzhen city, however this is speculation as reliable data regarding the ship traffic are not available. There were no large scale oil power generation plants [37]. Therefore, control of oil combustion sources was likely very challenging, since all of these sources were probably small. Some striking differences include an approximate eleven-fold increase for Pb and an eight-fold difference for Zn for $PM_{2.5}$ during uncontrolled periods at the PU site. Similarly, an approximate eight-fold difference was observed in the concentrations of Pb and a three-fold difference for Zn during uncontrolled periods at the LG site. Overall, the concentrations of most of the elements were higher at the LG site relative to the PU site. This may be because the LG site is close to significant point sources.

Water-soluble ions were quantified in the $PM_{2.5}$ collected at the two sampling sites during controlled and uncontrolled periods and are presented in Figure 3. Elevated levels of SO_4^{2-} were observed in $PM_{2.5}$ during uncontrolled periods at both sites. Sources like coal power plants and industries may be playing an important role in the emissions of sulfur dioxide, which leads to sulfate. NO_3^-/SO_4^{2-} ratio can be used as an indicator of the type of anthropogenic activity [38]. If the ratio is >1, it indicates greater NO_x emissions, indicating vehicular emissions are likely dominant. If the ratio is <1, it indicates greater SO_2 emissions, and that stationary sources are dominant [38,39]. The average ratio at the PU site was 0.071 ± 0.003 and 0.12 ± 0.02 during controlled and uncontrolled periods respectively, whereas the average ratio at the LG site was 0.15 ± 0.01 and 0.13 ± 0.01 during controlled and uncontrolled periods, respectively. At the two sampling sites, both during controlled and uncontrolled periods, the ratio is lower than 1, indicating that stationary sources like industry emissions and power plants are important contributors to $PM_{2.5}$ emissions in Shenzhen.

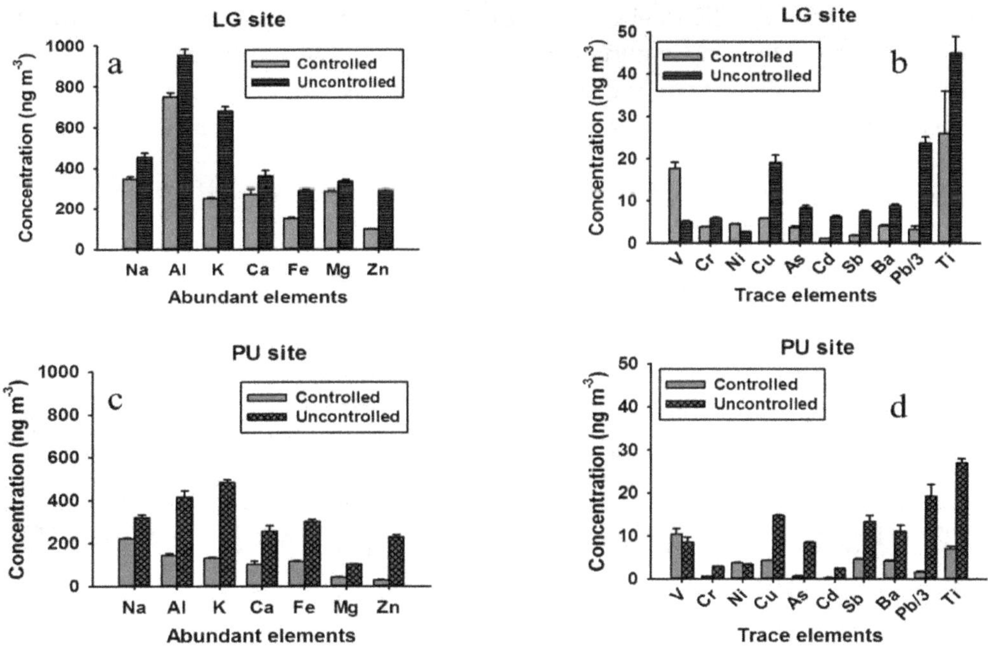

Figure 2. Average (ng· m^{-3}) abundant and trace elements observed in $PM_{2.5}$ at the LG site (**a**,**b**) and PU site (**c**,**d**) during controlled and uncontrolled periods.

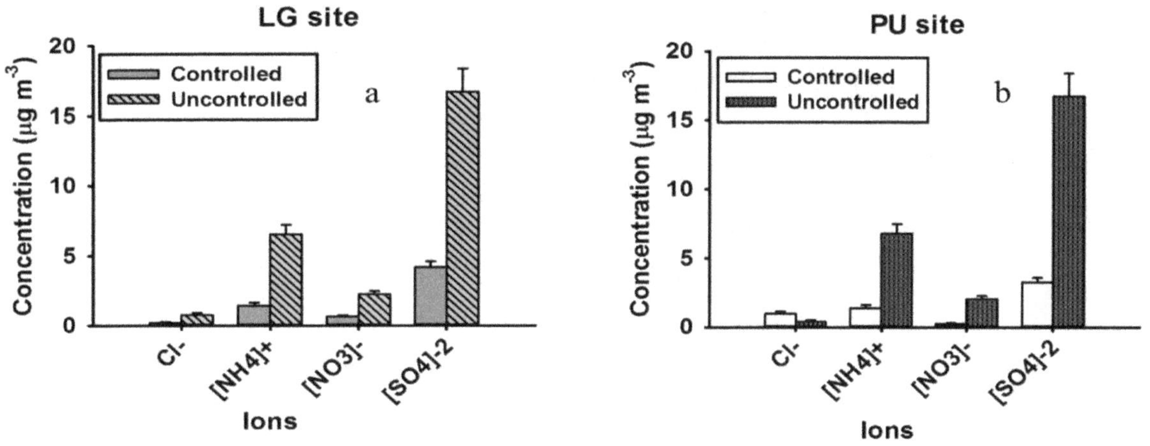

Figure 3. Average (µg· m^{-3}) water-soluble ions observed in $PM_{2.5}$ at the LG site (**a**) and PU site (**b**) during controlled and uncontrolled periods.

3.2. Crustal Enrichment Factors

Enrichment factors (EF) are used to assess whether the elements have a major crustal component [40]. EF of the elements was calculated by first normalizing the measured elemental concentrations in the sample with aluminum (Al), and then dividing by the Upper Continental Crust (UCC) ratio [41,42]. EF is calculated using the following formula:

$$EF_{element} = \frac{(\text{Concentration of element in sample/ Concentration of Al in sample})}{(\text{Concentration of element in crust / Concentration of Al in crust})} \quad (1)$$

EF is close to unity for the elements related to the reference, Al (marker for crustal emissions). A high EF (>>10) suggests that particular elements are enriched relative to the crust and thus are anthropogenically derived [43]. The dashed line (EF = 10) on the plots shown in Figure 4a,b represents the level above which the element is considered to have a major anthropogenic source. The error bar represents the standard deviation of the 13 samples each during the controlled and uncontrolled periods at the two sites. The dots represent the average EF of the 13 samples for each element.

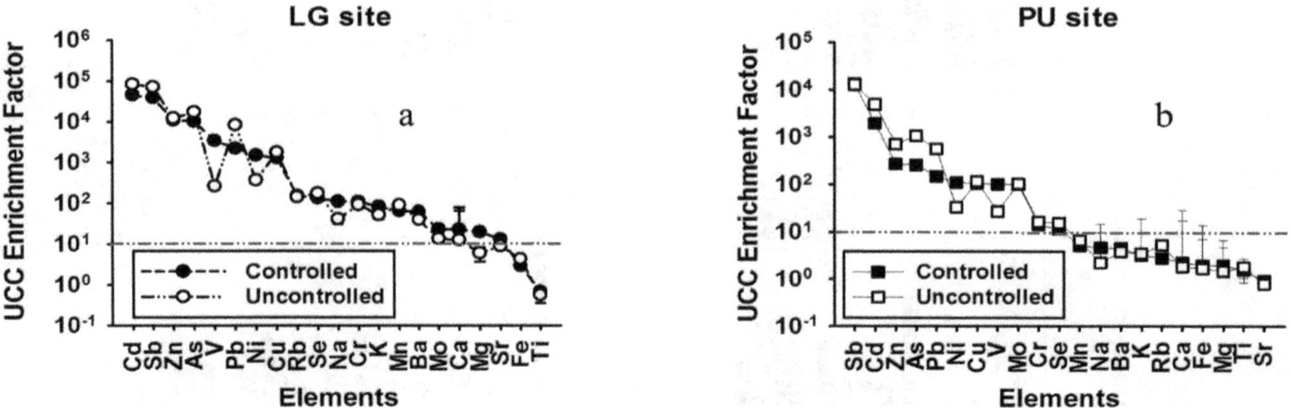

Figure 4. Enrichment factor (EF) for $PM_{2.5}$ collected at: (**a**) LG site during controlled and uncontrolled periods; and (**b**) PU site during controlled and uncontrolled periods. The error bar represents the standard deviation of the samples.

As observed in Figure 4, almost every measured element appears to have some anthropogenic source at the LG site. Conversely, at the PU site, mainly the industrially-sourced elements have high EF. Specifically, As, Cd, Cr, Cu, Ni, Pb, Sb, Se, Pb, V, and Zn, as well as K and Na were highly enriched at the LG site during both the controlled and uncontrolled periods (Figure 4a). Similarly, the elements As, Cd, Cr, Cu, Mo, Ni, Pb, Sb, Se, Pb, V, and Zn were associated with an anthropogenic source at the PU site during both the controlled and uncontrolled periods (Figure 4b). The fact that these elements are highly enriched is consistent with many other studies in urban areas [44,45]. It is also important to note that, while the concentrations were increased during uncontrolled periods, the EF essentially did not change. Significant differences in EF, however, were observed between the two sites.

3.3. Soluble Fe Oxidation State Analysis

As most atmospheric Fe is crustally-derived, Fe(III) dominates the major part of total iron in the PM but its relative importance also depends on local sources and the size fraction [46]. For example, crustal Fe is primarily in the Fe(III) oxidation state and shows a solubility of <1% [47], while ambient urban Fe shows solubility ~10%–20%, with a mixture of Fe(II) and Fe(III) [46], and Fe emitted directly from vehicles has been shown to be up to 70% water-soluble, being mostly Fe(II) sulfate [48]. Consequently, different locations exhibit different Fe solubility, depending on the dominant sources [46]. Our study shows Fe solubility of 7% and 15% during the controlled and uncontrolled periods, consistent with other urban sources. The oxidation state of the soluble Fe is shown in Figure 5. The majority of the soluble fraction was comprised of Fe(II) at the two sites both during controlled and uncontrolled periods. As Fe(III) is the dominant form of soluble Fe under oxidizing conditions [49], this implies that the PM contained other compounds which allowed the Fe to be stabilized in the reduced state. Possibilities include small chain organic acids, or even polycyclic aromatic hydrocarbons [50–52].

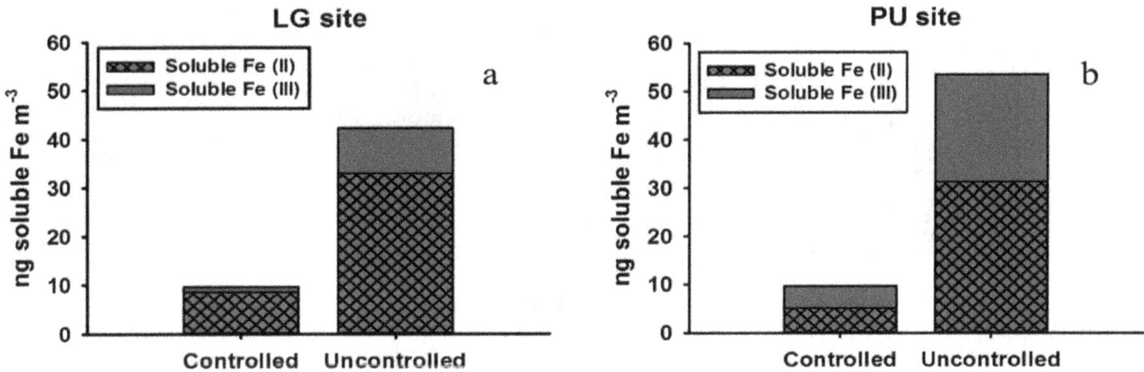

Figure 5. Soluble fractions of Fe(II) and Fe(III) at: (**a**) LG site; and (**b**) PU site.

At the LG site, the percent soluble Fe(II) of total iron collected on the filter was 5.9% during controlled periods and 12.2% during uncontrolled periods. Similarly, at the PU site, the percent soluble Fe(II) of total iron collected on the filter was 5.4% during controlled periods and 12.6% during uncontrolled periods. Of the soluble Fe, Fe(II) was 82% during controlled periods and 86% during uncontrolled periods at the LG site. Similarly, at the PU site, Fe(II) was 56% of the total soluble, during controlled periods and 61% during uncontrolled periods. Significant differences in relative Fe(II)/Fe(III) solubility between the PU and LG site indicates different sources between sites. However, similarities in Fe(II)/Fe(III) solubility at each site suggest that the sources are similar during the controlled and uncontrolled periods.

3.4. Stable Pb Isotope Ratios

Anthropogenic activities like mining, industry, and utilization of fossil fuels and tetraethyl lead in gasoline significantly affect the Pb levels in the environment to varying degrees. In this study, the greatest differences in elemental concentrations were Pb (Figure 2). Thus, we focus on Pb isotope ratios to help determine if its origin was different during controlled and uncontrolled periods. Pb has four stable isotopes; ^{204}Pb, ^{206}Pb, ^{207}Pb, and ^{208}Pb with the radioactive decay of ^{238}U, ^{235}U, and ^{232}Th eventually producing ^{206}Pb, ^{207}Pb, and ^{208}Pb, respectively. Depending on the geological history, different sources of Pb possess specific Pb isotopic signatures and these ratios do not fractionate during any chemical, physical, or biological process [53]. Therefore, Pb isotopic ratios are useful in distinguishing natural Pb from anthropogenic Pb and its origin in different ecosystems [53,54]. Stable Pb isotope ratios (^{207}Pb/^{206}Pb) in the range 0.7952–0.8405 can be used as a tracer species to identify natural sources of PM whereas, ^{207}Pb/^{206}Pb ratio in the range of 0.8504–0.9651 can be used to identify anthropogenic sources contributing to airborne PM [55,56].

Ratios of ^{207}Pb/^{206}Pb *versus* ^{206}Pb/^{204}Pb are presented in Figure 6 for the airborne $PM_{2.5}$ collected at the PU and LG sites during controlled and uncontrolled periods, with these ratios depending on local geology, rainfall, wind direction, and traffic [57,58]. The average ^{207}Pb/^{206}Pb ratios of the $PM_{2.5}$ collected at the LG site during the controlled and uncontrolled periods are 0.8599 (range, 0.8567–0.8590) and 0.8550 (range, 0.8536–0.8563), respectively. Similarly, at the PU site, the average ^{207}Pb/^{206}Pb ratios during the controlled and uncontrolled periods are 0.8564 (range, 0.8525–0.8591) and 0.8567 (range, 0.8539–0.8576), respectively. These ratios suggest that $PM_{2.5}$ has an anthropogenic Pb source [55], which is also in agreement with the high UCC EF. The ^{206Pb}/^{204}Pb ratios at the PU and the LG sites during the controlled and uncontrolled periods were similar (t-test, $p < 0.05$), indicating similar Pb sources. The ^{206}Pb/^{204}Pb ratios between the two sites during the controlled and uncontrolled periods were also similar (t-test, $p < 0.05$).

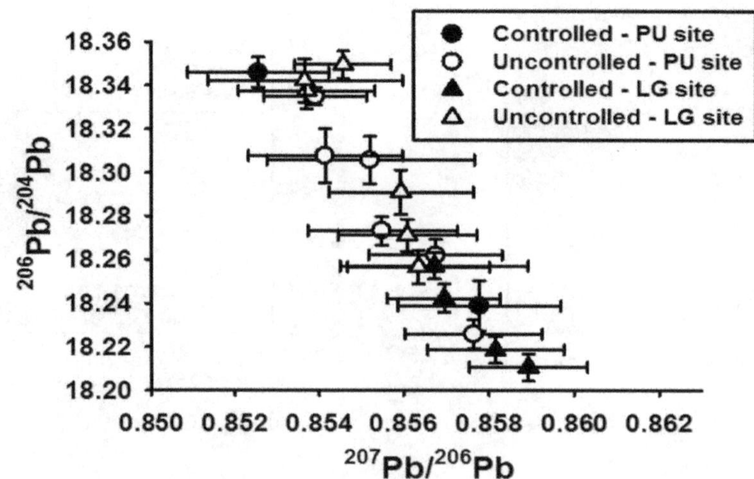

Figure 6. Isotopic composition of the $PM_{2.5}$ collected at LG and PU sites during controlled and uncontrolled periods. The error bars represent the standard deviation of the Pb isotopic ratios for all samples.

Industry emissions, coal combustion, and vehicle exhaust are considered to be the three main sources of Pb pollution in China [59,60]. Previous studies have determined the $^{207}Pb/^{206}Pb$ ratios ranging from 0.850 to 0.872 for wide varieties of Chinese coal [61]. Leaded gasoline, which was phased out in the early 1990s [62], showed an average $^{207}Pb/^{206}Pb$ ratio of 0.901, whereas unleaded exhaust showed the ratio of 0.872 in the PRD region [63]. The average ratio of $^{206}Pb/^{204}Pb$ observed for coal combustion and cement factories in Beijing ware 18.09 (range, 17.873–18.326) and 18.05 (range, 17.729–18.365), respectively [62]. There are about 26 coal power plants located in Guangdong Province, China [37]. Shenzhen is a major city in the Guangdong Province. These power plants are within 97 km of the sampling sites. A map of these power plants relative to the sampling sites is shown in Figure S1. Overall, the $PM_{2.5}$ Pb ratios are similar to the Pb ratios observed for coal varieties, which implies that coal combustion may be the primary Pb source(s) between the sampling sites in this study and the previous study [61]. In addition, the Pb isotope ratios were significantly similar at each site during both controlled and uncontrolled periods. This suggests that local and regional sources affect these two sites in a similar manner.

3.5. Source Identification Using Principal Component Analysis

PCA was used to identify major sources of $PM_{2.5}$. The results of the PCA for the combination of the two sites during controlled periods and uncontrolled periods are presented in the supplemental information (Tables S1 and S2). There were three factors contributing to the $PM_{2.5}$ in Shenzhen during controlled periods whereas five factors were present during uncontrolled periods at the two sites. The difference in reported factors between two periods is determined on the basis of inflection point in the scree plots and, we have selected the interpretable factors.

At the two sampling sites, during controlled periods, the first principal component has elevated loadings of Pb, Zn, K^+, Al, Na^+, Se, and OC, as shown in Table S1. The major sources of $PM_{2.5}$ are categorized in several groups and these are: coal combustion (Se and Pb), biomass burning (K^+ and OC), and vehicular abrasion (Zn and Sb) [3,23,64]. Therefore, we associate this factor with a mixture of combustion sources. We also note a potentially confounding correlation between Al and Se (r^2 = 0.858), which is present at both sites and during both the controlled and uncontrolled periods. For the second factor, characteristic values for V and Ni are the highest, which are tracers of heavy oil combustion [65]. Apart from oil-fired power plants and industries, ship emissions may be a prominent source of such combustion in Shenzhen [66]. Also, the second factor has prominent values for NH_4^+ and SO_4^{2-}, indicating the presence of secondary inorganic aerosols [67]. Sulfur dioxide would be emitted along with Ni and V during oil combustion. The third factor has high loadings of Mg^{2+}, Na^+, NH_4^+ SO_4^{2-}, and Cl^-, signifying this source was chiefly associated with partially aged sea salt [23], which is consistent with the fact that Shenzhen is a coastal city.

Table S2 represents the sources during the uncontrolled periods at the PU and LG sites. The first principal component shows high loadings of NH_4^+, SO_4^{2-}, and Sb, which are associated with secondary inorganic aerosols and potentially brake wear [68]. The second factor has elevated Al, Se, and K^+. Al is marker of crustal emissions and Se and K^+ are associated with coal combustion and biomass burning, respectively. Therefore, second factor is undetermined. High values of Na^+, Mg^{2+}, and Cl^- are associated with sea salt spray for the third factor indicating that sea spray is a contributing factor to $PM_{2.5}$ in Shenzhen regardless of wind direction. The fourth factor shows high values of V and Ni, which are tracers of heavy oil combustion emissions. The fifth factor has high values of Pb and OC, which is associated with combustion emission sources containing Pb. This is consistent with the Pb isotope data that the source of Pb is similar at the two sites during controlled and uncontrolled periods.

4. Conclusions

The results of this study provide insights into the effects of pollution restrictions in Shenzhen, China. The average $PM_{2.5}$ concentrations at the PU and LG sites during controlled periods were lower than 24-h mean ambient air quality standard of People's Republic of China of 75 $\mu g \cdot m^{-3}$ [25]. Surprisingly, both Ni and V, markers for the residual oil combustion, had lower concentrations relative to other trace elements during uncontrolled periods at the two sampling sites, suggesting that oil combustion emissions were not controlled at all by the restrictions imposed during the Universiade event.

While it is possible that the soluble Fe(II)/Fe(III) ratio depends, to some extent, on the original Fe phases present in the PM, the equilibrium speciation in solution is primarily dependent on the immediate redox environment in the extract solution [49]. Thus, the presence of source-specific organic compounds [50] and ions [69] are the major determinants of Fe speciation and solubility. While this is the first study to measure Fe speciation and solubility in airborne $PM_{2.5}$ in Shenzhen, China, it possesses similarities to previous Fe speciation studies. For instance, at the PU site, soluble Fe(II) and soluble Fe(III) were approximately equal, which was similar to the percentage of soluble Fe(II) at Waukesha, WI, USA [46]. Similarly, at the LG site, soluble Fe(II) was far greater than soluble Fe(III), which was similar to the results reported for Los Angeles, CA, USA [46] and Denver, CO, USA [70].

The $^{206}Pb/^{204}Pb$ ratios measured at the two sites during the controlled and uncontrolled periods, and between the two sites during the controlled and uncontrolled periods, were similar (t-test, $p < 0.05$), representing a common anthropogenic Pb source. This suggests that airborne $PM_{2.5}$ is dominated by local or regional combustion sources (as evidenced by the high EF), which was in agreement with the principal component analysis.

Previous studies have presented elemental and water-soluble ion concentrations for airborne $PM_{2.5}$ in southwest China [1,3]. The source apportionment based on positive matrix factorization (PMF) and chemical mass balances (CMB) revealed that coal combustion, secondary inorganic aerosols, biomass burning, metal industries, crustal dust, and sea spray were common sources in southwest China. The impact of control measures implemented before and during 2008 Olympics in Beijing showed 33% reduction in BC emissions [27] and controls implemented during the Universiade showed 30% reduction in traffic [71,72] and 50% reduction in $PM_{2.5}$ [26]. Although the relative importance of each component varied, the major sources at the two sites during controlled and uncontrolled periods were identified as residual oil combustion, secondary inorganic aerosols, combustion, and sea spray which is also in agreement with the previous studies.

In our study, however, every metric was consistent (e.g., Pb isotopes, PM mass trends, EC/OC trends, and individual element trends) between sites. The $PM_{2.5}$ levels in Shenzhen were mainly dominated by anthropogenic emissions. Reductions in emissions from point sources were observed, but it is unclear if this was due to the restrictions or from changes in meteorological conditions.

Supplementary Materials: The following are available online at http://www.mdpi.com/2073-4433/7/4/57/s1, Figure S1: Map showing the location of the PU and LG sampling sites (shown as red stars) relative to the location of power plants in Guangdong Province; Table S1: Principal component loadings of selected elements and ions for PU and LG sites during controlled periods; Table S2: Principal component loadings of selected elements and ions for PU and LG sites during uncontrolled periods.

Acknowledgments: The authors thank Elizabeth Stone and Ibrahim Al-Naghemah for their help and guidance in water soluble ion measurements. We also thank Kate Smith at the Wisconsin State Lab of Hygiene for her assistance with the Pb isotopic ratios measurement. This study was supported by the National Natural Science Foundation of China (Grant No. 41375131 and 21307129) and the Key Research Program of Chinese Academy of Sciences (Grant No. KJZD-EW-TZ-G06-01-0). We thank the anonymous reviewers for their helpful input to improve the paper.

Author Contributions: For research articles with several authors, a short paragraph specifying their individual contributions must be provided. All authors conceived and designed the experiments; all authors performed the experiments; Nitika Dewan and Brian J. Majestic analyzed the data; all authors contributed reagents/materials/analysis tools; Nitika Dewan and Brian J. Majestic wrote the paper." Authorship must be limited to those who have contributed substantially to the work reported.

References

1. Hagler, G.S.; Bergin, M.H.; Salmon, L.G.; Yu, J.Z.; Wan, E.C.H.; Zheng, M.; Zeng, L.M.; Kiang, C.S.; Zhang, Y.H.; Lau, A.K.H.; *et al.* Source areas and chemical composition of fine particulate matter in the Pearl River Delta region of China. *Atmos. Environ.* **2006**, *40*, 3802–3815. [CrossRef]
2. Yang, F.; Tan, J.; Zhao, Q.; Du, Z.; He, K.; Ma, Y.; Duan, F.; Chen, G.; Zhao, Q. Characteristics of $PM_{2.5}$ speciation in representative megacities and across China. *Atmos. Chem. Phys.* **2011**, *11*, 5207–5219. [CrossRef]
3. Tao, J.; Gao, J.; Zhang, L.; Zhang, R.; Che, H.; Zhang, Z.; Lin, Z.; Jing, J.; Cao, J.; Hsu, S.C. $PM_{2.5}$ pollution in a megacity of southwest China: Source apportionment and implication. *Atmos. Chem. Phys.* **2014**, *14*, 8679–8699. [CrossRef]
4. Dockery, D.W.; Pope, C.A.; Xu, X.; Spengler, J.D.; Ware, J.H.; Fay, M.E.; Ferris, B.G., Jr.; Speizer, F.E. An association between air pollution and mortality in six U.S. cities. *N. Engl. J. Med.* **1993**, *329*, 1753–1759. [CrossRef] [PubMed]
5. Englert, N. Fine particles and human health—A review of epidemiological studies. *Toxicol Lett.* **2004**, *149*, 235–242. [CrossRef] [PubMed]
6. Davidson, C.I.; Phalen, R.F.; Solomon, P.A. Airborne particulate matter and human health: A review. *Aerosol. Sci. Tech.* **2005**, *39*, 737–749. [CrossRef]
7. Charlson, R.J.; Schwartz, S.E.; Hales, J.M.; Cess, R.D.; Coakley, J.A., Jr.; Hansen, J.E.; Hofmann, D.J. Climate forcing by anthropogenic aerosols. *Science* **1992**, *255*, 423–430. [CrossRef] [PubMed]
8. Lee, Y.L.; Sequeira, R. Water-soluble aerosol and visibility degradation in Hong Kong during autumn and early winter, 1998. *Environ. Pollut.* **2002**, *116*, 225–233. [CrossRef]
9. Deng, X.J.; Tie, X.X.; Wu, D.; Zhou, X.J.; Bi, X.Y.; Tan, H.B.; Li, F.; Hang, C.L. Long-term trend of visibility and its characterizations in the Pearl River Delta (PRD) region, China. *Atmos. Environ.* **2008**, *42*, 1424–1435. [CrossRef]
10. Chameides, W.L.; Yu, H.; Liu, S.C.; Bergin, M.; Zhou, X.; Mearns, L.; Wang, G.; Kiang, C.S.; Saylor, R.D.; Luo, C.; *et al.* Case study of the effects of atmospheric aerosols and regional haze on agriculture: an opportunity to enhance crop yields in China through emission controls? *Proc. Natl. Acad. Sci.* **1999**, *96*, 13626–13633. [CrossRef] [PubMed]
11. Tsai, F.C.; Apte, M.G.; Daisey, J.M. An exploratory analysis of the relationship between mortality and the chemical composition of airborne particulate matter. *Inhal. Toxicol.* **2000**, *12*, 121–135. [CrossRef] [PubMed]
12. Verma, V.; Polidori, A.; Schauer, J.J.; Shafer, M.M.; Cassee, F.R.; Sioutas, C. Physicochemical and toxicological profiles of particulate matter in Los Angeles during the October 2007 Southern California wildfires. *Environ. Sci. Technol.* **2009**, *43*, 954–960. [CrossRef] [PubMed]
13. Valavanidis, A.; Fiotakis, K.; Bakeas, E.; Vlahogianni, T. Electron paramagnetic resonance study of the generation of reactive oxygen species catalysed by transition metals and quinoid redox cycling by inhalable ambient particulate matter. *Redox Rep.* **2005**, *10*, 37–51. [CrossRef] [PubMed]

14. Shi, T.M.; Schins, R.P.F.; Knaapen, A.M.; Kuhlbusch, T.; Pitz, M.; Heinrich, J.; Borm, P.J.A. Hydroxyl radical generation by electron paramagnetic resonance as a new method to monitor ambient particulate matter composition. *J. Environ. Monitor.* **2003**, *5*, 550–556. [CrossRef]
15. Prahalad, A.K.; Soukup, J.M.; Inmon, J.; Willis, R.; Ghio, A.J.; Becker, S.; Gallagher, J.E. Ambient air particles: Effects on cellular oxidant radical generation in relation to particulate elemental chemistry. *Toxicol. Appl. Pharm.* **1999**, *158*, 81–91. [CrossRef] [PubMed]
16. Alloway, B.J. *Heavy Metals in Soils*; Blackie Academic & Professional: Glasgow, UK, 1990.
17. Lee, C.S.; Li, X.D.; Shi, W.Z.; Cheung, S.C.; Thornton, I. Metal contamination in urban, suburban, and country park soils of Hong Kong: A study based on GIS and multivariate statistics. *Sci. Total Environ.* **2006**, *356*, 45–61. [CrossRef] [PubMed]
18. Nriagu, J.O. A silent epidemic of environmental metal poisoning. *Environ. Pollut.* **1988**, *50*, 139–161. [CrossRef]
19. Thompson, C.M.; Markesbery, W.R.; Ehmann, W.D.; Mao, Y.X.; Vance, D.E. Regional Brain Trace-Element Studies in Alzheimers-Disease. *Neurotoxicology* **1988**, *9*, 1–8. [PubMed]
20. Bocca, B.; Alimonti, A.; Petrucci, F.; Violante, N.; Sancesario, G.; Forte, G.; Senofonte, O. Quantification of trace elements by sector field inductively coupled plasma mass spectrometry in urine, serum, blood and cerebrospinal fluid of patients with Parkinson's disease. *Spectrochim Acta Part B At. Spectrosc.* **2004**, *59*, 559–566. [CrossRef]
21. Cao, J.J.; Lee, S.C.; Ho, K.F.; Zhang, X.Y.; Zou, S.C.; Fung, K.; Chow, J.C.; Watson, J.G. Characteristics of carbonaceous aerosol in Pearl River Delta Region, China during 2001 winter period. *Atmos. Environ.* **2003**, *37*, 1451–1460. [CrossRef]
22. Cao, J.J.; Lee, S.C.; Ho, K.F.; Zou, S.C.; Fung, K.; Li, Y.; Watson, J.G.; Chow, J.C. Spatial and seasonal variations of atmospheric organic carbon and elemental carbon in Pearl River Delta Region, China. *Atmos. Environ.* **2004**, *38*, 4447–4456. [CrossRef]
23. Dai, W.; Gao, J.; Cao, G.; Ouyang, F. Chemical composition and source identification of $PM_{2.5}$ in the suburb of Shenzhen, China. *Atmos. Res.* **2013**, *122*, 391–400. [CrossRef]
24. Air quality Guidelines for Particulate Matter, Ozone, Nitrogen Dioxide and Sulfur Dioxide. Available online: http://apps.who.int/iris/bitstream/10665/69477/1/WHO_SDE_PHE_OEH_06.02_eng.pdf (accessed on 10 February 2016).
25. HORIBA Technical Reports—The Trends in Environmental Regulations in China. Available online: http://www.horiba.com/uploads/media/R41E_05_010_01.pdf (accessed on 10 February 2016).
26. Wang, Y.Q.; Zhang, Y.X.; Zhang, Y.; Li, Z.Q.; He, L.Y.; Huang, X.F. Characterization of carbonaceous aerosols during and post-Shenzhen UNIVERSIADE period. *China Environ. Sci.* **2014**, *34*, 1622–1632. (In Chinese)
27. Wang, X.; Westerdahl, D.; Chen, L.C.; Wu, Y.; Hao, J.M.; Pan, X.C.; Guo, X.B.; Zhang, K.M. Evaluating the air quality impacts of the 2008 Beijing Olympic Games: On-road emission factors and black carbon profiles. *Atmos. Environ.* **2009**, *43*, 4535–4543. [CrossRef]
28. China. Available online: https://en.wikipedia.org/wiki/China (accessed on 13 April 2016).
29. Zhang, Y.; Zhang, Y.; Liu, H.; Wang, Y.; Deng, J. Design and application of a novel atmospheric particle sampler. *Environ. Monit. China* **2014**, *30*, 176–180.
30. NIOSH Manual of Analytical Methods. Available online: http://www.cdc.gov/niosh/docs/2003-154/pdfs/5040.pdf (accessed on 2 February 2016).
31. Rolph, G.D. *Real-Time Environmental Applications and Display System (READY)*; NOAA Air Resources Laboratory: Silver Spring, MD, USA, 2003.
32. Jayarathne, T.; Stockwell, C.E.; Yokelson, R.J.; Nakao, S.; Stone, E.A. Emissions of fine particle fluoride from biomass burning. *Environ. Sci. Technol.* **2014**, *48*, 12636–12644. [CrossRef] [PubMed]
33. Stookey, L.L. Ferrozine—A new spectrophotometric reagent for iron. *Anal. Chem* **1970**, *42*, 779–781. [CrossRef]
34. Majestic, B.J.; Schauer, J.J.; Shafer, M.M.; Turner, J.R.; Fine, P.M.; Singh, M.; Sioutas, C. Development of a wet-chemical method for the speciation of iron in atmospheric aerosols. *Environ. Sci. Technol.* **2006**, *40*, 2346–2351. [CrossRef] [PubMed]
35. Schaug, J.; Rambaek, J.P.; Steinnes, E.; Henry, R.C. Multivariate-analysis of trace-element data from moss samples used to monitor atmospheric deposition. *Atmos. Environ. Part A Gen. Top.* **1990**, *24*, 2625–2631. [CrossRef]

36. Elliott, A.C.; Woodward, W.A. *IBM SPSS by Example: A Practical Guide to Statistical Data Analysis*; SAGE Publications: Thousand Oaks, CA, USA, 2015.
37. List of major power stations in Guangdong. Available online: https://en.wikipedia.org/wiki/List_of_major_power_stations_in_Guangdong (accessed on 1 March 2016).
38. Arimoto, R.; Duce, R.A.; Savoie, D.L.; Prospero, J.M.; Talbot, R.; Cullen, J.D.; Tomza, U.; Lewis, N.F.; Ray, B.J. Relationships among aerosols constituents from Asia and the North Pacific during PEM-West A. *J. Geophys. Res.* **1996**, *101*, 2011–2023. [CrossRef]
39. Yao, X.; Chan, C.K.; Fang, M.; Cadle, S.; Chan, T.; Mulawa, P.; He, K.; Ye, B. The water-soluble ionic composition of $PM_{2.5}$ in Shanghai and Beijing, China. *Atmos. Environ.* **2002**, *36*, 4223–4234. [CrossRef]
40. Reimann, C.; Caritat, P.D. Intrinsic flaws of element Enrichment Factors (EFs) in environmental geochemistry. *Environ. Sci. Technol.* **2000**, *34*, 5084–5091. [CrossRef]
41. Buat-Menard, P.; Chesselet, R. Variable influence of the atmospheric flux on the trace metal chemistry of oceanic suspended matter. *Earth Planet. Sci. Lett.* **1979**, *42*, 399–411. [CrossRef]
42. Taylor, S.R.; Mclennan, S.M. The geochemical evolution of the continental-crust. *Rev. Geophys.* **1995**, *33*, 241–265. [CrossRef]
43. Cheung, K.; Daher, N.; Kam, W.; Shafer, M.M.; Ning, Z.; Schauer, J.J.; Sioutas, C. Spatial and temporal variation of chemical composition and mass closure of ambient coarse particulate matter ($PM_{10-2.5}$) in the Los Angeles area. *Atmos. Environ.* **2011**, *45*, 2651–2662. [CrossRef]
44. Clements, N.; Eav, J.; Xie, M.J.; Hannigan, M.P.; Miller, S.L.; Navidi, W.; Peel, J.L.; Schauer, J.J.; Shafer, M.M.; Milford, J.B. Concentrations and source insights for trace elements in fine and coarse particulate matter. *Atmos. Environ.* **2014**, *89*, 373–381. [CrossRef]
45. Jiang, S.Y.N.; Yang, F.; Chan, K.L.; Ning, Z. Water solubility of metals in coarse PM and $PM_{2.5}$ in typical urban environment in Hong Kong. *Atmos. Pollut. Res.* **2014**, *5*, 236–244. [CrossRef]
46. Majestic, B.J.; Schauer, J.J.; Shafer, M.M. Application of synchrotron radiation for measurement of iron red-ox speciation in atmospherically processed aerosols. *Atmos. Chem. Phys.* **2007**, *7*, 2475–2487. [CrossRef]
47. Cartledge, B.T.; Marcotte, A.R.; Herckes, P.; Anbar, A.D.; Majestic, B.J. The impact of particle size, relative humidity, and sulfur dioxide on iron solubility in simulated atmospheric marine aerosols. *Environ. Sci. Technol.* **2015**, *49*, 7179–7187. [CrossRef] [PubMed]
48. Oakes, M.; Ingall, E.D.; Lai, B.; Shafer, M.M.; Hays, M.D.; Liu, Z.G.; Russell, A.G.; Weber, R.J. Iron solubility related to particle sulfur content in source emission and ambient fine particles. *Environ. Sci Technol* **2012**, *46*, 6637–6644. [CrossRef] [PubMed]
49. Stumm, W.; Morgan, J.J. *Aquatic Chemistry: An Introduction Emphasizing Chemical Equilibria in Natural Waters*, 2nd ed.; Wiley: New York, NY, USA, 1981.
50. Pehkonen, S.O.; Siefert, R.; Erel, Y.; Webb, S.; Hoffman, M.R. Photoreduction of iron oxyhydroxides in the presence of important atmospheric organic compounds. *Environ. Sci Technol* **1993**, *27*, 2056–2062. [CrossRef]
51. Barbas, J.T.; Sigman, M.E.; Buchanan, A.C.; Chevis, E.A. Photolysis of substituted naphthalenes on SiO_2 and Al_2O_3. *Photochem. Photobiol.* **1993**, *58*, 155–158. [CrossRef]
52. Paris, R.; Desboeufs, K.V. Effect of atmospheric organic complexation on iron-bearing dust solubility. *Atmos. Chem. Phys.* **2013**, *13*, 4895–4905. [CrossRef]
53. Dickin, A.P. *Radiogenic Isotope Geology*, 2nd ed.; Cambridge University Press: Cambridge, UK, 2005.
54. Kendall, C.; McDonnell, J.J. *Isotope Tracers in Catchment Hydrology*, 1st ed.; Elsevier B.V.: Philadelphia, PA, USA, 1998; pp. 51–86.
55. Komerek, M.; Ettler, V.; Chrastny, V.; Mihailovic, M. Lead isotopes in environmental sciences: A review. *Environ. Int.* **2008**, *34*, 562–577. [CrossRef] [PubMed]
56. Dewan, N.; Majestic, B.J.; Ketterer, M.E.; Miller-Schulze, J.P.; Shafer, M.M.; Schauer, J.J.; Solomon, P.A.; Artamonova, M.; Chen, B.B.; Imashev, S.A.; *et al.* Stable isotopes of lead and strontium as tracers of sources of airborne particulate matter in Kyrgyzstan. *Atmos. Environ.* **2015**, *120*, 438–446. [CrossRef]
57. Monna, F.; Lancelot, J.; Croudace, I.W.; Cundy, A.B.; Lewis, J.T. Pb isotopic composition of airborne particulate material from France and the southern United Kingdom: Implications for Pb pollution sources in urban areas. *Environ. Sci. Technol.* **1997**, *31*, 2277–2286. [CrossRef]
58. Simonetti, A.; Gariepy, C.; Carignan, J. Pb and Sr isotopic compositions of snowpack from Quebec, Canada: Inferences on the sources and deposition budgets of atmospheric heavy metals. *Geochim. Cosmochim. Acta* **2000**, *64*, 5–20. [CrossRef]

59. Chen, J.M.; Tan, M.G.; Li, Y.L.; Zhang, Y.M.; Lu, W.W.; Tong, Y.P.; Zhang, G.L.; Li, Y. A lead isotope record of shanghai atmospheric lead emissions in total suspended particles during the period of phasing out of leaded gasoline. *Atmos. Environ.* **2005**, *39*, 1245–1253. [CrossRef]
60. Lee, C.S.L.; Li, X.D.; Zhang, G.; Li, J.; Ding, A.J.; Wang, T. Heavy metals and Pb isotopic composition of aerosols in urban and suburban areas of Hong Kong and Guangzhou, South China—Evidence of the long-range transport of air contaminants. *Atmos. Environ.* **2007**, *41*, 432–447. [CrossRef]
61. Bollhofer, A.; Rosman, K.J.R. Isotopic source signatures for atmospheric lead: The Northern Hemisphere. *Geochim. Cosmochim. Acta* **2001**, *65*, 1727–1740. [CrossRef]
62. Widory, D.; Liu, X.D.; Dong, S.P. Isotopes as tracers of sources of lead and strontium in aerosols (TSP & $PM_{2.5}$) in Beijing. *Atmos. Environ.* **2010**, *44*, 3679–3687.
63. Zhu, L.M.; Tang, J.W.; Lee, B.; Zhang, Y.; Zhang, F.F. Lead concentrations and isotopes in aerosols from Xiamen, China. *Mar. Pollut. Bull.* **2010**, *60*, 1946–1955. [CrossRef] [PubMed]
64. Thurston, G.D.; Ito, K.; Lall, R. A source apportionment of U.S. fine particulate matter air pollution. *Atmos. Environ.* **2011**, *45*, 3924–3936. [CrossRef] [PubMed]
65. Clayton, J.L.; Koncz, I. Petroleum Geochemistry of the Zala Basin, Hungary. *Aapg. Bull.* **1994**, *78*, 1–22.
66. Mueller, D.; Uibel, S.; Takemura, M.; Klingelhoefer, D.; Groneberg, D.A. Ships, ports and particulate air pollution—An analysis of recent studies. *J. Occup. Med. Toxicol.* **2011**, *6*, 1–6. [CrossRef] [PubMed]
67. Zhang, R.; Jing, J.; Tao, J.; Hsu, S.C.; Wang, G.; Cao, J.; Lee, C.S.L.; Zhu, L.; Chen, Z.; Zhao, Y.; *et al.* Chemical characterization and source apportionment of $PM_{2.5}$ in Beijing: Seasonal perspective. *Atmos. Chem. Phys.* **2013**, *13*, 7053–7074. [CrossRef]
68. Garg, B.D.; Cadle, S.H.; Mulawa, P.A.; Groblicki, P.J. Brake wear particulate matter emissions. *Environ. Sci. Technol.* **2000**, *34*, 4463–4469. [CrossRef]
69. Oakes, M.; Weber, R.J.; Lai, B.; Russell, A.; Ingall, E.D. Characterization of iron speciation in urban and rural single particles using XANES spectroscopy and micro X-ray fluorescence measurements: investigating the relationship between speciation and fractional iron solubility. *Atmos. Chem. Phys.* **2012**, *12*, 745–756. [CrossRef]
70. Cartledge, B.T.; Majestic, B.J. Metal concentrations and soluble iron speciation in fine particulate matter from light rail activity in the Denver-Metropolitan area. *Atmos. Pollut. Res.* **2015**, *6*, 495–502. [CrossRef]
71. SZ News. Available online: http://sztqb.sznews.com/html/2011-08/08/content_1694922.htm (accessed on 8 February 2016). (In Chinese).
72. GD.Xinhuanet. Available online: http://www.gd.xinhuanet.com/newscenter/2011-06/01/content_22914683.htm (accessed on 8 February 2016). (In Chinese)

Permissions

All chapters in this book were first published in MDPI; hereby published with permission under the Creative Commons Attribution License or equivalent. Every chapter published in this book has been scrutinized by our experts. Their significance has been extensively debated. The topics covered herein carry significant findings which will fuel the growth of the discipline. They may even be implemented as practical applications or may be referred to as a beginning point for another development.

The contributors of this book come from diverse backgrounds, making this book a truly international effort. This book will bring forth new frontiers with its revolutionizing research information and detailed analysis of the nascent developments around the world.

We would like to thank all the contributing authors for lending their expertise to make the book truly unique. They have played a crucial role in the development of this book. Without their invaluable contributions this book wouldn't have been possible. They have made vital efforts to compile up to date information on the varied aspects of this subject to make this book a valuable addition to the collection of many professionals and students.

This book was conceptualized with the vision of imparting up-to-date information and advanced data in this field. To ensure the same, a matchless editorial board was set up. Every individual on the board went through rigorous rounds of assessment to prove their worth. After which they invested a large part of their time researching and compiling the most relevant data for our readers.

The editorial board has been involved in producing this book since its inception. They have spent rigorous hours researching and exploring the diverse topics which have resulted in the successful publishing of this book. They have passed on their knowledge of decades through this book. To expedite this challenging task, the publisher supported the team at every step. A small team of assistant editors was also appointed to further simplify the editing procedure and attain best results for the readers.

Apart from the editorial board, the designing team has also invested a significant amount of their time in understanding the subject and creating the most relevant covers. They scrutinized every image to scout for the most suitable representation of the subject and create an appropriate cover for the book.

The publishing team has been an ardent support to the editorial, designing and production team. Their endless efforts to recruit the best for this project, has resulted in the accomplishment of this book. They are a veteran in the field of academics and their pool of knowledge is as vast as their experience in printing. Their expertise and guidance has proved useful at every step. Their uncompromising quality standards have made this book an exceptional effort. Their encouragement from time to time has been an inspiration for everyone.

The publisher and the editorial board hope that this book will prove to be a valuable piece of knowledge for researchers, students, practitioners and scholars across the globe.

List of Contributors

Yun-Chia Liang, Yona Maimury and Josue Rodolfo Cuevas Juarez
Department of Industrial Engineering and Management, Yuan Ze University, Taoyuan City 320, Taiwan

Angela Hsiang-Ling Chen
Department of Industrial and Systems Engineering, Chung Yuan Christian University, Taoyuan City 320, Taiwan

Xinyue Mo and Lei Zhang
College of Atmospheric Sciences, Lanzhou University, Lanzhou 730000, China
Collaborative Innovation Center for Western Ecological Safety, Lanzhou 730000, China

Huan Li
School of Information Science and Engineering, Lanzhou University, Lanzhou 730000, China

Zongxi Qu
School of Management, Lanzhou University, Lanzhou 730000, China

Ji Hoon Seo
Department of Health & Environmental Science, Korea University, Seoul 02841, Korea
Department of Public Health Sciences, BK21 FOUR R&E Center for Learning Health Systems, Graduate School, Korea University, Seoul 02841, Korea

Ji Soo Kim, Jinho Yang, Hyunjun Yun, Hyun Woo Jeon and Joung Sook Choi
Department of Health and Safety Convergence Science, Korea University, Seoul 02841, Korea

Minwoo Roh
Department of Environmental Science and Ecological Engineering, Korea University, Seoul 02841, Korea

Jung Woon Kim, Songhun Yu and Na Na Jeong
Department of Health & Environmental Science, Korea University, Seoul 02841, Korea

Jong Ryeul Sohn
Department of Health & Environmental Science, Korea University, Seoul 02841, Korea
Department of Public Health Sciences, BK21 FOUR R&E Center for Learning Health Systems, Graduate School, Korea University, Seoul 02841, Korea
Department of Health and Safety Convergence Science, Korea University, Seoul 02841, Korea

Rohit Chakraborty and Martin Mayfield
Department of Civil and Structural Engineering, The University of Sheffield, Sheffield S1 3JD, UK

James Heydon
School of Sociology and Social Policy, The University of Nottingham, Nottingham NG7 2RD, UK

Lyudmila Mihaylova
Department of Automatic Control and Systems Engineering, The University of Sheffield, Sheffield S1 3JD, UK

Afifa Aslam, Muhammad Ibrahim, Abid Mahmood, Muhammad Kashif Irshad, Ghazala and Muhammad Tariq
Department of Environmental Sciences & Engineering, Government College University Faisalabad, Faisalabad 38000, Pakistan

Imran Shahid
Environmental Science Centre, Qatar University, Doha, Qatar

Muhammad Yamin
Department of Farm Machinery & Power, Faculty of Agricultural Engineering & Technology, University of Agriculture, Faisalabad 38040, Pakistan

Redmond R. Shamshiri
Leibniz Institute for Agricultural Engineering and Bioeconomy, Max-Eyth-Allee 100, 14469 Potsdam-Bornim, Germany

Steven G. Brown
Sonoma Technology Inc., 1455 N. McDowell Blvd., Suite D, Petaluma, CA 94954, USA
Department of Atmospheric Science, Colorado State University, Fort Collins, CO 80523, USA

Taehyoung Lee
Department of Environmental Science, Hankuk University of Foreign Studies, Yongin 427-010, Korea

Paul T. Roberts
Sonoma Technology Inc., 1455 N. McDowell Blvd., Suite D, Petaluma, CA 94954, USA

Jeffrey L. Collett Jr.
Department of Atmospheric Science, Colorado State University, Fort Collins, CO 80523, USA

Xingcheng Lu
Division of Environment, Hong Kong University of Science & Technology, Clear Water Bay, Hong Kong, China

Jimmy C. H. Fung
Division of Environment, Hong Kong University of Science & Technology, Clear Water Bay, Hong Kong, China
Department of Mathematics, Hong Kong University of Science & Technology, Clear Water Bay, Hong Kong, China

Pavla Dagsson-Waldhauserova
Faculty of Environmental Sciences, Agricultural University of Iceland, Keldnaholt, Reykjavik 112, Iceland
Faculties of Physical and Earth Sciences, University of Iceland, Reykjavik 101, Iceland
Faculty of Environmental Sciences, Czech University of Life Sciences, Prague 165 21, Czech Republic

Agnes Ösp Magnusdottir
Faculty of Environmental Sciences, Agricultural University of Iceland, Keldnaholt, Reykjavik 112, Iceland

Haraldur Olafsson
Faculties of Physical and Earth Sciences, University of Iceland, Reykjavik 101, Iceland
Icelandic Meteorological Office, Reykjavik 108, Iceland

Olafur Arnalds
Faculty of Environmental Sciences, Agricultural University of Iceland, Keldnaholt, Reykjavik 112, Iceland

Jeffrey K. Bean, Cameron B. Faxon and Lea Hildebrandt Ruiz
McKetta Department of Chemical Engineering, The University of Texas at Austin, Austin, TX 78712, USA

Yu Jun Leong, Henry William Wallace, Basak Karakurt Cevik and Robert J. Griffin
Department of Civil and Environmental Engineering, Rice University, Houston, TX 77005, USA

Stephanie Ortiz, Sascha Usenko and Rebecca J. Sheesley
Department of Environmental Science, Baylor University, Waco, TX 76798, USA

Manjula R. Canagaratna
Aerodyne Research Inc., Billerica, MA 01821, USA

Frauke Hennig, Dorothea Sugiri, Lilian Tzivian and Kateryna Fuks
Working group of Environmental Epidemiology of Cardiovascular Aging and Allergies, IUF-Leibniz Research Institute for Environmental Medicine, Auf'm Hennekamp 50, Düsseldorf 40225, Germany

Susanne Moebus and Karl-Heinz Jöckel
Institute for Medical Informatics, Biometry and Epidemiology, University Hospital, University Duisburg-Essen, Essen 45141, Germany

Danielle Vienneau and Kees de Hoogh
Department of Epidemiology and Public Health, Swiss Tropical and Public Health Institute, Socinstr. 57, Basel 4051, Switzerland
University of Basel, Basel 4003, Switzerland

Thomas A.J. Kuhlbusch
IUTA e.V., Air Quality & Sustainable Nanotechnology Unit, Duisburg, Germany

Michael Memmesheimer and Hermann Jakobs
Rhenish Institute for Environmental Research (RIU), Aachenerstr. 209, 50931 Köln, Germany

Ulrich Quass
IUTA e.V., Air Quality & Sustainable Nanotechnology Unit, Duisburg, Germany

Barbara Hoffmann
Working group of Environmental Epidemiology of Cardiovascular Aging and Allergies, IUF-Leibniz Research Institute for Environmental Medicine, Auf'm Hennekamp 50, Düsseldorf 40225, Germany
Heinrich Heine University of Düsseldorf, Medical Faculty, Deanery of Medicine, Moorenstraße 5, Düsseldorf 40225, Germany

Hongmei Zhao
Key Laboratory of Wetland Ecology and Environment, Northeast Institute of Geography and Agroecology, Chinese Academy of Sciences, Changchun 130102, China

Daniel Q. Tong and Hyuncheol Kim
U.S. NOAA Air Resources Laboratory, College Park, MD 20740, USA
UMD/Cooperative Institute for Climate and Satellites, College Park, MD 20740, USA

Pius Lee and Hang Lei
U.S. NOAA Air Resources Laboratory, College Park, MD 20740, USA

Amanda S. Willoughby, Andrew S. Wozniak and Patrick G. Hatcher
Department of Chemistry and Biochemistry, Old Dominion University, Norfolk, VA 23529, USA

Nitika Dewan and Brian J. Majestic
Department of Chemistry and Biochemistry, University of Denver, Denver, CO 80208, USA

Yu-Qin Wang
College of Resources and Environment, University of Chinese Academy of Sciences, Beijing 100049, China

Yuan-Xun Zhang
College of Resources and Environment, University of Chinese Academy of Sciences, Beijing 100049, China
Huairou Eco-Environmental Observatory, Chinese Academy of Sciences, Beijing 101408, China

Yang Zhang
College of Resources and Environment, University of Chinese Academy of Sciences, Beijing 100049, China

Ling-Yan He and Xiao-Feng Huang
Key Laboratory for Urban Habitat Environmental Science and Technology, School of Environment and Energy, Peking University Shenzhen Graduate School, Shenzhen 518055, China

Index

P

R

S

T

U

V

W

Printed in the USA
CPSIA information can be obtained
at www.ICGtesting.com
JSHW050201070524
62621JS00005B/91